ADVANCED TOPICS ON ASTROPHYSICAL AND SPACE PLASMAS

Proceedings of the Advanced School on Astrophysical and
Space Plasmas
held in Guarujá, Brazil, June 26–30, 1995

Edited by

ELISABETE M. DE GOUVEIA DAL PINO

University of São Paulo, Instituto Astronômico e Geofísico, Brazil

ANTHONY L. PERATT

*Los Alamos National Laboratory, United States Department of Energy,
Germantown, MD, U.S.A.*

GUSTAVO A. MEDINA TANCO

University of São Paulo, Instituto Astronômico e Geofísico, Brazil

and

ABRAHAM C.-L. CHIAN

National Institute for Space Research, São José dos Campos, Brazil

Reprinted from *Astrophysics and Space Science*
Volume 242, Nos. 1–2, 1996

SPRINGER SCIENCE+BUSINESS MEDIA, B.V.

A C.I.P. Catalogue record for this book is available from the Library of Congress.

ISBN 978-94-010-6299-2   ISBN 978-94-011-5466-6 (eBook)
DOI 10.1007/978-94-011-5466-6

Printed on acid-free paper

TABLE OF CONTENTS

Preface   1–2

List of Participants   7–8

Institutes   9–11

M. BIRKINSHAW / Instabilities in Astrophysical Jets   17–91

ANTHONY L. PERATT / Advances in Numerical Modeling of Astrophysical
    and Space Plasmas   93–163

DIETER BISKAMP / Magnetic Reconnection in Plasmas   165–207

D.B. MELROSE / Particle Acceleration and Nonthermal Radiation in Space
    Plasmas   209–246

A.C.-L. CHIAN / Nonlinear Wave-Wave Interactions in Astrophysical and
    Space Plasmas   249–295

Advanced Topics on Astrophysical and Space Plasmas

PREFACE

During the 1994 International Conference on Plasma Physics held in Foz do Iguaçu, Brazil, a group of scientists, motivated by the recent considerable growth of the Brazilian community interest in Plasma Astrophysics and Space Plasma Physics, decided to organize an Advanced School on Astrophysical and Space Plasmas. Seven months later, from June 26 to 30, 1995, the event took place in a nice seaside resort in Guarujá, an island off the coast of Brazil, between the cities of São Paulo and Rio de Janeiro.

The aim of this School was to bring together researchers of related areas of Plasma Astrophysics and Space Plasma Physics, providing a forum for the discussion of subject matters, existing models and techniques of common interest. It was agreed that the best way to meet that objective was to focus on a few subjects of wide interest and to invite some renowned scientists to present the reviews.

The main topics covered included Particle Acceleration, Nonthermal Radiation, MHD Instabilities, Magnetic Field Reconnection, and Nonlinear Dynamics in astrophysical and space plasmas. Lectures of four and half hours on each of those subjects were presented at the School and this book brings together most of those valuable reviews. Dieter Biskamp of the Max-Planck Institute for Plasma Physics reviewed the Magnetic Field Reconnection and MHD Instabilities, Mark Birkinshaw of the Harvard-Smithsonian Center for Astrophysics and the University of Bristol reviewed the Instabilities in Cosmic Jets, Abraham Chian of the National Institute for Space Research reviewed the Nonlinear Wave-Wave Interactions in Space and Astrophysical Plasmas, Nelson Fiedler-Ferrara of the University of São Paulo reviewed the Chaos and Nonlinear Dynamics in Plasmas, Donald Melrose of the University of Sydney reviewed the Particle Acceleration and Nonthermal Radiation, and Anthony Peratt of the Los Alamos National Laboratory reviewed the Numerical Methods on Astrophysical and Space Plasmas.

About 60 researchers and graduate students from Brazil and Argentina attended the School. This event has been financially supported by the following Brazilian science foundations: FAPESP, CNPq, and CAPES, and also by the Instituto Astronômico e Geofísico of the University of São Paulo (IAG-USP) and DEDALUS of Brazil.

2

The organizing committee was composed by Abraham Chian (National Institute for Space Research), Elisabete M. de Gouveia Dal Pino (University of São Paulo) (chairperson), Gustavo A. Medina Tanco (University of São Paulo), and Irina Potapenko (State University of Rio de Janeiro).

We are indebted to Ms. Marina de Freitas, Marcia Pina Albe, Maria do Carmo Correêa and Mr. Julio Klafke for their excellent logistic support during the organization of the School.

We are also indebted to the Brazilian Agency FAPESP for providing financial support for the publication of this volume.

Elisabete M. de Gouveia Dal Pino
Anthony L. Peratt
Gustavo A. Medina Tanco
Abraham C.-L. Chian

List of participants on the Conference Picture

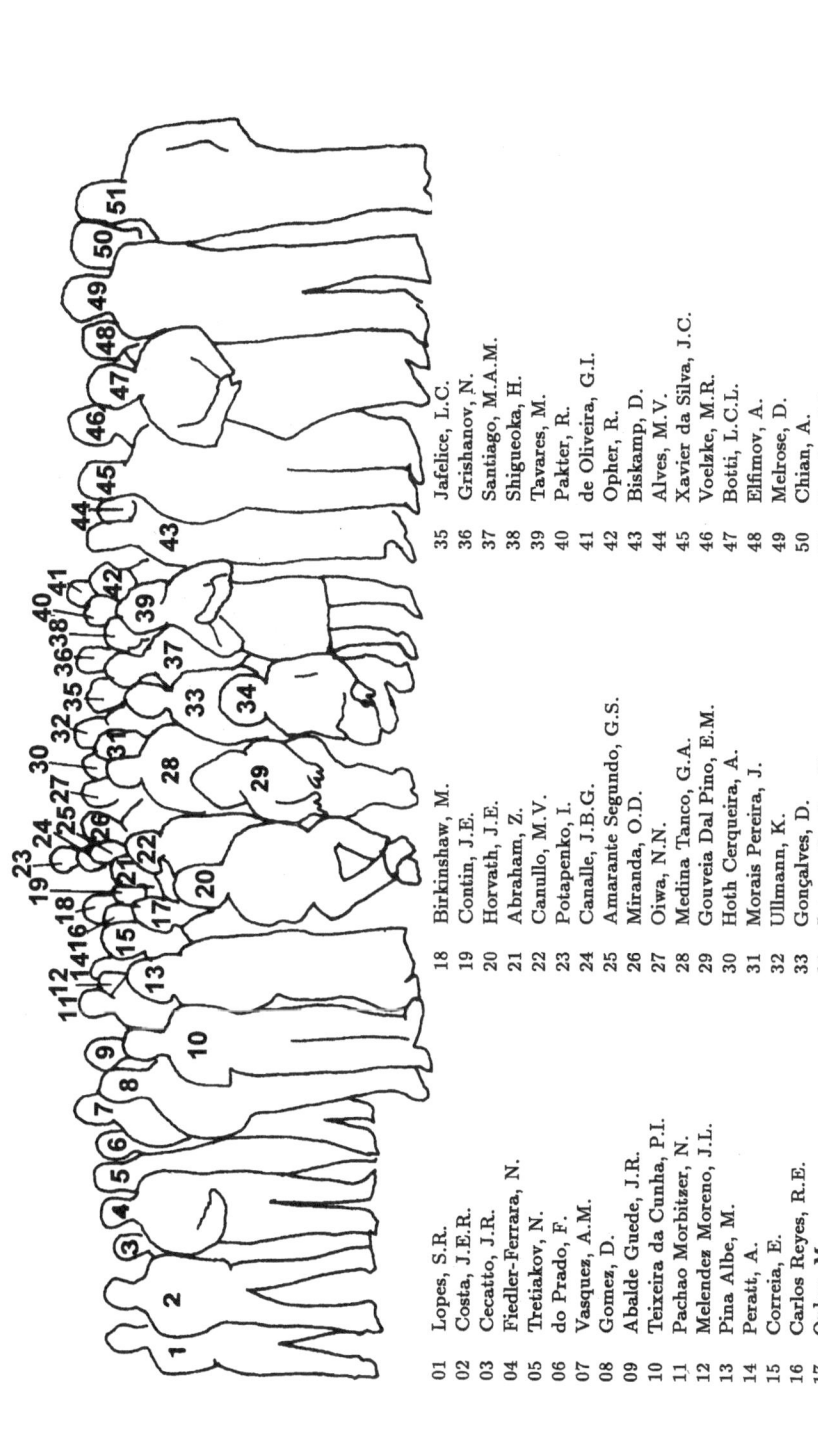

01 Lopes, S.R.
02 Costa, J.E.R.
03 Cecatto, J.R.
04 Fiedler-Ferrara, N.
05 Tretiakov, N.
06 do Prado, F.
07 Vasquez, A.M.
08 Gomez, D.
09 Abalde Guede, J.R.
10 Teixeira da Cunha, P.I.
11 Pachao Morbitzer, N.
12 Melendez Moreno, J.L.
13 Pina Albe, M.
14 Peratt, A.
15 Correia, E.
16 Carlos Reyes, R.E.
17 Opher, M.

18 Birkinshaw, M.
19 Contin, J.E.
20 Horvath, J.E.
21 Abraham, Z.
22 Canullo, M.V.
23 Potapenko, I.
24 Canalle, J.B.G.
25 Amarante Segundo, G.S.
26 Miranda, O.D.
27 Oiwa, N.N.
28 Medina Tanco, G.A.
29 Gouveia Dal Pino, E.M.
30 Hoth Cerqueira, A.
31 Morais Pereira, J.
32 Ullmann, K.
33 Gonçalves, D.
34 Jatenco Pereira, V.

35 Jafelice, L.C.
36 Grishanov, N.
37 Santiago, M.A.M.
38 Shigueoka, H.
39 Tavares, M.
40 Pakter, R.
41 de Oliveira, G.I.
42 Opher, R.
43 Biskamp, D.
44 Alves, M.V.
45 Xavier da Silva, J.C.
46 Voelzke, M.R.
47 Botti, L.C.L.
48 Elfimov, A.
49 Melrose, D.
50 Chian, A.
51 Ferreira, J.L.

List of participants

Name	Institute	E-mail address
José R. Abalde Guede	INPE	abalde@das.inpe.br
Zulema Abraham	IAG-USP	zulema@vax.iagusp.usp.br
Maria Virginia Alves	INPE	alves@plasma.inpe.br
Gesil S. Amarante Segundo	IF-USP	gesil@if.usp.br
Walter M. Andrade	IF-UERJ	canalle@vmesa.uerj.br
Mark Birkinshaw	CfA	mbl@cfa.harvard.edu
Dieter Biskamp	MPIPP	dfb@ipp-garching.mpg.de
Joáo B. G. Canalle	IF-UERJ	canalle@vmesa.uerj.br
Maria Victoria Canullo	IAFE	canullo@iafe.uba.ar
José R. Cecatto	INPE	jrc@das.inpe.br
Adriano H. Cerqueira	IAG-USP	adriano@vax.iagusp.usp.br
Abraham C. L. Chian	INPE	achian@das.inpe.br
Julia E. Contin	IFP	julia@lfpba.edu.ar
Emilia Correia	CRAAE	ecorreia@fox.cce.usp.br
Joaquim E. R. Costa	CRAAE	jercosta@fox.cce.usp.br
Pedro I. T. da Cunha	ICEX-UFMG	dpfcunha@oraculo.lcc.ufmg.br
Murilo da Silva Baptista	IF-USP	murilo@if.usp.br
José C. X. da Silva	IF-UERJ	canalle@vmesa.uerj.br
Vivian C. de Almeida Lopes	INPE	vivian@das.inpe.br
José de Morais Pereira	IF-USP	jmpereira@if.usp.br
Glaucius I.. de Oliveira	IF - UFRGS	gio@if.ufrgs.br
Wilson de Oliveira Lavras	IAG-USP	wilson@vax.iagusp.usp.br
Fábio do Prado	INPE	fprado@plasma.inpe.br
Artour Elfimov	IF-UERJ	elfimov@bruerj.bitnet
José L. Ferreira	UnB	leo@lccfis.unb.br
Nelson Fiedler-Ferrara	IF-USP	nferrari@if.usp.br
Erik Figueiredo Freire	IAG-USP	freire@vax.iagusp.usp.br
Daniel Gomez	IAFE	gomez@iafe.uba.ar
Denise R. Gonçalves	IAG-USP	denise@vax.iagusp.usp.br
Elisabete M. de Gouveia Dal Pino	IAG-USP	dalpino@astrol.iagusp.usp.br
Tatiana Grishahova	IF-UERJ	grishano@bruerj.bitnet
Nikolay Grishanov	IF-UERJ	grishano@bruerj.bitnet
Jorge E. Horvath	IAG-USP	foton@vax.iagusp.usp.br
Luiz C. Jafelice	UFRN	lcj@ncc.ufrn.br
Vera Jatenco-Pereira	IAG-USP	jatenco@vax.iagusp.usp.br
Konstantin G. Kostov	UnB	kostov@sunsl.fis.unb.br
Luiz C. Lima Botti	CRAAE	lclbotti@fox.cce.usp.br
Sérgio R. Lopes	INPE	lopes@pgrad.inpe.br
Marcos A. Matos Santiago	IF-UFF	gfimsan@vmhpo.uff.br
Gustavo A. Medina Tanco	IAG-USP	gustavo@vax.iagusp.usp.br
Jorge L. Melendez Moreno	INPE	jorge@das.inpe.br
Donald Melrose	Sydney Univ.	melrose@physics.usyd.edu.au
Oswaldo D. Miranda	IAG-USP	oswaldo@vax.iagusp.usp.br
Nestor N. Oiwa	IF-USP	oiwa@if.usp.br
Merav Opher	IAG-USP	merav@vax.iagusp.usp.br
Reuven Opher	IAG-USP	opher@vax.iagusp.usp.br
Nelson Pachao Morbitzer	Balseiro	pachao@df.edu.ar
Renato Pakter	IF-UFRGS	pakter@if.ufrgs.br
Anthony Peratt	LANL/DOE	alp@lanl.gov
Irina Potapenko	IF-UERJ	irina@bruerj.bitnet

Astrophysics and Space Science **242**: 7–8, 1997.

8

List of participants

Name	Institute	E-mail address
Rafael E . C . Reyes	IAG-USP	rafael@vax.iagusp.usp.br
Hisataki Shigueoka	IF-UFF	gfihisa@bruff.bitnet
Marilia Tavares	IF-UFF	gfimari@vmbpo.uff.br
Nikolai Tretiakov	UFG	nicolai@ufg.br
Vladimir Tsypin	IF-UERJ	tsypin@vmesa.uerj.br
Kai Ullmann	IF-USP	kai@if.usp.br
Alberto M. Vasquez	IAFE	plasma@iafe.uba.ar
Marcos R. Voelzke	IAG-USP	voelzke@astrol.iagusp.usp.br
Luis C. Yamamoto	IAG-USP	yamamoto@vax.iagusp.usp.br

Institutes

Center for Astrophysics (CFA)
60 Garden Street
Cambridge, 02138-1596 MA, USA
Telefone: (001) (617) 4957146
Telefax: (001) (617) 4957356

Centro de Radioastronomia e Aplicações Espaciais (CRAAE)
Escola Politécnica da USP
Departament o de Engenharia de Transportes
Caixa Postal 61548
05424-970 São Paulo SP
Telefone: (011) 8155936
Telefax: (011) 8156289

Instituto de Astronomia e Fisica del Espacio (IAFE)
Casilla de Correo 67
Sucursal 28
1428 Buenos Aires
Argentina
Telefone: (0054) (1) 7816755
Telefax: (0054) (1) 7868114

Instituto Astronômico e Geofísico da Universidade de São Paulo (IAG-USP)
Av. Miguel Stéfano, 4200
04301-904 São Paulo SP
Telefone: (011) 5778599
Telefax: (011) 5778599 ou 2763848

Instituto Balseiro
Centro Atomico Bariloche
8400 Rio Negro
Argentina
Telefone: (0054) (944) 61002
Telefax: (0054) (944) 61006

Instituto de Ciências Exatas da Universidade Federal de Minas Gerais (ICEX-UFMG)
Departamento de Física
Caixa Postal 702 30161-970 Belo Horizonte MG
Telefone: (031) 4485620

Telefax: (031) 4485600

Instituto de Física del Plasma
Ciudad Universitaria - Pab. 1
1428 Buenos Aires
Argentina
Telefone: (0054) (1) 7848104
Telefax: (0054) (1) 7872712

Instituto de Física da Universidade Estadual do Rio de Janeiro (IF-UERJ)
Rua São Francisco Xavier, 524 - Sala 3002-D
20550-013 Rio de Janeiro RJ
Telefone: (021) 2848322
Telefax: (021) 5674541

Instituto de Física da Universidade Federal Fluminense (IFOFF)
Av. Litoranea, s/n
Praia de Boa Viagem
24000-000 Niterói RJ
Telefone: (021) 7196735
Telefax: (021) 7174553

Instituto de Física da Universidade Federal do Rio Grande do Sul (IF-UFRGS)
AV. Bento Gonçalves, 9500
91500-900 Porto Alegre RS
Telefone: (051) 3368399
Telefax: (051) 3361762

Instituto de Física da Universidade de São Paulo (IFUSP)
Caixa Postal 66318
05389-970 São Paulo SP
Telefone: (011) 8186841
Telefax: (011) 8140503

Instituto Nacional de Pesquisas Espaciais (INPE)
Caixa Postal 515
12201-970 São José dos Campos SP
Telefone: (0123) 418977
Telefax: (0123) 218743

Max-Planck Institute for Plasma Physics
85748 Garching, Germany
Telefone: (0049) (89) 32991181
Telefax: (0049) (89) 32992126

United States Department of Energy
Office of Research and Development
DP-16, B-302
Germantown, MD 20874 USA
Telefone: (301) 903 9697
Telefax: (301) 903 9743

Universidade de Brasília (UnB)
Departamento de Física
Campus Universitário Asa Norte
70900-000 Brasília DF
Telefone: (061) 3482185
Telefax: (061) 2721053

Universidade Federal de Goiás (UFO)
Departamento de Física
Caixa Postal 131
74001-970 Goiânia GO
Telefone: (062) 2051000
Telefax: (062) 2051327

Universidade Federal do Rio Grande do Norte (UFRN)
Departamento de Física
Caixa Postal 1641
59072-970 Natal RN
Telefone: (084) 2319586
Telefax: (084) 2319749

University of Sydney
School of Physics
NSW 2006 Sydney
Australia
Telefone: (0061) (2) 3514234
Telefax: (0061) (2) 6602903

12

Figure 1. The Chair, Advanced School on Astrophysical and Space Plasmas. From left to right, Professors Gustavo A. Medina Tanco, Elisabete M. de Gouveia Dal Pino, and Abraham Chian.

Figure 2. Gustavo, holding a *Caipirinha* discusses magnetic merging with Dieter, holding a *Bier*.

Figure 3. Official school dance: The Samba.

Figure 4. Bete, giving samba lessons to Tony and Mark: *step-and-cut, step-and-cut.*

14

Figure 5. Advanced school in session.

Figure 6. Reuven Opher summarizes the findings presented by the lecturers.

Figure 7. View of Guarujá towards the Atlantic.

Figure 8. Return trip from Guarujá to São Paulo.

Mark Birkinshaw

Instabilities in Astrophysical Jets

M. Birkinshaw *

Department of Physics, University of Bristol, Tyndall Avenue, Bristol BS8 1TL, UK

Abstract. The observed properties of astrophysical jets are reviewed, and the techniques used to estimate the parameters of the underlying beams are described. This information is then used in a theoretical treatment of the Kelvin-Helmholtz instability of the flows, and the relevance of this instability to the persistence of the observed structures is emphasised.

Key words: Jets, Beams, Magnetohydrodynamics, Radio galaxies

1. Observed properties of astrophysical jets

I have chosen to start this review of astrophysical jet stability and instability by giving some of the essential observational background on real astronomical sources, and on the results of large-scale computer codes that attempt to explain what we are seeing in terms of a flow model. In later sections I will describe the methods used to try to get at the physical parameters of the beams that underlie the observed jets, and end with a discussion of the important issue of stability: whether the underlying beams can propagate to large distances without disrupting.

1.1. INTRODUCTION TO JETS: EXTRAGALACTIC, GALACTIC

M 87 provided the first example of an astrophysical jet (Curtis, 1918). The detail that is visible in modern optical and radio images of M 87 (Biretta & Meisenheimer, 1993) reinforces the strong impression of *flow* that we get when looking at them: how else, it might reasonably be asked, can a linear structure exist in a chaotic object like an elliptical galaxy? However, the modern idea that quasi-continuous flows from the nuclei of elliptical galaxies may appear as jets of this type was not developed until the 1970s, when the idea that these jets trace the flow of matter and energy into the outer parts of radio galaxies (such as the radio lobes of Cygnus A; Section 1.2.2) was expounded and elaborated (Rees, 1971; Scheuer, 1974; Blandford & Rees, 1974).

Jets are now recognised as an important phenomenon in both extragalactic radio sources (such as M 87, Cygnus A, and NGC 6251) and

---

* Also Smithsonian Astrophysical Observatory, 60 Garden Street, Cambridge, MA 02138, USA

Astrophysics and Space Science **242**: 17–91, 1997.

galactic objects (such as SS 433, HH 34, and GRO J1655-40). We see jets with a wide range of linear sizes and luminosities, and believe that jets are a *common* attribute of energetic objects. I will not be discussing the origins of jets—but whatever the controlling mechanisms are, they are effective from stellar to quasar luminosities.

The theoretical description of jet flows cannot be divorced from the observational basis of the phenomena: the limitations of astronomical data are such that many of the fundamental physical parameters of jets are not known. Often this looseness in the data gives us too much theoretical freedom of maneuver, but conversely, a theoretical understanding of jet flows should help us to measure the physical parameters of these interesting astronomical objects.

It will help us avoid confusion if we carefully distinguish between *jets* and *beams* in what follows. I will always use the term *jet* to describe an astrophysical structure which looks as if it corresponds to a mostly-linear, collimated flow. It should always be remembered that what we see as a jet may only be loosely related to the underlying structure (see Section 2.1). I will use the term *beam* to refer to the underlying flow that is supposed to result in the observed jet. That is, the beam is the theoretical flow that is supposed to result in the observed jet structure. The beam model is now quite well accepted, but observational studies of jets continue to test it.

A good reference on the properties of astrophysical jets and their interpretation in terms of the beam model is Hughes (1991).

1.2. OBSERVED STRUCTURES OF JETS

The easiest way to explain what is meant by a jet is to show some examples. In this section I discuss some of the more famous jets, and recent data that shows how much more detail of jet structures we are beginning to see.

1.2.1. *M 87 (NGC 4486; 3C 274)*

Recent radio and optical images of the jet of M 87, which is one of the closest and best-studied extragalactic jets, appear in Biretta & Meisenheimer (1993). On the scale of a few kpc, we see a knotty, somewhat bent, jet which has an angular width that generally increases at increasing distance from the center of the galaxy, and which is embedded in a large-scale "halo" of diffuse emission. At low radio frequencies, and on large angular scales, the jet is inconspicuous: the steep-radio-spectrum emission from the halo has a higher apparent surface brightness. A detailed examination of the maps shows limb brightening in the regions between the knots (Owen *et al.*, 1989).

One attraction of the study of M 87 is that the galaxy is relatively nearby (in the Virgo cluster), so that arcsecond-scale resolution from ground-based optical telescopes and interferometers corresponds to physical structure with about 100 pc scales. The highest resolution imaging, from VLBI (e.g., Junor & Biretta 1995) shows jet-like structures on scales < 0.1 pc, continuing very close to the bright unresolved radio component associated with the core of the galaxy. A high quality VLBI map (Reid *et al.*, 1989) shows that M 87 contains a jet which is already very linear on the scale of a few pc.

Recent HST data (Ford *et al.*, 1994; Harms *et al.*, 1994) show that the centre of M 87 contains a disk of gas and dust, which is rotating at $\sim$ 550 $\mathrm{km\,s}^{-1}$ only 20 pc from the centre of the galaxy. If this is interpreted as the Keplerian motion of cold material in the gravitational field of the nucleus, then the nucleus may contain a supermassive black hole, with mass $M \sim 3 \times 10^9 \ M_\odot$. A favourite prejudice is that it is the presence of such objects that provides the ultimate (gravitational) energy source that powers the jet flow: however, I do not intend to discuss the question of the power source, or the initial generation of jets.

On the smallest (sub-milliarcsec) scales, the jet does not vary significantly with time: the pattern of emission knots and extended emission that is seen appears to be static, with $v < 0.05c$ (Junor & Biretta, 1995). On the milli-arcsec scale, bright knots in the jet are moving out from the most compact component in the source with speed $(0.28 \pm 0.08)c$ (Reid *et al.*, 1989). On the arcsec scale of the large-scale jet, the knots appear to be expanding from the core of the source with speeds of about $0.5c$, with some compact features in the knots moving at several c (Biretta & Meisenheimer, 1993). Clearly this is indicative of a relativistic phenomenon, but the interpretation of an apparently accelerating flow is unclear: is it the fluid that is moving relativistically and accelerating (and if so, how is the acceleration produced at distances of order 1 kpc from the nucleus of M 87)? Or is what we are seeing merely a pattern velocity, which may be moving much faster than the flow itself (see Section 2)?

The VLBI source is known to be variable on month time-scales (e.g., Morabito *et al.*, 1988). Such variability of brightness may correspond to variability of input of energy into the radio jet, although on the arcsecond-scale of the large jet, where the length scales are larger than in the VLBI jet, it is likely that these variations are somewhat smoothed out if they correspond to flow variations.

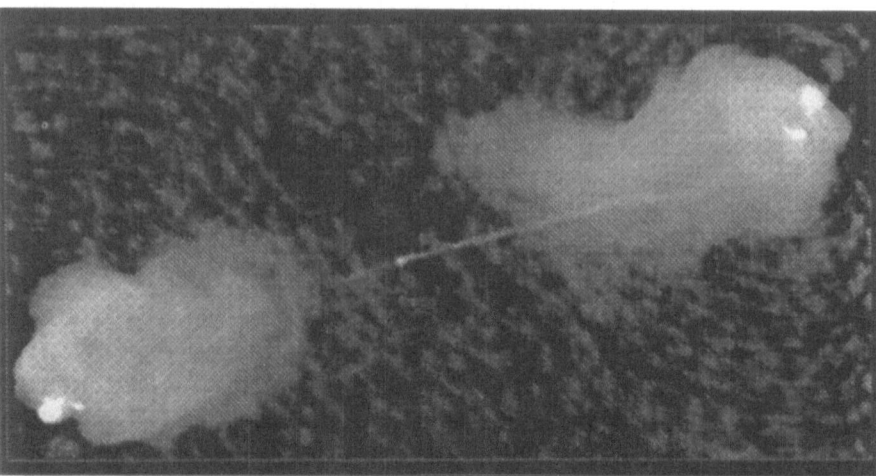

Figure 1. A 6-cm radio image of Cygnus A, from the VLA, with a logarithmic grey-scale. The dominant features of this image are the two bright radio lobes to the NW and SE of the radio core, with their embedded hot spots, but this high dynamic range image also shows linear features extending from the core towards the brightest parts of the lobes. These features are the jet (oriented to the NW) and counter-jet (to the SE). Figure supplied by Chris Carilli, CfA.

1.2.2. *Cygnus A (3C 405)*

The closest of the "classical double" radio sources is Cygnus A, 3C 405, with redshift $z = 0.0565$ (Figure 1.2.2). Its large scale structure is that of two bright, steep-spectrum, radio lobes, with prominent "hot spots" towards the outer edges of the lobes. Near the middle of the source lies a compact radio component with angular size < 1 arcsec, and a flat spectrum. With the excellent dynamic range that can be obtained with present-day imaging instruments, such as the VLA, and image processing techniques, the source can also be seen to posess faint linear features "jets" which form almost straight connections from the central component to the brightest parts of the source (Perley *et al.*, 1984b; Fig. 1.2.2). The NW jet is considerably brighter than the SE "counterjet". Both jets have length > 30 arcsec (50 kpc), so that they extend far from the center of the parent giant elliptical galaxy, and into the atmosphere of the poor(ish) cluster of galaxies that contains it (Arnaud *et al.*, 1984).

A close examination of the structure of the jets on Figure 1.2.2 shows that these are slightly misaligned, with the SE jet lying 7° to the south of the backward extrapolation of the NW jet. The NW jet shows clear internal structures. Where it is clearly separated from the lobe, it shows several bright knots, which are themselves elongated almost along the line of the jet, but in fact inclined at about 6° to this line, and which appear at regular intervals of about 10 arcsec. About 30 arcsec from the core, the NW jet seems to split into two, and bend significantly, before it becomes lost against the bright lobe emission. Multifrequency maps show that the jet has a spectral index of 0.85 ± 0.20 (Carilli *et al.*, 1991), but that the knot closest to the core may have a steeper spectrum, with spectral index ~ 1.4.

At very high (milliarcsec) angular resolution, a small jet source is also seen in the central component: this jet is closely aligned with the NW jet (Linfield, 1985). The core is variable—about 15 per cent variability is seen over time scales of 1 to 2 years (Carilli *et al.*, 1991). This variability is presumably related to the high-speed, $(0.4-1)c$, motions seen in the VLBI (Carilli *et al.*, 1994).

One particularly interesting set of results from the mapping of Cygnus A relates to the polarization of the radio emission in the lobes (Dreher *et al.*, 1987). The polarization pattern reflects a magnetic field that traces the edge of the source and the bright "loops" of emission at high frequency, suggesting that magnetic fields are being sheared at the edge of the source and define the loop structures. At lower frequency, the effect of Faraday rotation (inside or outside the source) is to cause the polarization angle of the radiation to change. The amplitude of this rotation as a function of frequency is measured by the rotation measure, which can be mapped across the source and which measures the product of density and parallel (to the line of sight) magnetic field in the Faraday screen. For Cygnus A, huge rotation measures and rotation measure gradients are found. These are interpreted as evidence for magnetic field structures and density structures in the gas surrounding the lobe. A particularly sharp rotation measure feature, seen near one of the hot-spots in the NW lobe (Carilli *et al.*, 1988), is interpreted in terms of compressed gas and/or magnetic fields in the medium ahead of a strong shock.

A final element of the polarization data is the depolarization induced by magnetic fields and gas within the lobe. Differential Faraday rotation between the front and the back of the lobe will reduce the polarization below the value expected from the synchrotron formulae. It is difficult to disentangle this effect from the beam depolarization caused by the large rotation measure gradients, but it appears that the internal electron density in the lobes is less than about 3×10^{-10} m^{-3} (Dreher *et al.*, 1987) if the lobes are interpreted as containing a conventional electron/proton plasma.

1.2.3. *NGC 6251*

NGC 6251 is a relatively nearby ($z = 0.0243$) and apparently unexceptional elliptical galaxy which came to prominence in 1977, when Waggett *et al.* mapped it in the radio. The source has an overall angular size of ~ 1 deg and a generally double structure, with a central region containing an exceptionally clear-cut radio jet more than 4.5 arcmin (190 kpc, if in the plane of the sky) long. This jet has been well mapped by Perley *et al.* (1984a), and a low-frequency map of it is shown in Fig. 1.2.3 (Birkinshaw *et al.*, 1995). The radio spectra of all parts of the jet are remarkably similar, so that this 330-MHz map looks almost exactly like a 5-GHz map with the same resolution.

Within the radio core Cohen & Readhead (1979) found a bright one-sided jet that is almost aligned with the large-scale structure and which has an angular size of a few milliarcsec (about 2 pc) and a very small radius. No sign of the counterjet is apparent on the VLBI image, although a counterjet with a few per cent of the brightness of the main jet has been detected (Perley *et al.*, 1984a).

Returning to the larger scale, NGC 6251 has been mapped in the X-ray (Birkinshaw & Worrall, 1991) and shows itself to be dominated by a bright small-scale source (with angular size less than 15 arcsec), which may be associated with the active nucleus of the galaxy—either as non-thermal emission or as the X-radiation from a cooling inflow. However, this bright emission is surrounded by a faint halo of low surface brightness, which is interpreted as the X-ray emission of 90-kpc scale group gas.

Detailed maps of the jet show a radio polarization pattern that reflects a principally longitudinal field in the inner part of the jet. Further out the edges of the jet show a longitudinal field, while the centre of the jet shows a perpendicular field (Perley *et al.*, 1984a). The polarization from which this field pattern was inferred is generally high, so that there is no evidence for internal thermal material (which would cause a differential Faraday rotation between the front and back of the jet, and cause the polarization to fall), but the position angle of the polarization was found to show signs of a foreground Faraday screen in the inner part of the galaxy in the large Rotation Measure variations that it displays. The gradients of rotation measure are presumably caused by a lumpy material (in either density or magnetic field, or both) which appears superimposed on the jet.

If the usual minimum energy arguments are applied in this case, then the jet internal pressures are found to be well above the external pressure that can be provided by the X-ray emitting gas. This is, presumably, saying something about the way in which a radio source can affect the state of the surrounding medium—it is likely that the gas in

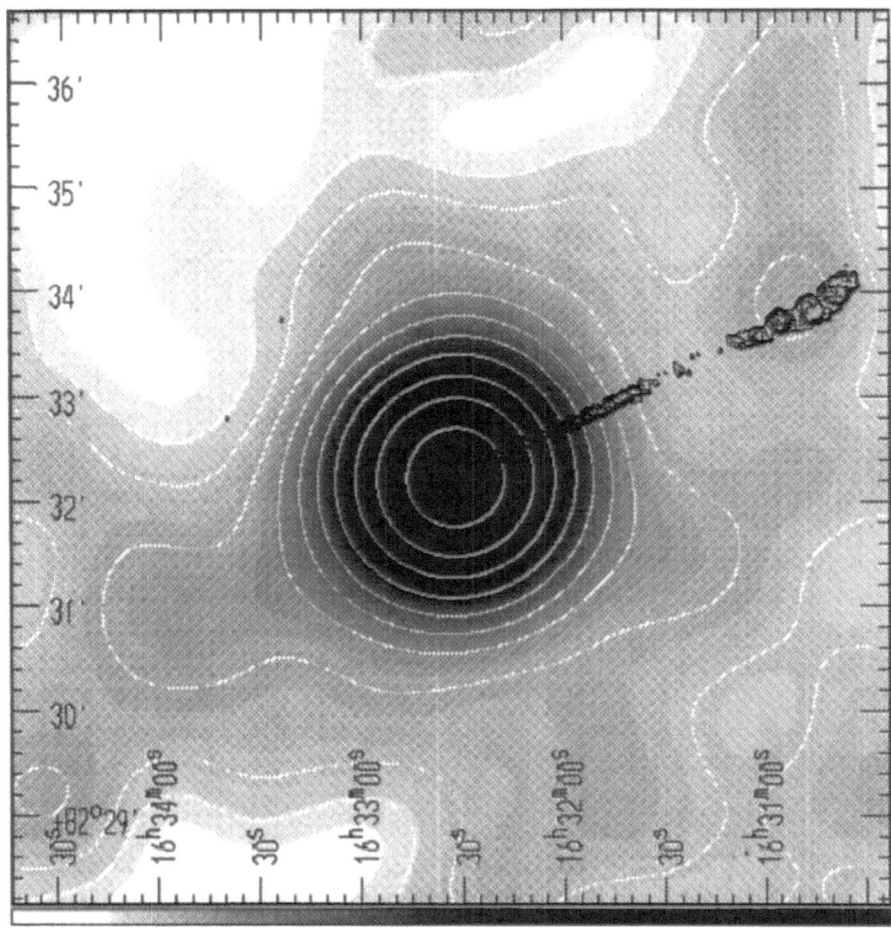

Figure 2. A contour plot of a 330-MHz radio image of NGC 6251, from the VLA (Birkinshaw *et al.* 1995) superimposed on the ROSAT PSPC 0.2-2.4 keV X-ray image (Birkinshaw & Worrall 1993).

direct contact with the jet has been shocked, and hence is at higher pressure than the general run of ambient gas. This also suggests that the radio source is non-static, since such a pressure imbalance implies that the source is changing structure on a sound crossing time in the ambient and jet gas.

1.2.4. *3C 449*

A 20-cm map of the inner part of 3C 449 appears in Perley *et al.* (1979), which is perhaps the paper that first demonstrated the effectiveness of the VLA for mapping radio jets. However, the source extends far beyond the well-mapped jet regions when mapped with higher surface brightness sensitivity, before terminating in low-surface brightness lobes. The inner jets show strong deflections, with the jet broadening significantly at about the location of the first major bends, and converting into plumes of emission that extend far beyond the optical galaxy. This is very unlike Cygnus A, for example, where the contrast between the jet brightness and the extended emission brightness is quite the other way round. Furthermore, 3C 449 shows nothing like the bright hot spots seen in Cygnus A.

These morphological differences between sources like 3C 449, which are fainter at the edge than in the centre, and Cygnus A, which is fainter at the centre than the edge, correspond well to radio power differences: 3C 449 has a radio power $P_{1\ \mathrm{GHz}} \sim 10^{25}$ W Hz^{-1}, whereas Cygnus A has a power $P_{1\ \mathrm{GHz}} \sim 10^{29}$ W Hz^{-1}. The morphology represented by 3C 449 is referred to as FR 1, and that represented by Cygnus A as FR 2, after the original description of the morphology/power correspondence by Fanaroff & Riley (1974).

1.2.5. *NGC 1265 (3C 83.1B)*

Lest you think that all radio jets are straight, consider the case of the FR 1 radio source 3C 83.1B, which is associated with NGC 1265 in the Perseus cluster (O'Dea & Owen, 1986). At low resolution the source appears to be a long trail of low surface-brightness radio emission, extending from the optical galaxy, which is located at the brightest part of the source. At higher resolution the brightest part of the source is seen to consist of two bright radio jets which emerge from a bright core roughly EW, and then curve to the north. Both jets are somewhat knotty, and end by expanding into large plumes that have the appearance of smoke trails seen from a factory chimney. The source is sometimes referred to as an example of a Narrow Angle Tail radio galaxy.

Very detailed mapping of the jets shows that they are brighter on their lower (convex) edges, and that the jets overlie lower-surface brightness emission with larger width but similar curvature. The centre lines of the jets "wiggle" from side to side about the smooth curvature that dominates. Polarization mapping suggests that the magnetic fields in the radio emitting plasma is predominantly aligned along the radio jets, at least in their inner parts (O'Dea & Owen, 1986). No strong spectral changes are seen over the jets, except that the knots are somewhat flatter than the extended emission.

1.2.6. *3C 273*

Extragalactic jets are seen in quasars as well as radio galaxies, with one of the best-studied being the quasar 3C 273 (Conway *et al.*, 1993). Although 3C 273 is somewhat special in showing both optical and X-ray jets in addition to the radio jet (Thomson *et al.*, 1993; Harris & Stern, 1987), the existence of the jet does make the point that there are some similarities in behaviour between radio galaxies and quasars. As seen in the Conway *et al.* radio map, the jet is around 80 kpc in projected size (its redshift $z = 0.158$). The general appearance of the jet is unlike the classical FR 1 or FR 2 jets and there is no detached hot spot of very high prominence. Instead the radio jet is of high surface brightness, which gradually increases towards the end. The core of the source is very bright relative to the jet.

Further information on the core is available from VLBI—3C 273 was the first source mapped with VLBI that showed definite evidence of superluminal expansion (at $v_{app} \approx 8c$; for $H_0 = 50$ km s^{-1} Mpc^{-1}; Pearson *et al.*, 1981), and hence relativistic flow on scales of a few pc.

1.2.7. *SS 433*

The star SS 433 was first reported as a bright Hα emission-line object. It is now realized that it is intensely interesting—it lies near the centre of the supernova remnant W 50, at a distance of 5 kpc, and displays prominent emission lines with a redshift which varies with a period of about 164 days. The amplitude of these variations corresponds to red- or blue-shifts of about 0.17 (see review by Margon 1984).

Such large redshifts force us to accept that relativistic ejection is taking place, and direct evidence for this ejection is available from radio mapping using the VLA and VLBI (Vermeulen *et al.*, 1993). Bright radio knots appear sporadically in the centre of the source, and move outwards at a speed of about 20 milliarcsec per day, brightening for about the first 10 days, and then fading in flux density. On the larger scale a continuation of this motion is seen (Hjellming & Johnston, 1988).

Both the motion of the individual radio brightness peaks and the periodic Doppler shifts are fitted well by a kinematic model in which blobs move out from the central source at a speed of $0.26c$. The direction of these ejections then precesses about an axis inclined at 79° to the line of sight, with the cone half-angle of the precession being about 20°. This causes the side-to-side wiggles in the locations of the blobs. The precession is driven by a binary companion, which has a 13 day period (seen in small Doppler shifts of the near-stationary line system).

The fact that a simple kinematic model for the radio peaks fits the structure on both the VLBI and VLA scales in SS 433 suggests that

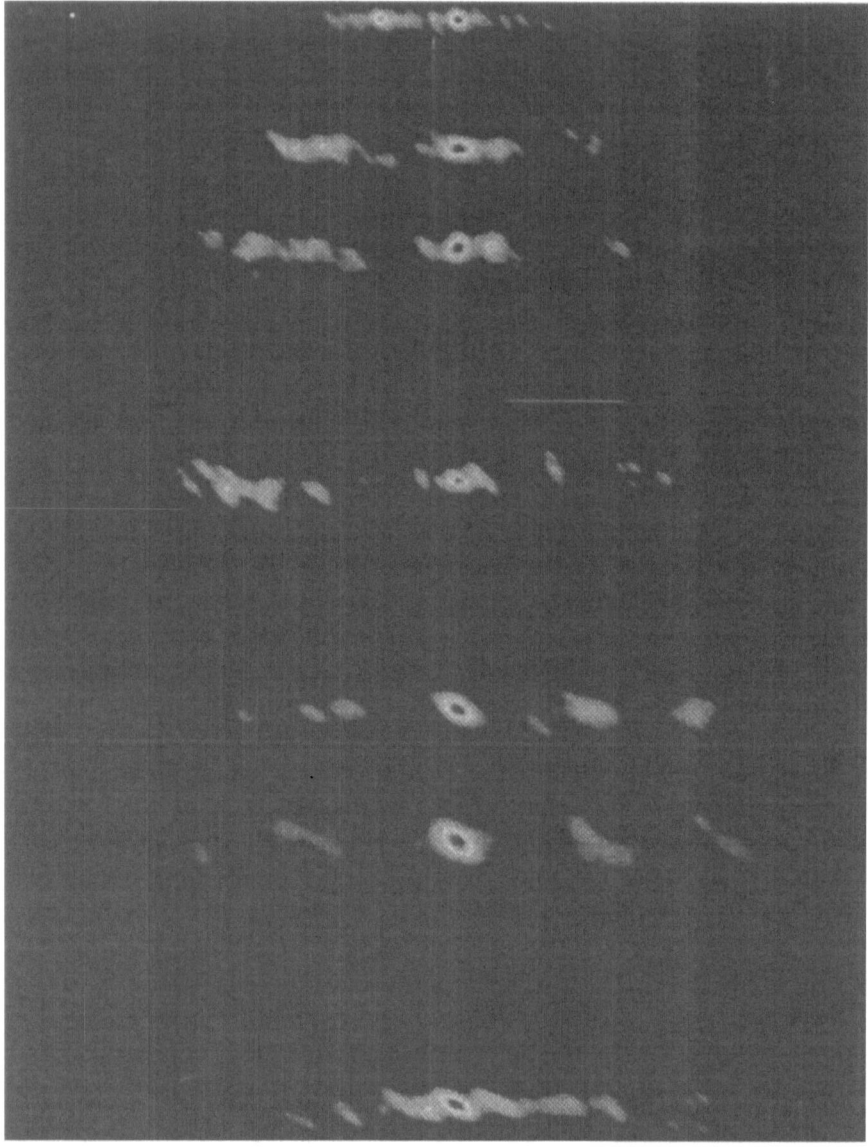

Figure 3. The object GRO J1655-40, as imaged by the VLBA between Aug 18 and Sep 21, 1994 (Hjellming& Rupen 1995). The pattern of knots on either side of the core appears to expand at an almost constant rate: of 54 milliarcsec/day (on the NE side of the core) and 45 milliarcsec/day (on the SW of the core). At the distance of this object (about 3 kpc), this corresponds to a velocity of about 0.9c.

the ejected material is dense, and little affected by the properties of the gas through which it flows (the hot gas in W 50). An X-ray image of W 50 (Seward *et al.*, 1980; Watson *et al.*, 1983) shows enhanced X-ray emission from the neighbourhood of the jets and where the jets should hit the edge of W 50.

1.2.8. *GRO J1655-40*

The Compton Gamma Ray Observatory's Burst and Transient Source Experiment detected the X-ray source GRO J1655-40 on 27 July 1994. It was quickly found to be a strong radio source. When mapped with the Very Long Baseline Array (VLBA), it was seen to be a long, thin, jet-like object (Fig. 1.2.6; Hjellming & Rupen, 1995) which varied in both structure and flux density over over a period of a few days. As the source was followed for about a month, it became clear that several episodes of knot ejection from the core were occurring, at the times that the total radio flux density of the source flared. Hjellming & Rupen found that these knots are moving away from the nucleus at about 0.05 arcsec per day: at a distance of 3 kpc for GRO J1655-40, derived from the HI absorption spectrum, this corresponds to an apparent speed of about $0.9c$.

The appearance of the jet at different times (Fig. 1.2.6) clearly shows its asymmetry about the core of the source. It appears that individual ejection episodes cause bright blobs of radio-emitting plasma to appear near the core, and then to move away from the core at almost constant speed, suggesting a ballistic motion rather than conventional fluid flow. Thus GRO J1655-40 appears to be an extreme example of objects like SS 433: faster, but similarly shooting out dense blobs. The optical object is bright enough that detailed spectroscopy is possible: so far it has been shown that the parent object is a binary with about a 2.6 day period, and a mass function $\approx 3M_\odot$ (Bailyn *et al.*, 1995a; Bailyn *et al.*, 1995b). A possible eclipse seen during optical monitoring (Bailyn *et al.*, 1995a) may cause any moving line system associated with the jet to be hard to see.

1.2.9. *HH 34*

It is not only energetic non-thermal objects that produce interesting jets. Star-forming regions also display jets. A well-known example is the HH 34 system. Herbig-Haro objects are bright, fuzzy, optical objects that appear on the dark backgrounds of molecular gas. They are thought to be a consequence of low-mass star formation, and frequently show high proper motions (a few hundred $km\,s^{-1}$ being not unusual). HH 34 is a particularly spectacular optical jet (Reipurth *et*

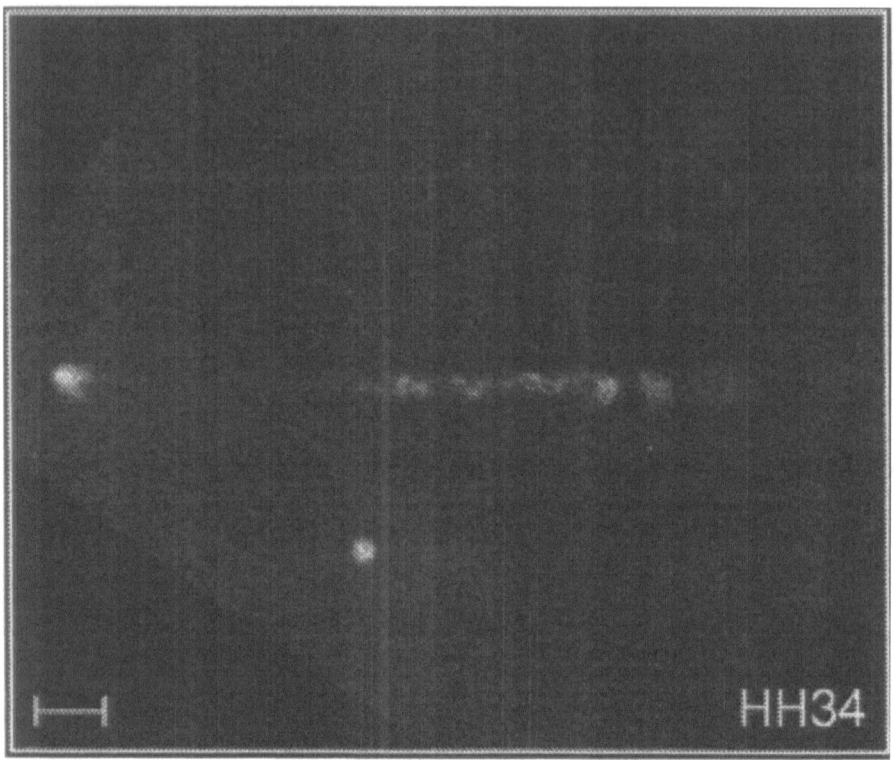

Figure 4. The object HH34, as seen by the HST WFPC-2 camera (Hester *et al.*
1995). A regular series of knots is ejected from the originating stellar source to the
left (North) of the image. The scale bar corresponds to 10^3 AU.

al., 1986), which points back towards a star which is thought to be its
source.

A recent HST image of HH 34 is shown in Fig. 1.2.9 (Hester *et
al.*, 1995): we see a remarkably regular structure with distinct blobs of
emission separated by a few hundred AU. Like the non-thermal sources
SS 433 and GRO J1655-40, these blobs appear to be the results of
distinct ejection episodes, "bullets" of high-speed gas that are shot out
from the star. It is possible that there is a counter-jet of bullets on
the other side of the star too, but these might be hidden by the dust
and gas in which the jet is embedded. The absence of a counterjet is a
little unusual, though, since the visual absorption of the stellar source
is only about 5 mag (Reipurth *et al.*, 1986), and it requires a somewhat
special circumstances for one side of the jet to be exactly pointing out
of the cloud while the other points into it.

The optical spectra of knots in the jet show emission lines that are produced by shocked gas and little or no continuum radiation. The knots are moving with a radial velocity of about -60 km s^{-1} relative to the parent star: once again we see evidence for motion, with a speed which is likely to be several times the sound speed in the medium in which it is embedded. The "head" of the flow, located well beyond the jet imaged by HST, is also moving rapidly relative to the core—and shows several distinct knots with significantly different radial velocities. Studies of the spectrum also suggest that the material emitting the line radiation has a density which is somewhat lower than that of the molecular gas near the jet: of course, this doesn't necessarily say anything about the non-radiating material, nor whether the jet is embedded in a locally lower-density medium.

In this case there is a second indication of motion, from CO observations which show the presence of a "bipolar flow" centered on the HH 34 jet (Chernin & Masson, 1995). Chernin & Masson's map of high-velocity molecular gas shows blueshifted material on the side where the jet lies and redshifted emission on the opposite side. The CO data can be used to estimate the mass and momentum in high-speed gas that has been generated by the star: it is found that roughly 0.003 $M_\odot$ is involved, and the momentum of the flow is about $0.006 M_\odot$ km s^{-1}. The jet is expected to be at least 10^4 years old (Bally & Devine, 1994), and it is plausible that the molecular outflow is material that has been accelerated by a beam of material from the star.

1.2.10. *L 1551*

Finally, another example of a molecular flow that has been associated with a jet, the L 1551 outflow. This was the first bipolar molecular outflow found, and remains one of the best studied. It is associated with the young stellar object IRS 5 and several Herbig-Haro objects. On the largest scale, molecular line observations have found redshifted and blueshifted high-velocity gas distributed in distinct lobes, each about 15 arcmin (0.7 pc) in size, lying on either side of IRS 5. The general structure of the source consists of two well-separated lobes of high-velocity gas seen in the ^{12}CO $J = 1 - 0$ line (Snell & Schloerb, 1985). At high resolution the brightest parts of the SW (blueshifted) lobe break up into a number of bright filaments that may represent hydrodynamical shocks (Barsony *et al.*, 1993).

Snell & Schloerb estimate the total mass of high-velocity gas to be about 0.9 $M_\odot$, and believe that it has been swept up by the expanding cavity that is defined by the lobes. The corresponding momentum is about 16 $M_\odot$ km s^{-1} (Moriarty-Schieven & Snell, 1988).

The entire system appears to be being driven by IRS 5 which lies between the lobes, and which has a jet-line radio and optical appearance (Mundt & Fried, 1983; Rodriguez et al., 1986; Fridlund & Liseau, 1994). Over a period of years, the knots in the IRS 5 optical structure have been seen to move apart, with an average tangiential velocity of about 180 km s^{-1}, while optical spectroscopy (Stocke et al., 1988) reveals radial velocities between -80 and -290 km s^{-1} for the knots. Not all the knots are travelling in the same direction: the spread in directions is $\sim 20°$. There is no clear pattern in the directions of motion: it is possible that the knot ejections have not been stable, although there is a pattern of one knot being ejected every 20 years (Fridlund & Liseau, 1994). The outflow momentum in the jets, calculated using knot masses estimated from their optical spectral lines, is much less than the momentum in the high-velocity molecular gas, which suggests that IRS 5 shows intermittent bursts of very high activity. Perhaps the jet seen today is the remnant of the last burst, or the precursor to the next.

1.3. THE BEAM MODEL

So what are the characteristics of the beam model? In its simplest form, the idea is that the wide variety of jet structures that are seen can be attributed to the interactions of a collimated flow of gas (and magnetic fields) with the ambient gas through which it moves. A cartoon of a simple beam model, identifying the important segments of the flow, is shown in Fig. 1.3.

The initial injection of the beam is a difficult problem, for which there are a wide variety of models ranging from largely electromagnetic effects to largely gas-dynamical effects, such as the suggestion of a de Laval nozzle by Blandford & Rees (1974). In the context of the large-scale structures of jets that we are dealing with here, I shall ignore the important question of how a high-speed collimated flow is produced and simply examine how it propagates. Thus the beginning of the beam model for us will be where the flow exits some nozzle and comes into contact with the surrounding gas.

Assuming that the beam is supersonic, it will drive a shock into the ambient gas in front of the head of the beam. This bow shock compresses, heats, and accelerates the ambient medium. Material flowing down the beam leaves the collimated flow through a shock at the head of the beam, where it becomes much hotter (and hence highly over-pressured) and encounters the shocked ambient medium. In the rest-frame of the beam head, a circulation pattern is set up where the beam fluid expands into a backflow parallel to the beam. The inter-

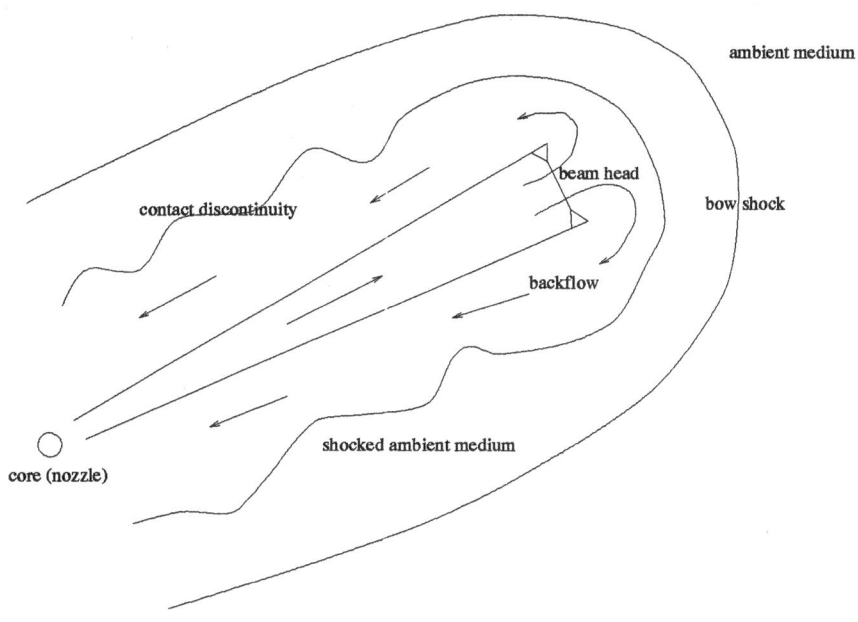

Figure 5. A cartoon of a supersonic beam model, showing the most important features of the flow.

face between the rapidly-backflowing beam material and the shocked ambient medium is the contact discontinuity.

Since the velocities of the shocked ambient gas and the backflowing beam material are very different, there is the potential for creating strong Kelvin-Helmholtz instabilities at the contact discontinuity. The contact discontinuity may even be erased entirely. The interface between the beam and the backflowing gas provides another site where this instability might be active: and it was long thought that this instability would have a severe effect on the properties of the jet, destroying the collimated flow and hence preventing the creation of well-developed structures.

If the beam material can radiate energy efficiently, then compressions provide sites where substantial energy loss can take place and the flow can be severely modified (as it is expected to be in galactic jets). In extragalactic jets, where the energy loss is expected to be less significant, the beam head shock serves as a site at which particle acceleration can take place (Drury, 1983; Blandford & Eichler, 1987), and

hence where synchrotron radiation might be possible. The beam head
is therefore identified with the hot spots in FR 2 radio galaxies.

Further shocks can be created in the beams themselves, either as
they pass through the structured cocoon that they have created, or
because the injection mechanism is not completely steady. The latter
suggestion has been used in the case of M 87, for example, to explain the
prominent knots, and it is clear from some of the galactic jets (such as
SS 433 and HH 34) that the unsteady injection of material is common
on the stellar scale.

How about the FR 1 galaxies, which have no hot-spots? Here we
can modify the previous picture by suggesting that the flows are not
strongly supersonic, but transsonic or, perhaps subsonic. Then there is
no strong shock at the head of the beam, and it is likely that the flow
will be less well collimated and more likely to be distorted by disturbances in the ambient medium. In this context we can see a possible
explanation for the very distorted jet in NGC 1265: it is likely that a
beam emerging from the nucleus of this galaxy is being swept back by
the ram-pressure of the ambient medium through which the galaxy is
moving (Begelman *et al.*, 1994). NGC 1265 is a particularly good case
for this explanation: it is located on the edge of the Perseus cluster,
and has an anomalously high radial velocity relative to the systemic
velocity of the cluster.

And what about the magnetic fields? If the fields are strong, then
it is possible that they might dominate the flow: perhaps helping to
collimate the beam (if circumferential), perhaps "guiding" the beam
(if axial). If the fields are weak, then they will be dragged around by
the flow, being sheared in some regions and tangled in others. But
understanding the field structures is difficult in the abstract, and it is
better to turn to full calculation to extract the details of the flow in
particular cases.

1.4. NUMERICAL MODELS

Perhaps the most convincing evidence that we can understand the
details of jets structure by a flow model is to examine numerical simulations of these flows. These simulations seek to solve the equations
of gas-dynamics (or magneto-gas-dynamics, in the more adventurous
codes) using computer models. Two cases are generally considered:

- adiabatic flows, where radiation from the beam is not significant
  in the evolution of the flow, and

- radiative flows, where the beam loses energy rapidly, through some
  radiation process.

Adiabatic flows are generally used to model extragalactic jets, since the physics involving the acceleration and transport of the relativistic particles that radiate energy is difficult, and it is expected that the observed radiation is only a small fraction of the kinetic luminosity of the beam. Radiative flows are used for the jets seen in star-forming regions, where the integrated radiation is a major coolant for the flow.

1.4.1. *Hypersonic, adiabatic flow*

Recent work by Loken *et al.* (1992) and Cioffi & Blondin (1992) considered the flow of a beam with very high speed and low density in a uniform ambient medium. Loken *et al.*'s 2D calculation involved a beam with density 0.01 of the ambient density, and a velocity 4650 times the sound speed in the ambient medium. The resulting flow pattern consists of a coherent high-speed beam, with a narrow forward-flowing cocoon next to the beam, a wider backward-flowing cocoon further out, then a further forward-flowing cocoon layer which is in contact with ambient gas which has passed through the bow shock. At the head, the beam material flows through a Mach disk into the cocoon, where it expands into a very low density medium which contains a complicated pattern of eddies. The beam itself carries significant density and pressure oscillations, and the beam head advances at a variable speed as the area of the head of the beam varies.

What does this tell us? It says that very fast, low density, flows can remain coherent as they propagate very many jet radii (more than 100 in this simulation), and that the cocoon that they inflate around themselves can be strongly over-pressured relative to the external medium. This is of interest because the apparent pressures within many jets (e.g., NGC 6251) are higher than the pressures in the X-ray emitting gas through which they move. The simulation by Loken *et al.* assures us that it is not the original pressure of the ambient gas that keeps the beam from expanding dramatically, but the pressure of the self-generated cocoon of shock-heated gas, which is itself expanding into the intergalactic medium.

1.4.2. *Supersonic, adiabatic flows*

A typical ambient medium is intracluster gas around a radio galaxy, with a sound speed ≈ 1000 km s^{-1}, so that the calculation by Loken *et al.* (1992) was certainly of a relativistic beam that cannot consistently be treated by non-relativistic gasdynamics. However, it is generally agreed that the existence of compact hot-spots in FR II sources is probably due to the presence of a strong shock at the end of a supersonic jet, so this is a particularly interesting case to consider.

Many authors have discussed the propagation of supersonic jets (e.g., Norman *et al.* 1982; Williams & Gull 1985; Hardee & Clarke 1992; see the review by Williams 1991). As indicated by the cartoon of Fig. 1.3, the jets inflate a substantial cocoon about themselves, with a very low density backflowing material in contact with the beam. Strong waves can be seen in the contact discontinuity: these are driven by oscillations in the development of the beam head, but are likely to be maintained and amplified by the Kelvin-Helmoltz instability and (especially in 2-D simulations) by vortical flow at the injection plane and the beam-head.

1.4.3. *Subsonic, adiabatic flows*
As the speed of the flow drops, so the width of the insulating cocoon around the jet decreases. Williams (1991) shows a representative calculation in this case. The principal difference from the supersonic, adiabatic flows is that the beam is now in direct contact with the ambient gas. There is no bow shock, and the ambient material can flow fairly smoothly into contact with the beam around the plume of material that develops at the head of the flow. The pressure enhancement at the head is also small, so that there is no particular reason to expect a strong hot-spot. This type of flow may be a good model for FR 1 radio galaxies—the slower speed also makes them susceptible to environmental disturbances, which may help to bend them into some of the shapes that are seen.

1.4.4. *Supersonic, radiative flow*
The effect that radiative losses can have on a supersonic jet flow is well illustrated by the simulations of Blondin *et al.* (1990) and Gouveia Dal Pino & Benz (1993). As might be expected, compressions of the gas (especially in the beam) cause a rapid energy loss through radiation, so that the beam tends to be strongly compressed by the ambient medium, and a cold, dense shell of material tends to form at the head of the beam (Fig. 1.4.4). This shell of material is very unstable, and breaks into distinct clumps and (for some beams) a dense plug of material that is advected with the flow. The dispersing dense shell bears some resemblance to the complicated patterns of knots seen in Herbig-Haro objects, and displays something of the wide range of velocities that these knots show.

1.4.5. *More complicated flows*
Most of the simulations in the past have considered the propagation of simple gas-dynamical beams through a uniform medium. With the continuing improvements in computer power, many more complicated

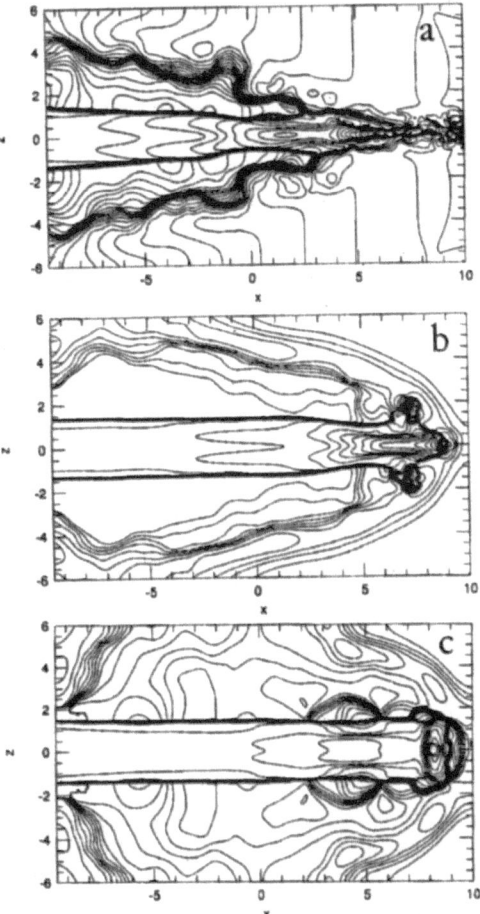

Figure 6. The density structure of a radiative beam with initial density 3 times greater than the density of the ambient medium and Mach number $\mathcal{M}_a = 24$ after propagating into an ambient medium with increasing external density (a), constant external density (b), and decreasing external density (c). From Gouveia Dal Pino *et al.*, (1995).

calculations are now being done. For example, Hardee *et al.* (1992) considered the propagation of a slab jet through a medium with a temperature and pressure structure: Gouveia Dal Pino *et al.* (1995) have redone this calculation with a radiative flow in three dimensions (Fig. 1.4.4), finding a number of structural effects that are amplified by the presence of pressure gradients. There have also been a number of calculations of the effect on a beam of crossing a sudden jump in density or pressure, such as may be encountered as a beam crosses a shock front between the interstellar and intergalactic medium (e.g., Wiita *et al.* 1990, Wiita & Norman 1992). Balsara & Norman (1992) have repeated the analysis by Williams & Gull (1984) of the ram-pressure distortion of NGC 1265's jet, elucidating many of the structures in the deformed jet. De Young (1991) has studied cloud-jet interactions, finding dramatic structural changes which may be relevant to the wide angle tail radio galaxies (such as 3C 465; Leahy 1984) or double hot-spot objects. There has also been progress on flows containing dynamically-significant magnetic fields (e.g., Clarke *et al.* 1986b), although many of the resulting flows look rather unlike most radio jets.

2. Jet parameters

In Section 1 some of the general observational properties of (principally radio) jets were discussed. In the present Section, I will describe the essence of the methods that have been used to extract the properties of the underlying flow from these data. In other words, I am going to attempt to move from the "jet" to the "beam". I will concentrate on the extragalactic jets, for which the emission has a synchrotron origin, although the non-thermal galactic jets (associated with SS 433 and GRO 1655-40, for example) may be understood in the same way.

The properties of thermal galactic jets are studied principally through their line spectra, although thermal continuum is also seen. Line diagnostics of the density and temperature of the line-emitting region, and velocity information from proper motions and line shapes and centres, provide an extensive set of data for the galactic jets which helps to set their parameters more clearly than is possible for extragalactic jets.

2.1. JET VISUALIZATION

What are we seeing in the views of extragalactic jets. The high brightness of the radiation, and the polarization, imply that we see the jets through their radio synchrotron radiation. The jets may account for either a small (as in Cyg A) or a large (as in 3C 449) part of the overall energy radiated from the sources.

How is the radiation pattern that we see related to the material that is emitting? In synchrotron jets, the radiation is coming from ultrarelativistic electrons, with energy $E_e \gg m_e c^2$, moving in a magnetic field, of strength B. The effect of the field is to cause the electrons to circulate with frequency ν_g/γ_e, where $\gamma_e = E_e/m_e c^2$ and ν_g is the electron gyrofrequency

$$\nu_g = \frac{e\,B}{2\pi\,m_e}.$$

(1)

The total power radiated by an electron is

$$P_{\text{tot}} = 2\,\sigma_{\text{T}}\,c\,\gamma_e^2\,u_{\text{B}}\,\sin^2\psi \tag{2}$$

where σ_{T} is the Thomson cross-section, $u_{\text{B}} = B^2/2\mu_0$ is the magnetic energy density, and ψ is the pitch angle of the particle's motion about the magnetic field line on which it moves. The energy at which this radiation is produced lies in a fairly narrow band (compared to the wide frequency range of synchrotron radiation) about the critical frequency

$$\nu_{\text{crit}} = \frac{3}{2}\gamma_e^2\,\nu_g\,\sin\psi \tag{3}$$

where the leading factor of γ_e^2 arises because the radiation is beamed into the direction of motion of the radiating electron (as seen in the observer's frame), and because the motion of the electron about the field line causes this beamed radiation pattern to be pulsed into the line of sight of the observer only for a short fraction of each gyration. We can also estimate the radiative lifetime of an electron, the time taken to radiate a large fraction of its energy, $\tau_e \approx E/P_{\text{tot}}$.

If we now consider an ensemble of electrons, with an isotropic distribution of pitch angles (as is expected in radio jets), and a number per unit volume described by a power-law energy distribution

$$n(E)dE = n_{\text{E}}\,E^{-\gamma}dE \tag{4}$$

with $E_{\text{L}} < E < E_{\text{U}}$, the averaged emission coefficient of the population of electrons is

$$j_\nu = \Lambda_j\,N_e\,(B\,\sin\theta)^{\alpha+1}\left(\frac{\nu}{c_\nu}\right)^{-\alpha} \tag{5}$$

where Λ_1 is a pure number which depends on α, θ is the angle between the magnetic field vector $\mathbf{B}$, and the line of sight, $c_\nu = 3e/4\pi m_e^3 c^4$, and the spectral index,

$$\alpha = \frac{\gamma - 1}{2}. \tag{6}$$

(5) applies only in the range of frequencies for which the radiating electrons have energies $E_{\text{L}} \ll E \ll E_{\text{U}}$. Above and below this band, the spectrum turns down. Examples of broad-band spectra with synchrotron fits are shown in Fig. 2.1.

If we assume that the radio jets that we observe are optically thin (as is usually the case), then the radio brightness at the surface of the jet is

$$I_\nu = \int_{-\infty}^{\infty} j_\nu\,ds. \tag{7}$$

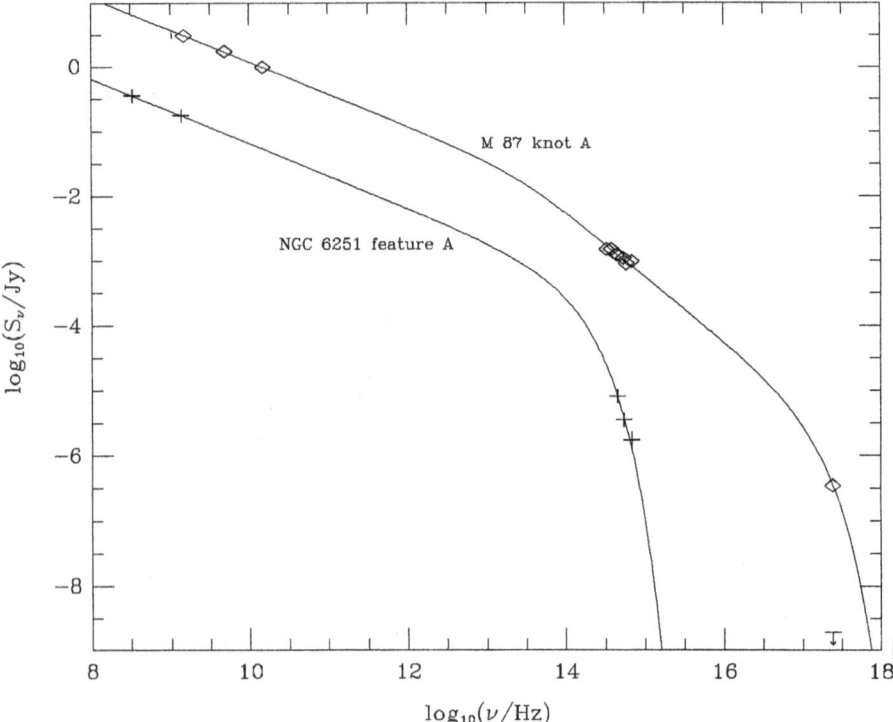

Figure 7. The radio to X-ray spectra of feature A in NGC 6251 (crosses and upper limit) and knot A in M 87 (diamonds), from Birkinshaw & Worrall (1991). For most of these points the errors are smaller than the sizes of the symbols. The solid lines represent rough fits to these spectra from the synchrotron emission of optically-thin model electron energy distribution functions. The fit for NGC 6251 feature A is based on a simple power-law electron energy distribution that is cut off at energies $E > E_U$, while the model distribution for M 87 knot A also posesses a break of unity in the electron energy index (and hence a break of 0.5 in the photon spectrum) at intermediate energy $0.02E_U$, which may be related to electron ageing, although the location and size of the break are poorly constrained. Neither spectrum shows any sign of a turn-down at low energies due to a lower energy cutoff of the electron spectrum.

so that the observable flux density from a radio jet is determined by the mix of magnetic field strengths, magnetic field orientations, and electron spectra along the line of sight. Thus variations in the brightness of a radio jet may reflect variations in the number of relativistic electrons per unit volume, variations in the magnetic field strength or its direction, changes in the fraction of the jet width that is occupied by radiating plasma, or all of these. Unique inversion of synchrotron data to obtain an accurate internal structure for a jet will clearly be difficult!

Some additional clues about the structure in the magnetic field within a jet can be gained through studies of the polarization. Within the frequency range that the spectrum appears power-law (i.e., the optically-thin regime between the frequency limits), the fractional linear polarization of the emission is

$$\Pi_{\mathrm{L}} = \frac{\alpha + 1}{\alpha + \frac{5}{3}} \tag{8}$$

which for sources with typical spectral indices, $\alpha \approx 0.5$, leads to a polarization of about 70 per cent. If the magnetic field is structured on the line of sight through a synchrotron plasma, the polarization can be dramatically reduced. If magnetic field lines viewed along the line of sight have a random orientation about that line of sight, the polarization will drop to zero (provided that the relativistic electron population in each region of magnetic field is the same). If the magnetic field lies in a single direction, but with many reversals, then the polarization will be at the value given by (8), since the polarization is independent of field *sense*. Laing (1980) and Hughes (1985) give more details on this *structural depolarization*.

In view of this effect, the large fractional linear polarizations seen in many jets (e.g., in M 87) indicate that the magnetic fields in the regions that are radiating are at least partially ordered and that there are not many random cells of magnetic field along the line of sight. Of course, this only relates to the regions where the relativistic electrons are concentrated: if the electrons can diffuse freely across the field lines, then this is a strong statement about the magnetic field structure over the entire jet. If the electrons cannot diffuse freely, or have radiative lifetimes which are small compared with the time taken to diffuse across the jet, then it is possible that we see only a small sample of the magnetic field structure of a jet in each polarization measurement.

The apparent polarization structure of a radio jet is also affected by Faraday rotation and Faraday depolarization. When a linearly polarized radio wave travels through a medium containing non-relativistic electrons and magnetic field, the fractional polarization remains unaltered, but the plane of polarization rotates by an angle

$$\chi = \mathcal{R} \lambda^2 \tag{9}$$

where the rotation measure

$$\mathcal{R} \propto n_{\mathrm{e,th}} B_{\parallel} \tag{10}$$

with $n_{\mathrm{e,th}}$ being the thermal electron density and $B_{\parallel}$ the component of the magnetic field associated with this thermal electron population and parallel to the line of sight. This means that if a radio jet, with a high fractional polarization, is located behind a screen containing magnetic field and electrons, the apparent position angle of the polarized flux will change as the wavelength of observation changes. At very short wavelengths, χ will be small and the plane of polarization indicates the intrinsic electric vector of radio waves emitted in the source. The magnetic field in the source is then 90° from the direction of the electric vector. Polarization maps in the literature are usually corrected for Faraday rotation as measured from multi-wavelength data and corrected to zero wavelength using (9).

More insidious is the possibility that thermal electrons are mixed with relativistic electrons in the radio source itself. Then the phenomenon of Faraday depolarization can arise: radiation from the rear of the jet and from the front of the jet pass through different amounts of thermal material, and therefore show different Faraday rotations. This means that the apparent degree of polarization of the radiation is reduced below the expected fraction (8). The wavelength-dependence of the effect can then produce a variety of patterns of variation in the apparent linear polarization fraction as a function of frequency (e.g., Cioffi & Jones 1980; Laing 1984). Relativistic electrons in the beam plasma are much less effective at producing Faraday rotation than thermal electrons (Jones & O'Dell, 1977), and if the beam fluid is composed of electrons and positrons only, there is no Faraday rotation since the particles respond to circularly-polarized radiation in equal but opposite ways.

A final, major, complication to the use of the observed synchrotron radiation as an indicator of the underlying flow is the question of what relation the emission that we see has to the flow itself. Is the plasma that is *radiating* the same plasma that is *carrying the energy and momentum* in the beam, or even a constant fraction of it? In other words, are the large brightness variations seen in radio jets (such as the appearance of bright knots in the M 87 jet) produced by gross changes in the flow, or merely small changes in a trace population of relativistic electrons?

The answer to this question lies in the still unknown relationship between the properties of the bulk flow and the acceleration of relativistic particles within the beam. It is very likely that the electrons (and, perhaps, positrons) that are emitting synchrotron radiation are accelerated within the beam, and close to the regions in which the are observed, since their radiative lifetimes τ_e are often shorter than the light travel time from the core of the source to the point at which they are detected, at least at high radio frequencies and in bright parts of jets. Thus the radio brightness is something close to a snapshot of the particle acceleration, and is, perhaps, showing us where in the flow the

Table I. The ambient medium

quantity	symbol	typical value		unit
		extragalactic	galactic	
sound speed	c_{Sa}	$300 - 1000$	$1 - 20$	$\mathrm{km\,s^{-1}}$
density	ρ_a	$10^{-25} - 10^{-22}$	$10^{-21} - 10^{-15}$	$\mathrm{kg\,m^{-3}}$
magnetic field	B_a	$0 - 1000$	$0 - 100$	nT
temperature	T_a	$10^6 - 10^8$	$10 - 10^4$	K
gas pressure	P_a	$10^{-15} - 10^{-10}$	$10^{-16} - 10^{-9}$	Pa
polytropic index	Γ_a	$\frac{5}{3}$	$\frac{7}{5}$ $\frac{5}{3}$	

strong shocks lie, since these shocks are a likely site for accelerations (e.g., Blandford & Eichler 1987; Kirk 1994).

Similar considerations are important for the galactic jets. Much of the line emission that is seen arises from shocked gas, so that the material that is radiating may be only a thin layer around the material that carries most of the momentum.

2.2. THE AMBIENT MEDIUM

In extragalactic jets, the advent of a new generation of X-ray satellites has allowed us to get a good view of the medium around the jets. Thus, for example, near NGC 4261 the X-ray emission as seen by ROSAT is shown in Fig. 2.2. This emission consists of three components: an X-ray core (which is either non-thermal emission associated with the active nucleus of this galaxy, or thermal emission from a small-scale galactic atmosphere with high density), a small X-ray halo (Worrall & Birkinshaw 1994), and a group-gas halo with scale ~ 25 kpc (Davis *et al.* 1995).

The X-ray data provide the temperatures of the extended gas. The density distribution of the gas near a radio source can be found either by fitting the image (usually to a symmetrical function; e.g., Worrall & Birkinshaw 1994) or by attempting a detailed deprojection. As a result, the distributions of pressure, density, and other gas properties in the ambient medium near the beam flow can be extracted reasonably well, although the data are rarely good enough for detailed temperature profiles to be measured. Table 2.2 presents the range of temperatures and densities that are appropriate for large-scale jets, based on the range of gas properties of clusters and groups of galaxies given in the reviews of Forman & Jones (1982) or Sarazin (1986).

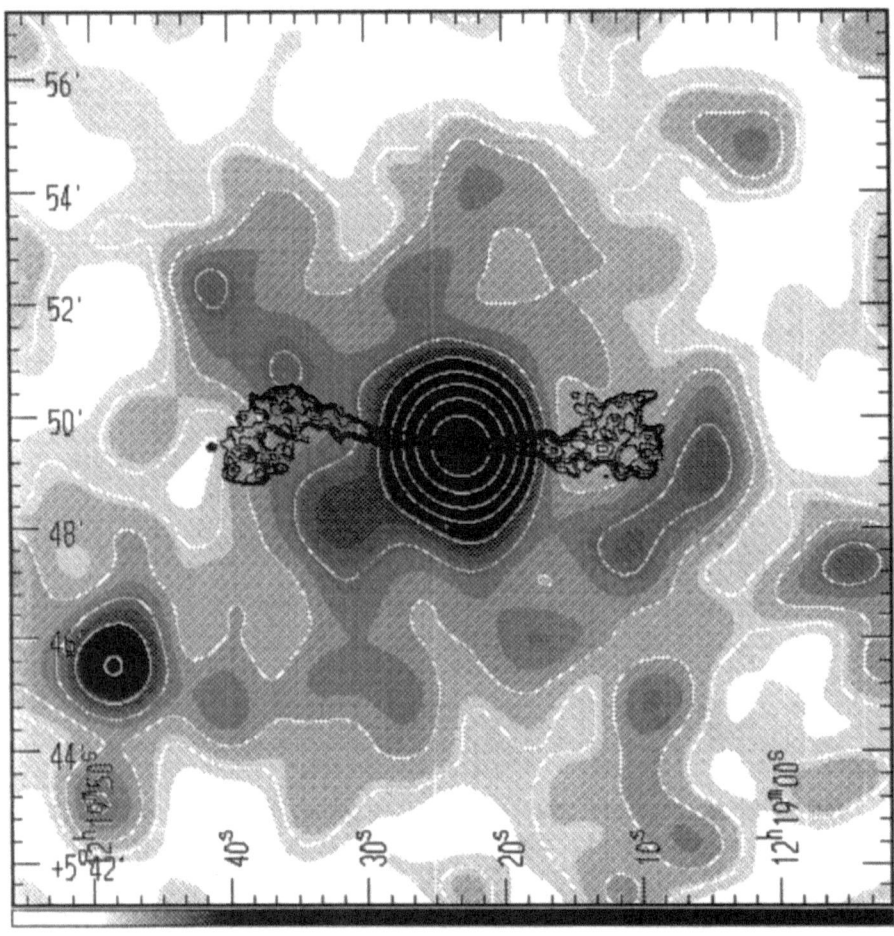

Figure 8. A soft X-ray image of NGC 4261 made using the *ROSAT* PSPC. About half of the counts from the core of the galaxy lie in an unresolved component associated with the radio core of NGC 4261. The remainder lie in a halo with an angular scale ~ 25 arcsec (4 kpc). In addition, there is clear evidence for larger-scale emission associated with the NGC 4261 group (Davis *et al.* 1995), which affects the dynamics of the outer parts of the radio source (3C 270).

The parameters recorded in Table 2.2 should be extended to lower temperatures ($T_a \sim 10^4$ K) and higher densities ($\rho_a \sim 10^{-12}$ kg m^{-3}) if the jets in the broad–line and narrow–line emission regions in the active nuclei are to be discussed. However, here the medium is likely to be very clumpy, and strongly affected by the presence of a high-power nuclear source that is emitting ionizing radiation. This makes it is difficult to make general statements about the environments of VLBI jets: each jet should be considered independently.

Similar remarks can be made about the environments of galactic jets. Here the properties of the gas can be found from the CO line emission from the molecular material in which the jet is embedded (or the HI line, if the jet is in a warmer environment; or the optical or X-ray emission in a few special cases, like SS 433). An example is L 1551, where the density in the cavity or the density in the surrounding molecular cloud can be obtained from the CO data. The inferred temperature of the ambient medium in which the jet lies is generally low (less than 30 K), but the densities may be high (for L 1551, $\sim 3 \times 10^9$ molecules m^{-3} before the cavity inflated, according to Snell & Schloerb 1985).

Just as for VLBI jets, the ambient medium in galactic jets is undoubtedly very clumpy. The widths of molecular lines from the surrounding molecular gas usually exceed the thermal line-widths predicted from the temperature (deduced from line strengths). This suggests that the surrounding gas is made up of denser cloudlets moving supersonically in a lower-density environment. Thus there is a significant probability that a beam in such an enviroment will hit a cloudlet, possibly with dramatic effects on the flow. Note also that galactic jets like SS 433, which is embedded in a supernova remnant, may be in a medium more like an extragalactic than a galactic jet.

In either the galactic or extragalactic cases the ambient medium is rarely uniform on the large scale. Thus, for example, the NGC 4261 jet has a size ~ 3 arcmin (35 kpc), comparable with the scale of the atmosphere in which it finds itself (Fig. 2.2). The external pressure therefore changes significantly on the scale of the jet. Thus the jets are never in a simple cylindrically-symmetrical flow, but are continually adjusting themselves to the properties of the medium that they flow through. This continuing adjustment may contribute to exciting the plasma that is flowing: perhaps helping it to generate the relativistic particles that cause the radio emission, or the collisionally-excited gas that we see from HH object complexes. It also suggests that a jet is never really in a steady state: beams are likely to change structure significantly on the timescales of sound crossings, and this may cause the jets that we see in our maps to be poor approximations to the mean flows that we want to understand.

In extragalactic jets the magnetic environments are now becoming better known through detailed rotation measure mapping. For example, the magneto–ionic halo around M 84 has been studied by Laing & Bridle (1987), who find magnetic fields $B_a \sim 0.1$ nT which are ordered on scales of a few kpc, and Kim $et\ al.$ (1986) have inferred large-scale fields of a similar strength in the Coma cluster. Larger fields are inferred from the enormous rotation measures seen in some sources (e.g., Cygnus A; Dreher *et al.* 1987). Even stronger external fields are necessary if jets are magnetically confined, and radial gradients of azimuthal field intensity are then expected: the upper limit to B_a in Table 2.2 reflects this possi bility. The evidence for magnetic fields in the medium around galactic jets comes from measurements of Zeeman splittings (e.g., Goodman *et al.* 1989) or from arguments about compression of the Galactic magnetic field.

2.3. THE BEAM FLUID

The parameters of the beam fluid are much less certain. Let's start with galactic jets. Here the appearance of line emission allows good estimates of the gas density and temperature in the line-emitting part of the flow. For HH 34, for example, the electron density near in this gas is estimated to be about 10^9 electrons m^{-3} (Reipurth *et al.*, 1986), based on the strength of the [SII] emission lines. This emission is, however, from the shocked and radiating material that is likely to be associated only with the bow shocks of discrete units of the flow. The gas that is carrying most of the momentum in the beam may be much denser: indeed, the appearance of the HH 34 jet (Sec. 1.2.9) is more consistent with discrete lumpy ejection of material than a quasi-steady beam flow. Thus we will assume that galactic beams have a density that is similar to, or higher than, the density of the medium through which they move.

Since the material in the flow is radiating, the temperature in the unshocked gas is likely to be less than the temperature $\sim 10^4$ K to which radiation regulates an HII region. Furthermore, since this part of the flow does not seem to be radiating, the temperature is likely to be substantially lower. Hence outside the line-emitting regions of shocked material, the temperature should be $10 - 3000$ K, with sound speed within a factor ~ 10 of the sound speed of the ambient medium.

The speeds of galactic flows can be seen directly, from the proper motions of the knots, as well as estimated from the velocities of the line-emitting gas relative to the velocity of the originating object: indeed, by measuring both speeds, the full 3-dimensional velocity of the knots can be measured, and the orientation of the jet to the line of sight can be found. The typical speeds of prominent flows in star-forming regions (such as the HH 34 jet) are a few hundred km s^{-1}, but it is likely that many weaker flows also exist.

The magnetic fields in these jets are essentially unknown, so we will assume values similar to those in the external medium, and rely on the results of detailed calculations (yet to be done: an excellent thesis project) to investigate the possible consequences of stronger fields.

Table II. The beam

quantity	symbol	typical value		unit
		extragalactic	galactic	
core radius	R	$0.1 - 1000$	$10^{-5} - 10^{-2}$	pc
core speed	v_b	$c_{Sa} \cdot 0.95c$	$c_{Sa} - 0.95c$	
core sound speed	c_{Sb}	$0.1 - 1000$	$0.1 - 10$	c_{Sa}
core density	ρ_b	$10^{-6} - 10^2$	$0.1 - 10^3$	ρ_a
core magnetic field	B_b	$0 - 1000$	$0 - 100$	nT
core fluid polytropic index	Γ_b	$\frac{4}{3} - \frac{5}{3}$	$\frac{5}{3}$	
sheared region width	δ	$0 - 2$	$0 - 2$	
sheared fluid polytropic index	Γ_s	$\frac{4}{3} - \frac{5}{3}$	$\frac{5}{3}$	

Of course, the discussions above relate only to the types of jet seen in star-forming regions. The other class of galactic jets, such as SS 433, is in some ways similar, but may contain much stronger magnetic fields (which, together with their relativistic particles, are responsible for the radio emission), and are certainly faster flows! A summary of the results for galactic beams is given in Table 2.3.

In extragalactic jets there is rarely any direct indication of the flow velocity. Where there is structural change, for example in VLBI jets, the apparent velocities are often much larger than the speed of light. This is explained as a consequence of relativistic motion in the source (Fig. 2.3): if the jet contains a knot that is moving with speed v_{knot}, and the jet lies at an angle θ to the line of sight, then the observer will see a motion across the line of sight with apparent velocity

$$v_{app} = v_{knot} \sin \theta \left(1 \pm \frac{v_{knot}}{c} \cos \theta \right)^{-1} \tag{11}$$

where the positive sign indicates an approaching knot. Then if $v_{knot} > c/\sqrt{2}$, v_{app} can exceed c for some orientations of the jet to the line of sight. For VLBI jets, where speeds $> 10c$ have been observed, it is clear that v_{knot} must be close to c. Interestingly, (11) has been used for GRO 1655-40 to obtain limits on both the velocity v_{knot} and on the angle θ, using the angular velocity difference on the two sides of the core (Hjellming & Rupen, 1995).

A major issue in the use of (11) is whether the observed motions in VLBI jets are indicators of the true motion of the bulk flow, or of a moving pattern of material within the flow. A simple example of a situation where the pattern speed can exceed the flow speed is shown in Fig. 2.3. If two oblique waves within the beam are converging from

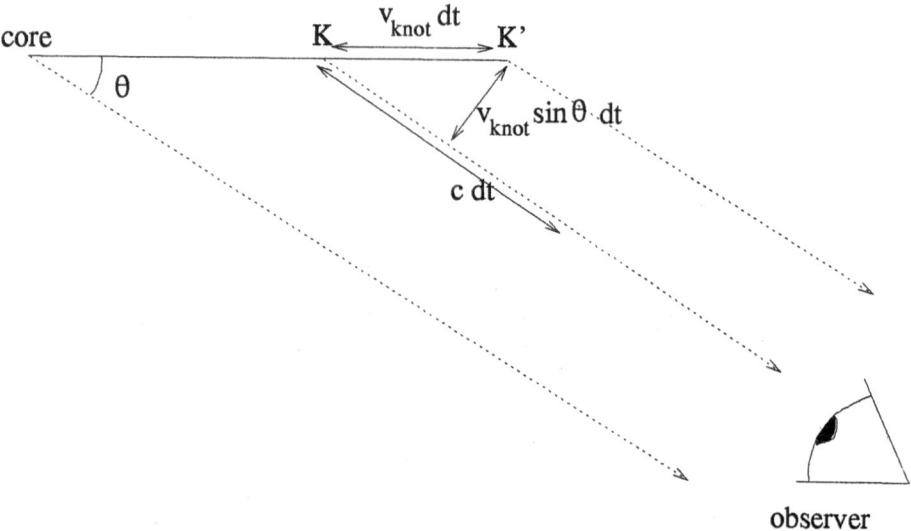

Figure 9. The geometry of superluminal motion occuring as a result of relativistic motion near the line of sight. In the interval dt, the knot K moves to K$'$, a projected distance $v_{knot} \sin \theta \, dt$ away, from the observer's point of view. In the same interval, light pulses from K and K$'$ are separated by $dt(c - v_{knot} \cos \theta)$, so that the apparent time taken for the motion from K to K$'$, as seen by the observer, is $dt(c - v_{knot} \cos \theta)/c$, and the apparent transverse velocity of the knot is therefore given by (11).

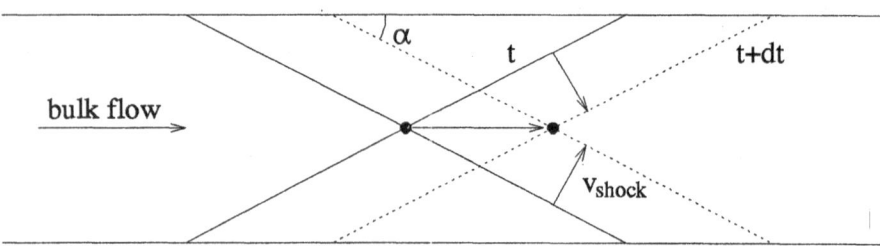

Figure 10. A simple example of a pattern speed that is different from the flow speed. Suppose that a radio knot is identified with the crossing point of two shocks which converge from the edge of the beam. Then in time dt, the shocks move distance $v_{shock} \, dt$ while the crossing point moves $v_{shock} \, dt / \sin \alpha$. The pattern speed of the crossing point is therefore $v_{shock} / \sin \alpha$, which may exceed the speed of the bulk flow.

the edges of the beam towards the center (perhaps as the result of an expansion of the beam), and we identify the region of brightest radio emission with the point at which these fronts cross (which is where the strongest particle acceleration is likely), then the speed with which the crossing point advances up the flow is $v_{shock}/\sin\alpha$. If we assume that the beam is heated so that the internal sound speed is comparable with the flow speed, then it is clear that the crossing point will advance faster than the flow speed even for large angles α, and may be much faster for small α.

For the larger-scale jets only indirect arguments exist for the beam velocities, and the results quoted in the literature are model-dependent and unreliable. In many cases it is not clear whether the flows are relativistic or non-relativistic. Some velocities are derived on a ballistic model for the jets, in which the flows are undecelerated (and, for 3C 449, for example, the speeds that are fitted must be a few times the escape speed from the host galaxy if interesting bends are to be formed; Blandford & Icke 1978). Other speeds are often based on a balance between the radiated energy from the radio source and the energy carried down the beam, after using a density estimate derived from the Faraday depolarization. Unfortunately, the measurement of Faraday depolarization is fraught with difficulties (see the review by Leahy 1991), and the resulting velocities are not reliable. Perhaps the best indications about beam speed are obtained by looking at simulations: to produce flows that *look* like FR 2 jets, light, supersonic flows are needed. For FR 1 jets, slower speeds may suffice. However, it is clear that the speeds are *at least* sufficient for the flow to escape from the host galaxy, and are hence a few times the sound speed. The entry in Table 2.3 is suitably cautious.

As mentioned above, there is little reliable information on the flow density either. If the Faraday depolarization is known, then some magnetic field can be guessed (perhaps from the minimum energy argument, see later), and the density estimated, but this always implies a model for the locations of the relativistic particles, fields, and energy-carrying material within the flow and we know (e.g., from M 87) that the distribution of radio emission can be complicated. Furthermore, we do not know whether the jet material is "conventional" astrophysical material, an electron/positron mixture, or some combination. It is only when the positive and negatively charged particles have different masses that the density can be obtained through Faraday depolarization. The best indication of density, then, is presumably again the simulation results, and I shall assume that the beams of extragalactic jets are generally very light.

The radii of extragalactic radio jets, R, vary from less than 1 parsec in VLBI jets to more than 1 kpc in low–power radio sources, or in the outer parts of high-power sources. In general, though, R is less than the scale on which we expect significant pressure structures in the external medium (except for shocks or dense blobs, which are most likely in the nuclear environment of VLBI jets).

Presumably the flowing beam is close to sideways dynamical equilibrium: that is, it is in contact with the surrounding gas and its internal pressure (from magnetic field, thermal gas, and relativistic particles) is close to being in balance with the external pressure. For fast flows this is not the pressure of the ambient medium, but rather the significant over-pressure of the cocoon (Loken *et al.*, 1992).

It is conventional to estimate the internal pressure of a beam using the minimum energy argument to estimate the magnetic field and the relativistic particle energy density, and then assuming some field geometry. The minimum energy argument consists of minimising the total energy density in some section of the radio source, while keeping the observed brightness (given by 7) constant. The result that is obtained is

$$B \propto ((1 + k)\, L_\nu)^{2/7} \tag{12}$$

Burbidge (1956), where k is the ratio of the energy densities in non-radiating particles and radiating electrons (and positrons), and L_ν is the synchrotron luminosity at some frequency. However, the existence of the "fiddle factor" k, which is essentially unknown, and the suspicion that the source may not be "efficient" in the sense of producing the maximum radio emission for a given relativistic electron population prevents the estimated pressures from being more than crude lower limits.

If we apply this argument, with $k = 0$ which implies that the non-relativistic component of the flow does not contribute to the pressure, then weak radio sources such as 3C 449 and NGC 6251 exhibit minimum pressures of 10^{-14} to 10^{-11} Pa, so that the thermal pressure of the ambient medium (up to 10^{-10} Pa; Table 2.2) can confine the jets (Fig. 2.3; but see Birkinshaw & Worrall 1991). For the most powerful jets in quasars and high–power radio galaxies, on the other hand, the assumption that the beams are confined implies that the external magnetic field or the jet cocoon is making a significant contribution to the pressure, since the ambient pressure is clearly insufficient to confine the beam.

The mixture of relativistic gas and thermal particles in the beam is generally treated as a single fluid, with no explicit account being taken of the transfer of energy between the constituents of the fluid. It is

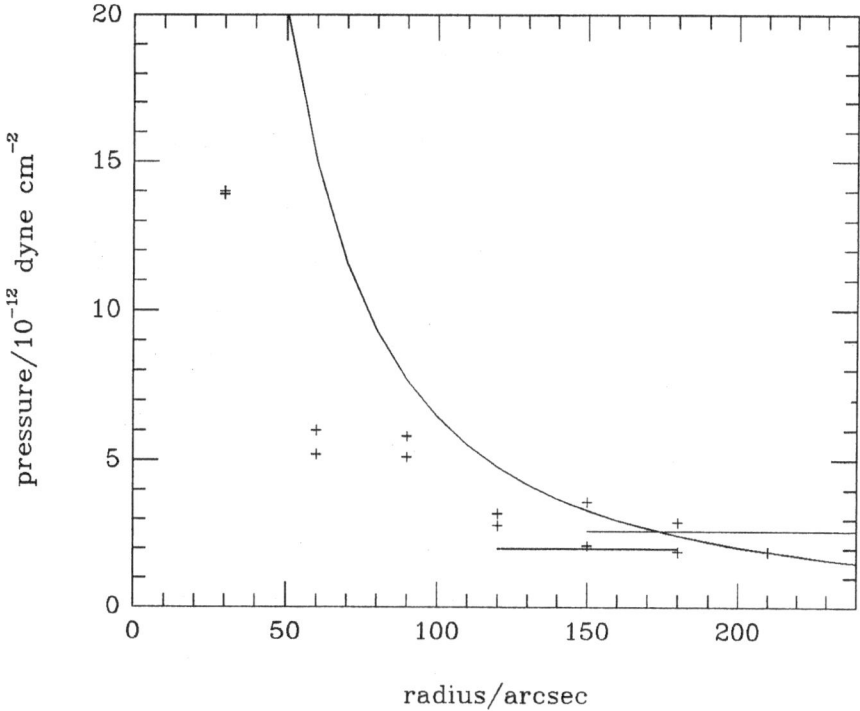

Figure 11. Pressures in and near the radio jets of NGC 4261, assuming that the source lies in the plane of the sky. Points indicate total pressures in the E and W jets, and the horizontal lines mark estimates of the total pressure in the radio lobes (assuming that only electrons and magnetic fields contribute). The external pressure, shown by the curve, exceeds the internal minimum pressure over the entire radio jet and is about equal to the lobe minimum pressure.

generally assumed that the beam fluid is a neutral mixture of electrons and protons, and this assumption will be made here, but the beams may consist of an electron–positron fluid (see, for example, Burns & Lovelace 1982).

Just as we are uncertain about the velocity, density, and pressure of a beam, we are uncertain about its internal structure. Until we can do real plasma diagnostics on jets there is little we can do about this: my approach here is to treat the beam core as being a uniform flow, and the ambient medium as a second uniform flow, and then put all the messy assumptions about structure into a "shear layer" dividing the two. In the notation of Table 2.3 (and Figure 2.3), this shear layer is given a scale δ, defined as

$$\delta = 2\frac{R_{\text{sa}} - R_{\text{bs}}}{R_{\text{sa}} + R_{\text{bs}}} \tag{13}$$

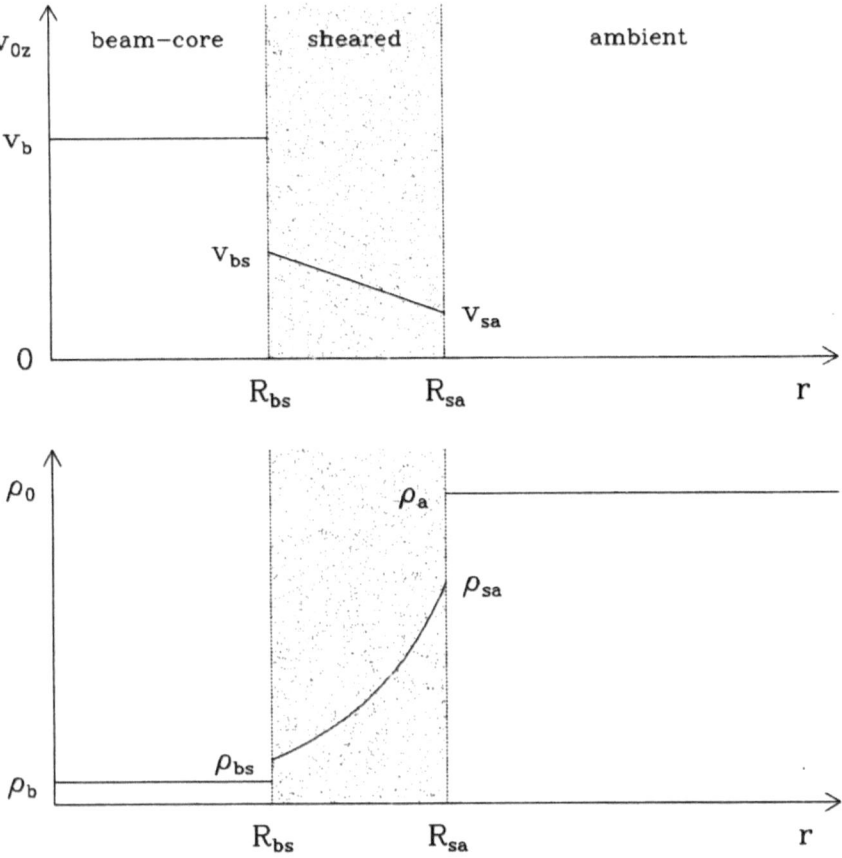

Figure 12. A schematic diagram of a beam structure including a uniform beam core (of radius R_{bs}, density ρ_b, and velocity v_b) and a uniform ambient medium (of radius R_{sa}, density ρ_a, and at rest) separated by a shear layer of dimensionless thickness δ (13) where the density varies from ρ_{bs} to ρ_{sa} as the velocity drops from v_{bs} to v_{sa}. The radius of the beam is identified with the mid-point of the shear layer, $R = (R_{sa} + R_{bs})/2$.

which varies from zero (no shear layer) to two (no beam core).

2.4. DIMENSIONLESS NUMBERS

Let's use the parameters in the Tables to derive some facts about the flows that we are discussing. First, let's consider the role of the magnetic field. A parameter that captures the relative importance of particle pressure and magnetic fields to the dynamics is

$$\beta_P = \frac{P}{u_B} \tag{14}$$

where P is the fluid pressure and u_B is the magnetic field energy density. For the ambient medium, far from the beam, it is likely that β_P is very large. For a typical cluster of galaxies, $\beta_P \gtrsim 10^3$, and the magnetic field is not dynamically important. If the larger field strengths in Table 2.2 or 2.3 are taken with the smaller fluid pressures, however, small values of β_P may result, and magnetic fields could dominate the flow. This can also be said in terms of the Alfvén speed,

$$c_A = \frac{B}{(\mu_0\,\rho)^{1/2}}.$$ (15)

Magnetic fields become important when the Alfvén speed approaches the sound speed in the gas.

In discussing the flow so far, we have been referring to the flow velocity as if it were a well-defined quantity. However, if the beam flow is turbulent, then we have been discussing only the *mean flow velocity* and other mean flow parameters. The Reynolds number describes whether an astrophysical beam can support turbulence: for magnetically–active plasmas we use the usual fluid Reynolds number

$$Re = \frac{\rho\,v\,l}{\eta},$$ (16)

and the magnetic Reynolds number

$$Re_m = v\,l\,\sigma\,\mu_0,$$ (17)

where v is the velocity scale and l is the length scale of the flow ($v \approx v_b$ and $l \approx R$). ρ, η, and σ are the density, viscosity, and electric conductivity of the fluid, respectively, and μ_0 is the permeability of free space. Using representative values for the temperature, magnetic field, density, velocity, and radius of a jet (from Table 2.3) and the expressions for σ and η given by Spitzer (1962), these Reynolds numbers may be calculated to be

$$Re \sim 4 \times 10^{28}$$ (18)

and

$$Re_m \sim 10^{28}$$ (19)

unless the velocity and velocity gradient in the flow are accurately parallel to the magnetic field.

The large value of Re_m indicates that magnetic field diffusion is much less important than the advection of the field by the flow (but note that this may not be true in thin *current sheets*, where large energy releases can take place). The large value of Re implies that turbulent motions are almost undamped, and astrophysical beams should be highly turbulent. Although much work on turbulent flows has been done under particular assumptions about the physics of the processes involved (Bicknell, 1984; Henriksen, 1987), the lack of definite knowledge about turbulence, particularly supersonic turbulence, makes it difficult to evaluate these results.

Perhaps the most important dimensionless number for describing these flows is their Mach number. Using the values in Tables 2.2 and 2.3, we can calculate an external Mach number

$$\mathcal{M}_a = \frac{v_b}{c_{Sa}} = \begin{cases} 1 - 10^5 & \text{extragalactic} \\ \\ 1 - 10^8 & \text{galactic} \end{cases} \tag{20}$$

where the extreme values for galactic jets arise in cases like GRO 1655-40. Thus in most cases we are dealing with supersonic or transsonic flows, and some hypersonic flows are also likely.

A parameter η is also in common use. η is the density ratio of the beam and the ambient medium: $\eta > 1$ is a dense flow, $\eta < 1$ is a light flow. The parameter range specified in Table 2.3 allows for either light or dense beams in extragalactic and galactic flows.

Finally, we should examine the continuity of the flows. Many jets, particularly the most powerful jets, are one–sided. This has been taken as evidence for Doppler preference, but also as evidence that the ejection is intrinsically one-sided (Burns, 1984), when the directions of jet ejection must vary on timescales much less than the flow time from the core to the lobes (Rudnick & Edgar, 1984). The ages of jets are easy to determine where jet motions are seen (in the galactic cases): star formation only produces jets over a small fraction of the time that the entire process takes. For GRO 1655-40, we believe that the jet started at about the time that the source was detected (Sec. 1.2.8). For the extragalactic sources, the ages are determined from spectral features which indicate which electrons have radiated all their energy: the age of the source is then related to the corresponding τ_e (Sec 2.1). However, no well-determined jet spectral feature exists that can be used to test the duty cycle or turn-on time of an extragalactic jet.

2.5. EQUATIONS OF MOTION OF BEAMS

In accordance with the arguments given above, we model the beam fluid flows as an inviscid, electrically–conducting (but not heat–conducting), compressible, relativistic gas moving in the absence of any gravitational field. The equations describing this motion are the equation of conservation of energy and momentum

$$T^{\mu\nu}{}_{,\nu} = 0. \tag{21}$$

The energy–momentum tensor $T^{\mu\nu}$ contains the energy and momentum of the fluid and the electromagnetic field $T^{\mu\nu} = T^{\mu\nu}_{\text{fluid}} + T^{\mu\nu}_{\text{EM}}$, where

$$T^{\mu\nu}_{\text{fluid}} = \left(\rho + \frac{P}{c^2}\right) u^\mu u^\nu + P \eta^{\mu\nu} \tag{22}$$

$$T^{\mu\nu}_{\text{EM}} = \frac{1}{\mu_0}\left(F^{\mu\alpha} F^\nu{}_\alpha - \frac{1}{4}F^{\alpha\beta} F_{\alpha\beta} \eta^{\mu\nu}\right). \tag{23}$$

The density of the fluid, ρ, includes both the rest–mass and the mass equivalent of the energy density of the fluid, P is the pressure of the fluid, and u^μ is the usual velocity four–vector. Greek indices are assumed to run from 0 to 3, and $\eta^{\mu\nu}$, the Minkowski metric tensor, is defined according to a spacelike convention. $F^{\mu\nu}$ is the electromagnetic field tensor. μ_0 is the permeability of free space, and the relative permittivity and permeability of the fluid, ϵ and μ, are assumed to be unity.

In addition to the equations of motion, an equation is needed to describe the presence of a gradient in the entropy due to the density and velocity shears and the generation of entropy by Joule heating. This entropy equation is

$$(n\, s\, u^\mu)_{,\mu} = -\frac{1}{T}\, u_\mu\, F^{\mu\nu}\, J_\nu, \tag{24}$$

where n is the baryon number density, s is the entropy per baryon in the fluid, T is the temperature of the fluid and J^μ is the charge–current four–vector.

The electromagnetic field tensor $F^{\mu\nu}$ obeys the Maxwell equations

$$F_{\alpha\beta,\gamma} + F_{\beta\gamma,\alpha} + F_{\gamma\alpha,\beta} = 0 \tag{25}$$

$$F^{\alpha\beta}{}_{,\beta} = \mu_0\, J^\alpha \tag{26}$$

which imply the equation of conservation of electric charge, $J^\mu{}_{,\mu} = 0$.

We also require a constitutive equation describing the relation between the charge current and the fields (the conduction equation),

$$J^\mu + \frac{u^\mu u_\nu}{c^2}\, J^\nu = \sigma\, F^{\mu\nu}\, u_\nu \tag{27}$$

where σ is the electrical conductivity (written as a scalar here, for simplicity).

Finally, we have equations describing the interrelation of the thermodynamic parameters n, ρ, P, s, and T. These equations are the equation of baryon conservation

$$(n\, u^\mu)_{,\mu} = 0, \tag{28}$$

two equations of state

$$P = P(n, s) \tag{29}$$
$$T = T(n, s), \tag{30}$$

and the first law of thermodynamics

$$d\rho = (\rho + \frac{P}{c^2}) \frac{dn}{n} + \frac{nT}{c^2} ds \tag{31}$$

(Landau & Lifshitz 1959; Misner *et al.* 1973; Königl 1980).

These equations can be written in terms of variables **E**, **B**, **j**, ρ_c, ρ, P, **v**, n, s and T, as

$$\frac{\partial}{\partial t} \left(\gamma^2(\rho + \frac{P}{c^2}) - \frac{P}{c^2} \right) + \nabla \cdot \left(\gamma^2(\rho + \frac{P}{c^2})\mathbf{v} \right) = \frac{\mathbf{E.j}}{c^2} \tag{32}$$

$$\gamma^2(\rho + \frac{P}{c^2}) \left(\frac{\partial \mathbf{v}}{\partial t} + (\mathbf{v}.\nabla)\mathbf{v} \right) + \nabla P + \frac{\mathbf{v}}{c^2} \frac{\partial P}{\partial t} = \rho_c \mathbf{E} + \mathbf{j} \times \mathbf{B} - \frac{\mathbf{v}}{c^2}(\mathbf{E.j}) \tag{33}$$

$$nT \left(\frac{\partial s}{\partial t} + \mathbf{v}.\nabla s \right) = (\mathbf{E.j} - \mathbf{v.j} \times \mathbf{B} - \rho_c \mathbf{v}) \tag{34}$$

$$\frac{\partial(\gamma n)}{\partial t} + \nabla.(\gamma n \mathbf{v}) = 0 \tag{35}$$

$$\nabla.\mathbf{E} = \frac{\rho_c}{\epsilon_0} \tag{36}$$

$$\nabla.\mathbf{B} = 0 \tag{37}$$

$$\nabla \times \mathbf{E} = -\frac{\partial \mathbf{B}}{\partial t} \tag{38}$$

$$\nabla \times \mathbf{B} = \mu_0 \mathbf{j} + \frac{1}{c^2} \frac{\partial \mathbf{E}}{\partial t} \tag{39}$$

$$\gamma(\mathbf{v.j} - \rho_c \mathbf{v.v}) = \sigma \, \mathbf{v.E} \tag{40}$$

$$\mathbf{j} + \frac{\gamma^2}{c^2} \mathbf{v}(\mathbf{v.j} - \rho_c c^2) = \gamma \sigma \, (\mathbf{E} + \mathbf{v} \times \mathbf{B}). \tag{41}$$

where

$$\gamma = \left(1 - \frac{\mathbf{v.v}}{c^2} \right)^{-1/2}. \tag{42}$$

These equations may be simplified for the highly–conducting plasmas of astrophysical beams, since the large value of the magnetic Reynolds number Re_m causes $F^{\mu\nu} u_\nu \approx 0$, or equivalently

$$\mathbf{E} = -\mathbf{v} \times \mathbf{B} \tag{43}$$

to an accuracy $O(Re_\mathrm{m}^{-1})$, or about one part in 10^{28}.

The assumption that the astrophysical plasma behaves as a continuum, which is essential if these equations are to apply, is excellent in the cases in which we are interested: the Debye length, $\lambda_D = (\epsilon_0 k_B T/ne^2)^{1/2}$, and the particle gyroradii are always smaller than the jet scales. Furthermore, we expect the fields and flows to change slowly compared with the time on which the plasma re–adjusts to charge neutrality. That is, the frequencies of interest are much lower than the plasma frequency, the ion gyrofrequency, and so on.

Formally, (29 – 43) describe the beam and the ambient medium together by a single set of velocity, density, pressure, and magnetic field functions. However, the problem is simplified by defining boundaries between regions of the flow which are described by different functions. One convenient choice, shown in Fig. 2.3, is to refer to the flow as three distinct regions (the beam core, the sheared flow, and the ambient medium), where only one of these regions (the sheared flow) contains gradients in the velocity or density. The functions describing the flow are then matched at the interfaces between these artificial domains.

2.5.1. *The relativistic, non-magnetic, equations of motion*

A simpler case of (29 – 43), which has been subjected to theoretical analysis, is that of relativistic, non–magnetic, flows, for which the equations reduce to the set

$$\frac{\partial}{\partial t}\left(\gamma^2(\rho + \frac{P}{c^2}) - \frac{P}{c^2}\right) + \nabla.\left(\gamma^2(\rho + \frac{P}{c^2})\mathbf{v}\right) = 0 \qquad (44)$$

$$\gamma^2(\rho + \frac{P}{c^2})\left(\frac{\partial\mathbf{v}}{\partial t} + (\mathbf{v}.\nabla)\mathbf{v}\right) + \frac{\mathbf{v}}{c^2}\frac{\partial P}{\partial t} + \nabla P = 0 \qquad (45)$$

$$\frac{\partial s}{\partial t} + \mathbf{v}.\nabla s = 0 \qquad (46)$$

$$s = s(P, \rho) \qquad (47)$$

which may be recognised as the usual equations of relativistic fluid flow. Codes that solve these equations in full generality are rare.

2.5.2. *The non-relativistic, magnetic flow*

The basic equations of fluid motion for an ideal, compressible, magnetic fluid in non–relativistic motion ($v \ll c$) can be derived from the general equations of motion as

$$\frac{\partial\rho}{\partial t} + \nabla.(\rho\mathbf{v}) = 0 \qquad (48)$$

$$\rho\left(\frac{\partial\mathbf{v}}{\partial t} + (\mathbf{v}.\nabla)\mathbf{v}\right) = -\nabla P + \frac{1}{\mu_0}(\nabla \times \mathbf{B}) \times \mathbf{B} \qquad (49)$$

$$\frac{\partial s}{\partial t} + \mathbf{v}.\nabla s = 0 \qquad (50)$$

$$s = s(P, \rho) \qquad (51)$$
$$\nabla . \mathbf{B} = 0 \qquad (52)$$
$$\frac{\partial \mathbf{B}}{\partial t} = \nabla \times (\mathbf{v} \times \mathbf{B}). \qquad (53)$$

These equations represent a considerable simplification, but the presence of the magnetic field causes additional difficulties over the non–magnetic case, although all relativistic effects have been ignored.

3. Kelvin-Helmholtz instabilities

If we are to accept the beam model (Section 2), so that jets are interpreted in terms of a directed flow from a compact object into a diffuse medium, then at least some beams must be sufficiently stable that they can preserve their identity and ability to transport energy over a distance of many jet radii (see the examples in Section 1).

This Section is concerned with the stability of beam flows. The subject is rather technical, but leads to some interesting conclusions, and some of the methods can be used in other stability problems. Generally, work on beam stability has used (analytical) linear instability theory (e.g., Chandrasekhar 1961; Drazin & Reid 1981). This methodology has been fairly successful at describing the stability of laboratory free shear flows (see the review of Ho & Huerre 1984): the extension to astrophysical flows needs to take account of the compressibility of the beam fluid, the presence of magnetic fields, and the possibility of energy loss through radiation.

3.1. LINEAR INSTABILITY THEORY

Plasma instabilities can be divided into the categories of *microinstabilities*, where the details of the distribution functions of particle species in the plasma are essential, and *macroinstabilities*, where only the bulk equations of motion of the fluid are necessary to describe the instability. Only macroinstabilities operate on large scales, such as the radius of a beam, and directly affect the gross pattern of flow. The Kelvin–Helmholtz instability is a classic example of a macroinstability: its natural length scale is a beam radius, and its time scale is a beam sound crossing time. By contrast, the two-stream instability, a well-known microscopic instability, operates at the characteristic timescale of plasma oscillations.

Kelvin–Helmholtz instabilities arise when two fluids are in relative motion on either side of some common boundary. Their origin may be traced to the Bernouilli equation—if a ripple develops at the interface, then fluid flowing faster to pass over that ripple exerts less pressure, and the ripple tends to grow. Kelvin–Helmholtz modes grow into striking wave structures (*e.g.*, Roberts *et al.* 1982).

The large-scale instability of a specific beam can be examined using numerical methods (e.g., Hardee & Clarke 1992), and numerical techniques are essential to check whether an instability grows to disrupt the beam or is stopped by some dissipative mechanism. However, analytical methods give an indication of stability for a variety of beam parameters without needing to run detailed numerical models for each case, and can check a wide range of linear scales in a problem, including scales which might be beyond the resolution limit of a particular numerical calculation.

What do we mean by "instability"? Generally, a fluid system is said to be unstable if small perturbations of that system grow. Note that large perturbations may be unstable even if small perturbations of all kinds are stable, and the flow that results from the growth of small perturbations may be stable (we say that the growth of the small perturbations *saturates*). An instability may disrupt the flow entirely, or it may cause the flow to readjust into a new configuration, which may itself be stable or unstable.

The simplest type of linear instability analysis revolves around a *dispersion relation*

$$\mathcal{D}(\omega, \mathbf{k}) = 0 \tag{54}$$

where perturbations $V^{(1)}$ of fluid variable V are assumed to be of the form

$$V^{(1)} \approx e^{i(\mathbf{k}.\mathbf{r} - \omega t)}. \tag{55}$$

The stability of the flow is then inferred from the behaviour of the *complex* angular frequency (ω) or wavevector ($\mathbf{k}$) roots of (54). If ω is assumed to be real, then roots of (54) with a negative imaginary part of $\mathbf{k}$ correspond to (spatially) growing perturbations in some direction: this is *spatial instability*. Alternatively, $\mathbf{k}$ can be assumed to be real, when the presence of roots with a positive imaginary part of ω indicates *temporal instability*.

The spatial instability formalism seems appropriate for beam flows, where an unstable disturbance propagates downstream as it grows. Nevertheless, the applicability of spatial instability theory is a subtle issue. A spatially growing mode will, according to (55), have infinite amplitude far downstream of its point of origin, and this contradicts the usual boundary condition for unstable modes—that they should tend to zero at a large distance from their source. The contradiction is usually more apparent than real, because it is usually the case that the growth of the mode to a large amplitude at far downstream points occurs *after* it grows to large amplitude at some fixed location—i.e., the relative amplitude does decline as a function of distance from the source. Clemmow & Dougherty (1969) find that spatial and temporal descriptions of instability are equivalent in cases where the real ω axis can be smoothly transformed to the real k, and that this is the case in free shear flows like the beams in which we are interested.

The survival of astrophysical beams is largely determined by their response to the Kelvin–Helmholtz instability. The study of Kelvin–Helmholtz instabilities on radio jets was introduced by Turland & Scheuer (1976) and Blandford & Pringle (1976) to test the feasibility of the transport of energy to the outer parts of radio galaxies. In this context, the presence of instabilities is *undesirable*, because they inhibit the effective transport of energy. The possibility that instabilities on the jet of M 87 might cause the optical knots was discussed by Stewart (1971), and Hardee (1979) has emphasized the possibility that instabilities on radio jets might cause knots and bends which resemble observed jet structures. Instabilities have also been invoked as causes of the distorted lobes of radio sources (Blake, 1972) and the shapes of comet tails by (Ershkovich, 1980; Kochhar & Trehan, 1988), and as a source of energy that might cause *in situ* particle acceleration (Ferrari *et al.* 1979). In this context, the presence of instabilities is seen as *desirable*. Overall the study of beam stability aims

- to discover whether there are beam structures capable of transporting energy to radio lobes without being wholly disrupted; and

- to determine whether particle acceleration and structural distortions may arise through the operation of instabilities which do not disrupt beams.

A dispersion relation such as (54) is derived under the assumption that the perturbations are sufficiently small that the fluid/plasma equations are linear in the perturbation variables. This assumption ceases to be true when the perturbations grow to observable sizes, so that linear analyses can predict *at most* the scales and types of structure that *might* develop. At this point different (often numerical) methods must be used to decide whether the instability develops to cause a detectable change in the flow. If the flow develops shocks, then they provide an opportunity for particle acceleration, and can cause abrupt changes in the direction or the physical properties of the beams. Alternatively, the instability might lead to the formation of *solitons*, non–linear disturbances that propagate with no (first order) change of shape, and which retain their identity in collisions with other solitons. The vortex rings developed around jets observed in the laboratory may be disturbances of this type. Lerche & Wiita (1980), Fiedler (1986), and Roberts (1987) have investigated this possibility. Instabilities seem likely to cause the beam flow to become make the transition to turbulence, with the turbulent energy cascade causing particle acceleration (Ferrari *et al.* 1979) and heating the beam material.

3.2. A MODEL OF THE STEADY-STATE FLOW

The discussion of beam stability begins with a description of its steady–state structure. Analytical studies are restricted to flows which are slowly–varying and near pressure equilibrium, so that the steady–state flow is relatively simple. The flow may be laminar or turbulent, but the lack of a convincing theoretical description of turbulence leads us to consider the development of instabilities on a laminar jet. Jets of circular section are of greatest interest for astrophysicists, and a plausible model structure is sketched in Fig. 2.3, where the velocity profile $v_z^{(0)}(r)$ and the density profile $\rho^{(0)}(r)$ of the unperturbed flow are defined for a model flow with a constant radius.

In Fig. 2.3, I have divided the model beam into three segments: a beam core (subscript b), the ambient medium (subscript a), and a sheared region between them (subscript s). The ambient medium will be assumed to be uniform, with no density, temperature, or magnetic field gradients that might cause the steady–state flow to be non–cylindrical. Where this is not the case, for example if the density of the ambient medium is decreasing in the direction of the flow, then it this structure must be explicitly modeled. Rapid variations in the properties of the ambient medium cause the instability problem to become intractable: for long–wavelength instability modes the nonuniformity of the ambient gas will limit the applicability of the model.

Analytical instability studies are restricted to flows near dynamical equilibrium, where the total internal pressure in the beam balances the external pressure of the ambient medium. If the jets are *free*, and not in effective contact with the ambient medium, then the instability analyses described here are inapplicable. It is also necessary that we take the unperturbed flows as being steady on the timescales of interest.

If we adopt the relativistic equations of motion developed in Sections 2.5.1 and 2.5.2, then a simple solution in cylindrical coordinates, (r, θ, z), which describes an unperturbed beam with significant structure is a steady, non–rotating, cylindrical beam flow with

$$\rho = \rho^{(0)}(r) \tag{56}$$

$$\mathbf{v} = (0, 0, v_z^{(0)}(r)) \tag{57}$$

$$\mathbf{B} = (0, B_\theta^{(0)}(r), B_z^{(0)}(r)) \tag{58}$$

where $\rho^{(0)}(r)$, $v_z^{(0)}(r)$, $B_\theta^{(0)}(r)$ and $B_z^{(0)}(r)$ are arbitrary functions of the distance from the beam axis, r. The pressure, $P = P^{(0)}(r)$, must then obey the equation of transverse force equilibrium

$$\frac{dP^{(0)}}{dr} = \frac{1}{\mu_0}\left(B_\theta^{(0)2}\frac{v_z^{(0)}}{c^2}\frac{dv_z^{(0)}}{dr} - \frac{1}{\gamma^{(0)2}}\frac{B_\theta^{(0)}}{r}\frac{d}{dr}(rB_\theta^{(0)}) - B_z^{(0)}\frac{dB_z^{(0)}}{dr}\right)$$

(59)

where $\gamma^{(0)}(r)$ is the local Lorentz factor,

$$\gamma^{(0)}(r) = \left(1 - \frac{v_z^{(0)2}}{c^2}\right)^{-1/2}.$$

(60)

(56 – 60) formally describe the beam, ambient medium, and shear layer by a single set of model functions. It is generally much more convenient to define boundaries between regions of the flow, and then use different model functions in each region, as has been done in Fig. 2.3. The functions describing the flow are then matched at the interfaces between these artificial domains.

3.3. SOME DISPERSION RELATIONS

The steady–state solutions (56 – 60) are written with suffix $^{(0)}$, as in the unperturbed pressure, $P^{(0)}$. To investigate the stability of this solution, the physical variables are expanded as sums of the unperturbed $^{(0)}$ parts and small $^{(1)}$ perturbations (for example, $P = P^{(0)} + P^{(1)}$). These forms for the variables are then substituted into the equations of motion (Sections 2.5.1 and 2.5.2), and the equations are linearized (expanded to first order in the perturbed quantities) by requiring that the perturbed flow variables are much smaller than the unperturbed variables (for example, $v_r^{(1)} \ll v_z^{(0)}$). The resulting equations are homogeneous in the perturbed quantities, but are too complicated for most purposes. Analyses of the stability of beam flows are based on simplifying assumptions about the functions $\rho^{(0)}$, $v_z^{(0)}$, $B_\theta^{(0)}$, and $B_z^{(0)}$.

For a cylindrical flow geometry, and a steady–state flow of the form (56 – 60), the homogeneity of the linearized equations allows the perturbations to be decomposed into wave–like eigenfunctions

$$P^{(1)} = P^{(1)}(r)\,e^{i(kz+n\theta-\omega t)}$$

(61)

$$\rho^{(1)} = \rho^{(1)}(r)\,e^{i(kz+n\theta-\omega t)}$$

(62)

$$\mathbf{v}^{(1)} = \mathbf{v}^{(1)}(r)\,e^{i(kz+n\theta-\omega t)}$$

(63)

$$\mathbf{B}^{(1)} = \mathbf{B}^{(1)}(r)\,e^{i(kz+n\theta-\omega t)}.$$

(64)

The spatial stability of the flow is described by the properties of the complex wave–number k, of the Kelvin–Helmholtz *normal modes* (61 · 64) for assumed real ω (Sec. 3.1). k and ω appear as eigenvalues of the solutions of the flow equations subject to the boundary conditions. General perturbations of the flow can be expressed as a superposition of these wavelike solutions provided that these solutions constitute a complete set of basis functions (see the discussion in Drazin & Reid 1981).

(61 – 64) imply that the Kelvin–Helmholtz normal modes are wavelike in z with wavenumber $\mathrm{Re}(k)$, wavelike in time with angular frequency ω, and develop $|n|$ oscillations around the circumference of the beam. The radial structure, specified by (for example) $P^{(1)}(r)$, dictates the extent to which these waves are localised near the beam. The azimuthal mode number, n, is a positive or negative integer, so that the eigenfunctions are single–valued, and the eigenfrequencies for positive and negative n are degenerate for the simplest beam configurations (non–rotating beams with only axial magnetic fields). For any given n, we expect the dispersion relation of (54) to describe a solution $k(\omega)$ (or a family of solutions) with similar azimuthal characteristics. $n = 0$ corresponds to a 'breathing' or 'pinching' mode of the beam, where the beam expands or contracts coherently at any z: only this mode causes the beam area to change. $n = \pm 1$ modes display sideways displacements of the beam with a helical pattern in z (these are the 'helical' or 'kink' modes): only these modes cause the beam centroid to be displaced. Modes with $|n| > 1$ (referred to as 'fluting' modes) produce distortions with ripples around the circumference of the beam—for $n = \pm 2$ there are two nodes on the beam surface.

The boundary conditions on the flow variables are imposed at $r \to \infty$ in the ambient medium, and at the boundaries between the ambient and sheared fluids (at $r = R_{sa}$ before perturbation) and the sheared and beam fluids (at $r = R_{bs}$ before perturbation). These conditions are

- all the flow variables should be finite at all r at finite t (including downstream from any z—see Sec. 3.1);

- all the flow variables should $\to 0$ as $r \to \infty$ at all finite time;

- the flow variables should be continuous functions of r;

- the radial phase velocity of the perturbations should be positive at $r \to \infty$ (so that perturbations are radiated from the beam);

— the radial displacements of the beam fluid and the ambient medium, $\xi_r^{(1)}$, where

$$\frac{d\xi^{(1)}}{dt} = v_r^{(1)}, \qquad (65)$$

should match at $r = R_{bs}$ and R_{sa};

— the total (fluid plus magnetic) pressures should match at $r = R_{bs}$ and R_{sa};

— the perpendicular (radial) component of the magnetic field should match at the boundaries; and

— the parallel component of the electric field should match at the boundaries.

For high Re_m, conditions (g) and (h) imply that the magnetic field is continuous at the boundary. Note the relevance of the problem of the divergence of the flow at downstream infinity (condition b), and the use of the Sommerfeld finiteness and radiation conditions (a, b, and d) for the perturbations.

3.3.1. *Relativistic, non-magnetic beams*

It is fairly easy to derive the eigenfunction equation for the pressure perturbation for a relativistic, non-magnetic beam with a general velocity and density structure. The flow variables obey the usual equations of relativistic fluid flow (44 – 47). If the equilibria (56 – 60) and perturbations (61 – 64) are substituted for the flow variables and the equations are linearised, they can be manipulated into the form

$$-i\omega\left(\gamma^{(0)2}(\rho^{(1)} + \frac{P^{(1)}}{c^2}) - \frac{P^{(1)}}{c^2} + 2\gamma^{(0)4}\frac{v_z^{(0)}v_z^{(1)}}{c^2}(\rho^{(0)} + \frac{P^{(0)}}{c^2})\right)$$

$$+\frac{1}{r}\frac{d}{dr}\left(r\gamma^{(0)2}(\rho^{(0)} + \frac{P^{(0)}}{c^2})v_r^{(1)}\right) + \frac{in}{r}\gamma^{(0)2}(\rho^{(0)} + \frac{P^{(0)}}{c^2})v_\theta^{(1)}$$

$$+ik\left(\gamma^{(0)2}v_z^{(0)}(\rho^{(1)} + \frac{P^{(1)}}{c^2}) + \gamma^{(0)4}(\rho^{(0)} + \frac{P^{(0)}}{c^2})(1 + \frac{v_z^{(0)2}}{c^2})v_z^{(1)}\right) = 0 \; (66)$$

$$i\gamma^{(0)2}(\rho^{(0)} + \frac{P^{(0)}}{c^2})(kv_z^{(0)} - \omega)v_r^{(1)} + \frac{dP^{(1)}}{dr} = 0 \; (67)$$

$$i\gamma^{(0)2}(\rho^{(0)} + \frac{P^{(0)}}{c^2})(kv_z^{(0)} - \omega)v_\theta^{(1)} + \frac{in}{r}P^{(1)} = 0 \; (68)$$

$$i\gamma^{(0)2}(\rho^{(0)} + \frac{P^{(0)}}{c^2})(kv_z^{(0)} - \omega)v_z^{(1)}$$

$$+\gamma^{(0)2}\left(\rho^{(0)}+\frac{P^{(0)}}{c^2}\right)v_{\mathrm{r}}^{(1)}\frac{dv_{\mathrm{z}}^{(0)}}{dr}+i\left(k-\frac{\omega v_{\mathrm{z}}^{(0)}}{c^2}\right)P^{(1)}=0 \ (69)$$

$$i(kv_{\mathrm{z}}^{(0)}-\omega)\left(\frac{P^{(1)}}{P^{(0)}}-\Gamma\frac{\rho^{(1)}}{\rho^{(0)}}\right)-\frac{\Gamma}{\rho^{(0)}}\frac{d\rho^{(0)}}{dr}v_{\mathrm{r}}^{(1)}=0 \ (70)$$

where the final equation results from a model for the beam fluid in which the entropy is given by

$$s=s_0+\frac{k_{\mathrm{B}}}{\Gamma-1}\ln P\rho^{-\Gamma} \tag{71}$$

with s_0 and Γ, the polytropic index, being constants. This form for $s(P,\rho)$ should be adequate when $nk_{\mathrm{B}}T<\rho c^2$, but is, nevertheless, not sufficiently general for a relativistic fluid (see Synge, 1957).

(66 – 71) can be manipulated to yield a single, second-order, differential equation for the pressure perturbation,

$$\frac{d^2P^{(1)}}{dr^2}+\frac{dP^{(1)}}{dr}\left(\frac{1}{r}+2\,\gamma^{(0)2}\frac{(k-\frac{\omega v_{\mathrm{z}}^{(0)}}{c^2})}{(\omega-kv_{\mathrm{z}}^{(0)})}\frac{dv_{\mathrm{z}}^{(0)}}{dr}-\frac{\frac{d\rho^{(0)}}{dr}}{\rho^{(0)}+\frac{P^{(0)}}{c^2}}\right)$$

$$+P^{(1)}\left(\gamma^{(0)2}\left(\frac{(\omega-kv_{\mathrm{z}}^{(0)})^2}{c_{\mathrm{S}}^2}-\left(k-\frac{\omega v_{\mathrm{z}}^{(0)}}{c^2}\right)^2\right)-\frac{n^2}{r^2}\right) \tag{72}$$

where $c_{\mathrm{S}}(r)$ is the local sound speed,

$$c_{\mathrm{S}}=\left(\frac{\Gamma P^{(0)}}{\rho^{(0)}}\right)^{1/2} \tag{73}$$

(Birkinshaw, 1984). Hardee (1984) has derived a similar equation for the pressure perturbation on an expanding beam.

Simple solutions of (72) for $P^{(1)}$ arise in the ambient medium and the beam core, where the density and velocity gradients vanish, since (72) then simplifies to Bessel's equation in αr, where the radial wavenumber

$$\alpha=\gamma^{(0)}\left(\frac{(\omega-kv_{\mathrm{z}}^{(0)})^2}{c_{\mathrm{S}}^2}-\left(k-\frac{\omega v_{\mathrm{z}}^{(0)}}{c^2}\right)^2\right)^{1/2} \tag{74}$$

The pressure perturbation in $r>R_{\mathrm{sa}}$ which satisfies the boundary conditions is then

$$P^{(1)}(r)=\epsilon_{\mathrm{a}}P^{(0)}\frac{H_n^{(1)}(\alpha_{\mathrm{a}}r)}{H_n^{(1)}(\alpha_{\mathrm{a}}R_{\mathrm{sa}})} \tag{75}$$

where $H_n^{(1)}(z)$ is a Hankel function of the first kind with order n. $\epsilon_a \ll 1$ is a measure of the smallness of the pressure perturbation relative to $P^{(0)}$, and α_a is given by (74) with $v_z^{(0)} = 0$, $\rho^{(0)} = \rho_a$, and $c_S = c_{Sa}$, the sound speed in the ambient medium. (75) represents an outward–going, decaying mode (a radiated sound wave) as $r \to \infty$, as required by boundary conditions (a – d).

In the unsheared beam core, the solution regular at $r = 0$ is

$$P^{(1)}(r) = \epsilon_b P^{(0)} \frac{J_n(\alpha_b r)}{J_n(\alpha_b R_{bs})} \tag{76}$$

where $J_n(z)$ is a Bessel function of the first kind and order n. $\epsilon_b \ll 1$ is a constant defining the smallness of the internal pressure perturbation, and the internal radial wavenumber, α_b, is given by (74) with $\rho^{(0)} = \rho_b$, $v_z^{(0)} = v_b$, and $c_S = c_{Sb}$, the sound speed in the core. The beam-core solution is a standing pressure wave.

Finally, the pressure perturbation must be found in the sheared region ($R_{bs} < r < R_{sa}$) between the beam core and the ambient medium. The simplest choice is to assume that the beam core and ambient medium are in direct contact, at $R_{bs} = R_{sa} = R$. In this case, Ferrari *et al.* (1978), Hardee (1979), and others match the radial fluid displacements

$$\xi_r^{(1)} = \frac{\frac{dP^{(1)}}{dr}}{\gamma^{(0)2}(\rho^{(0)} + \frac{P^{(0)}}{c^2})(\omega - kv_z^{(0)})^2} \tag{77}$$

and the perturbed pressures, $P^{(1)}$, at the interface (as required by the boundary conditions) to obtain the dispersion relation

$$\mathcal{D}(k,\omega) = \frac{H_n^{(1)'}(\alpha_a R)}{H_n^{(1)}(\alpha_a R)} \cdot \frac{\alpha_a}{\omega^2 (\rho_a + \frac{P^{(0)}}{c^2})}$$

$$- \frac{J_n'(\alpha_b R)}{J_n(\alpha_b R)} \cdot \frac{\alpha_b}{\gamma_b^2(\omega - kv_b)^2 (\rho_b + \frac{P^{(0)}}{c^2})}$$

$$= 0 \tag{78}$$

where the primes ($'$) denote differentiation with respect to argument, and γ_b is the beam Lorentz factor. The essential character of the dispersion relation for a conical beam with small opening angle is the same (Hardee 1984, 1986).

It is clear from (75, 76, and 78) that the Kelvin–Helmholtz modes are sound waves associated with the beam/ambient medium interface. The growth rate of the instability is related to the extent to which these waves are localised in the beam or efficiently radiated, i.e., to the match between the wave impedances of the beam and ambient medium fluids as a function of frequency (Payne & Cohn 1985).

The use of a vortex sheet interface is beguilingly simple, and (78) is susceptible to detailed mathematical analysis, but several important points are lost by this choice. The shear layer around a supersonic beam contains the *critical surface* at which the speed of a Kelvin–Helmholtz mode equals the speed of the fluid flow. The nature of this critical surface plays an important role in the development of the instability in laboratory jets, and it represents the location of a regular singularity in the differential equation for $P^{(1)}$. A second rationale for a study of a shear layer is as a test of the assertion that shear layers are transparent to long–wavelength modes, but suppress short wavelength instabilities. This view may be an oversimplification, since the presence of shear may *destabilise* some modes at sufficiently high Mach number $\mathcal{M}_b = v_b/c_{Sb}$ for shear layers which are not too wide.

The pressure perturbation in a linearly–sheared sheathing layer around a beam is

$$P^{(1)}(r) = \epsilon_{s1} \, P^{(0)} \frac{F_n^{(1)}(r)}{F_n^{(1)}(R_{bs})} + \epsilon_{s2} \, P^{(0)} \frac{F_n^{(2)}(r)}{F_n^{(2)}(R_{sa})} \qquad R_{sa} > r > R_{bs}$$

(79)

where $F^{(1)}$ and $F^{(2)}$ are independent solutions of (78) in the sheared layer, and can be expressed as infinite, convergent, series of terms or pairs of series (Birkinshaw, 1991). A dispersion relation $\mathcal{D}(\omega, k)$ similar to (78), but involving $F^{(1)}$ and $F^{(2)}$ as well as J_n and $H_n^{(1)}$, can be deduced by matching the pressures and fluid displacements at $r = R_{sa}$ and $r = R_{bs}$.

Bodo *et al.* (1989) have extended the theory by including the effects of uniform rotation of the steady-state beam (adding a velocity component $v_\theta^{(0)} \propto r$). Their work has discussed only non–relativistic, $n = 0$ modes, and indicates the existence of a new channel of instability.

3.3.2. *Non-relativistic, magnetic beams*

The basic equations of fluid motion for an ideal, compressible, magnetic fluid in non–relativistic motion ($v \ll c$) are (48 – 53). Interpreting the fluid variables as a superposition of the steady–state equilibria defined by (56 – 60) and perturbations of the form (61 – 64), they may be reduced to a set of eight non–redundant equations in the perturbations, but these eight equations are too complicated for useful manipulation if general $v_z^{(0)}(r)$, $\rho^{(0)}(r)$, $B_\theta^{(0)}(r)$ and $B_z^{(0)}(r)$ are assumed.

A variety of treatments of these equations in different limits, or with different forms for the magnetic field distributions have been discussed. Cohn (1983) considered the stability of a magnetically–pinched beam of this type with a vortex boundary layer at $r = R$, no density or velocity gradients, and an azimuthal field $B_\theta^{(0)} \propto \frac{1}{r}$ for $n = 0$ modes only. In this limit a solution for the fluid perturbations can be obtained in terms of confluent hypergeometric functions. Ray (1981) studied a cylindrical plasma beam with an incompressible equation of state and uniform velocity, density, and internal magnetic field $B_z^{(0)}$. Ferrari *et al.* (1981) have discussed the stability of cylindrical flows of the same type, but with a relativistic beam velocity. Following the same methods as in Sec. 3.3.1, it is simple to manipulate (48 – 53) to obtain the non–relativistic analogue of Ferrari *et al.*'s governing equation for the pressure perturbation,

$$\frac{d^2 P^{(1)}}{dr^2} + \frac{1}{r}\frac{dP^{(1)}}{dr} + P^{(1)} \left(\frac{\left(k^2 - \frac{(kv_z^{(0)}-\omega)^2}{c_S^2}\right)\left(\frac{(kv_z^{(0)}-\omega)^2}{c_A^2} - k^2\right)}{\left(k^2 - (kv_z^{(0)} - \omega)^2 \left(\frac{1}{c_A^2} + \frac{1}{c_S^2}\right)\right)} - \frac{n^2}{r^2}\right) = 0$$

(80)

which is a form of Bessel's equation. The solution for $P^{(1)}$ can be written

$$P^{(1)}(r) = \begin{cases} \epsilon_a P_a \frac{H_n^{(1)}(\alpha_a r)}{H_n^{(1)}(\alpha_a R_{sa})} & r > R_{sa} \\ \epsilon_b P_b \frac{J_n(\alpha_b r)}{J_n(\alpha_b R_{bs})} & r < R_{bs} \end{cases}$$

(81)

where the radial wavenumbers are given by

$$\alpha = \left(\frac{\left(1 - \left(\frac{\omega - kv_z^{(0)}}{kc_S}\right)^2\right)\left(\left(\frac{\omega - kv_z^{(0)}}{kc_A}\right)^2 - 1\right)}{1 - \left(\frac{\omega - kv_z^{(0)}}{kc_A}\right)^2 \left(1 + \left(\frac{c_A}{c_S}\right)^2\right)} \right)^{1/2}$$

(82)

with sound speed $c_S = c_{Sa}$ or c_{Sb}, Alfvén velocity $c_A = c_{Aa}$ or c_{Ab}, density $\rho^{(0)} = \rho_a$ or ρ_b, flow velocity $v_z^{(0)} = 0$ or v_b, and pressure $P^{(0)} = P_a$ or P_b for the ambient and beam wavenumbers, α_a and α_b, respectively. The condition of transverse pressure equilibrium is then

$$P_a + \frac{B_{za}^2}{2\mu_0} = P_b + \frac{B_{zb}^2}{2\mu_0},$$

(83)

and for a vortex layer boundary (associated with a current sheet) the dispersion relation becomes

$$\mathcal{D}(k,\omega) = \frac{H_n^{(1)'}(\alpha_a R)}{H_n^{(1)}(\alpha_a R)} \cdot \frac{\alpha_a}{\rho_a} \cdot \frac{1}{\omega^2 - k^2 c_{Aa}^2}$$

$$-\frac{J'_n(\alpha_b R)}{J_n(\alpha_b R)} \cdot \frac{\alpha_b}{\rho_b} \cdot \frac{1}{(\omega - kv_b)^2 - k^2 c^2_{Ab}}$$
$$= 0. \tag{84}$$

The Kelvin–Helmholtz modes are magnetosonic waves partially trapped in the beam cavity. Note that Ferrari et $al.$'s (1981) treatment is more general than that given above, since it includes the effects of relativistic beam motion.

Another version of (84), incorporating the effects of uniform beam rotation, was derived by Bodo et $al.$ (1989). Fiedler & Jones (1984) considered a beam carrying a uniform current density, with both axial and azimuthal fields, bounded by a vortex sheet, and with no gradients in density or velocity, and derived the properties of the unstable modes numerically (since no analytic dispersion relation could be found). A further modification of the theory by Trussoni et $al.$ (1988) considers the effect of allowing the fluid pressure to be anisotropic, with independent components parallel and perpendicular to an axial magnetic field inside the beam. No qualitatively-new features appear for fast beams, but flows with low Mach numbers tend to become unstable.

3.4. THE FORM OF THE INSTABILITY

A dispersion relation such as (78) or (84) is analysed from a spatial viewpoint by solving for complex k as a function of real ω. The wavelength and growth length of the instabilities are then

$$\lambda_0 = \frac{2\pi}{\mathrm{Re}(k)} \tag{85}$$

$$\lambda_c = -\frac{1}{\mathrm{Im}(k)} \tag{86}$$

and the pressure perturbation grows with distance down the beam, z, as e^{z/λ_e}. If many instability modes are active, then it is usually assumed that the fastest–growing mode (of wavelength λ_0^* and growth length λ_e^*) dominates the flow after a short incubation period (of length a few λ_e^*). Thus the dispersion relations are searched for the values of ω and n that produce the minimum value for λ_e^*, when the stability of the beam is described by this value, and the shape of the deformed beam is described by the corresponding eigenmode.

A spatial analysis of the shear-protected beam has been given by Birkinshaw (1991), and curves of $\lambda_e(\lambda_0)$ for several beams are shown in Figs 3.4 – 3.4, where the instability curves are drawn for azimuthal mode numbers $n = 0$ (the "pinching" modes, drawn with solid lines), $n = 1$ (the "helical" modes, drawn with dashed lines), and $n = 2$ (the first "fluting" modes, drawn with dot-dashed line). For each value of n the curve may show many branches. These branches have been labelled by N, the radial mode number, which represents the number of nulls in the pressure perturbation in $0 < r < R$. $N = 0$ modes are the 'Ordinary Modes' and $N > 0$ modes are the 'Reflection Modes' of Gill (1965) and Ferrari $et\ al$ (1981). Reflection modes appear only for sufficiently large beam velocity (if $v_b > c_{Sa} + c_{Sb}$; Payne & Cohn 1985); at smaller v_b only the ordinary mode is present. Analytical approximations for the values of λ_0^* and λ_e^* are given by Hardee (1987a, b) for low–n and low–N modes, and by Zaninetti (1986a, b) for short-wavelength modes in magnetic beams.

For beams which support reflection modes, Figs 3.4 to 3.4 show that many modes can be excited at sufficiently small wavelength, λ_0. For example, at $\lambda_0 = 10R$ in a beam with Mach number $\mathcal{M}_b = 10$ and $\rho_a = \rho_b$ (Fig. 3.4), modes with $n = 0$, $N = 0, 1, 2$, with $n = 1$, $N = 0, 1$, and with $n = 2$, $N = 0, 1$ are excited (as well as all modes with $n > 2$ and $N = 0$). If only modes of this wavelength and $n < 3$ are accessible to the beam then the $n = 2$, $N = 0$ mode, which has the smallest growth length, $\lambda_e \approx 7R$, would dominate the structure of the beam and should be visible to observers. Since there is no reason to suspect that $\lambda_0 = 10R$ is the only wavelength that is excited on the beam, Fig. 3.4 suggests that the fastest–growing mode tends to be that of smallest wavelength and largest n (the cylindrical analogue of the result obtained for a planar shear layer by Turland & Scheuer 1976). Even for a beam protected by a shear layer (Fig. 3.4), this theory does not predict an unambiguous mode which dominates the instability of a beam—rather we must appeal to other physical principles to set the ranges of ω and n which are accessible to instabilities, and then locate the values of λ_0^* and λ_e^* given this range. However, as n increases the ordinary ($N = 0$) mode has a smaller growth length than an increasing number of reflection ($N > 0$) modes. High–n and N modes are unlikely to have a significant individual effect on a beam, although their effective $continuum$ at small wavelengths may cause a smooth beam profile to develop by broadening the vortex layer to create a hot, sheared, sheathing layer that will inhibit the further growth of small–λ_0 instabilities.

Figs 3.4 to 3.4 demonstrate that the spatial instability of low–density beams is greater than that of high–density beams with the same velocity, and that the overall stability of a beam is increased as the Mach number $\mathcal{M}_b$ is increased, but the growth lengths of pinching ($n = 0$) modes increase faster then the growth lengths of helical ($n = 1$) modes (Ray, 1981). This suggests that beams tend to display knots at small Mach numbers and tend to twist at high Mach numbers. The figures

$$\lambda_e(\lambda_0)$$

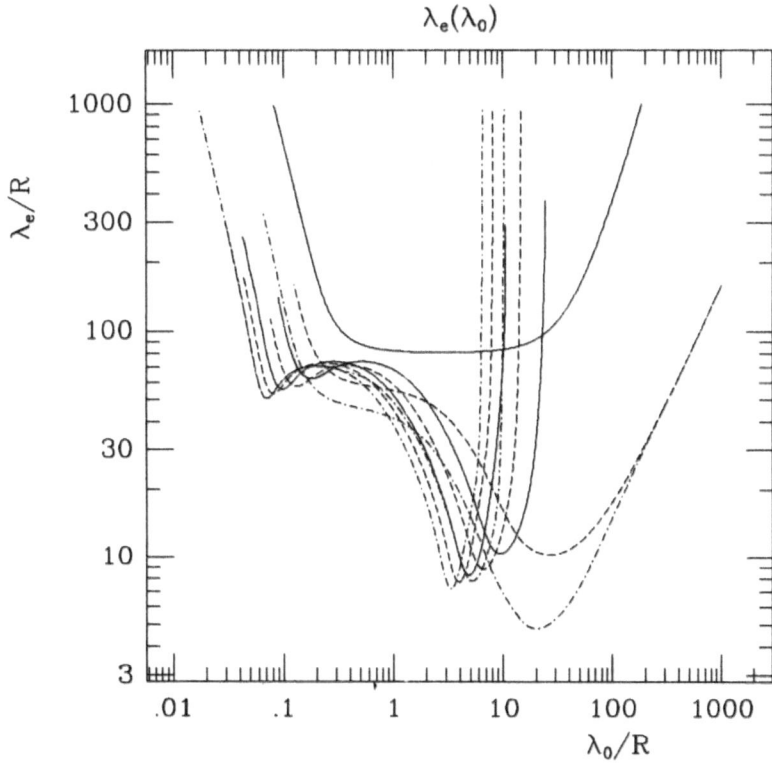

Figure 13. The mode structure for a beam velocity 3000 $km\,s^{-1}$, density equal to the density of the ambient medium, and internal source speed 300 $km\,s^{-1}$, bounded by a vortex sheet ($\delta = 0$). The ordinary modes extend to large λ_0, with the $n = 1$ and $n = 2$ modes degenerate at large λ_0 and the $n = 0$ mode more stable. The reflection modes appear at $\lambda_0 < 25R$, and have sharper minima than the ordinary modes. Many further reflection modes extend to shorter wavelengths, and are more unstable than the shortest-wavelength reflection mode appearing here (which has $n = 2$, $N = 2$).

also show that the growth lengths of the dominant instabilities are small unless the beam is supersonic both internally and externally (*i.e.*, $v_b > c_{Sa}$ and $v_b > c_{Sb}$). Thus the condition that the instability growth lengths $\lambda_e > R$ for wavelengths $\lambda_0 > R$ is just the condition that the beam is supersonic, and hence that its stability is dominated by reflection modes at short wavelengths and ordinary modes at long wavelengths (as in Fig. 3.4). It may be concluded that if the observed structures in radio jets are caused by the Kelvin–Helmholtz instability,

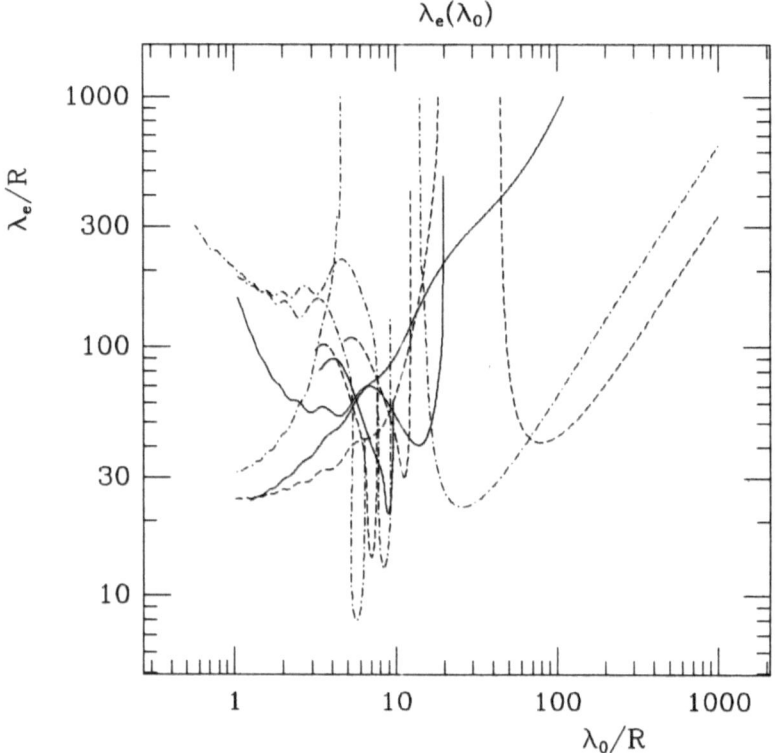

Figure 14. The mode structure for the beam shown in Fig. 3.4, but now bounded by a shear layer with $\delta = 1$. The reflection modes are similar, except that they show ripples at the short wavelengths, and the regions of peak instability are slightly sharper, because of resonances with shear-layer modes. The ordinary modes are more severely modified, with the $n = 1$ and $n = 2$ modes now showing regions of stability at $\lambda_0 \approx 20R$, while the $n = 0$ mode is destabilised at short wavelengths.

then the beams are supersonic, and that the influence of the shortest–wavelength modes on the gross structure of the beam has been slight.

If the density of a beam is increased whilst the kinetic energy flux and the internal pressure are held constant, the velocity, v_b, and the internal sound speed, c_{Sb}, of the flow decrease. The combined effect of these changes is that the growth lengths λ_e^* of the modes increase slightly, and that the wavelengths λ_0^* increase faster (if $\rho_b > \rho_a$) as functions of ρ_b. An exception arises for the $n = 0$, $N = 0$ mode, which is destabilised. If the density of the beam is increased whilst the momentum flux is held constant, only small changes in the values of

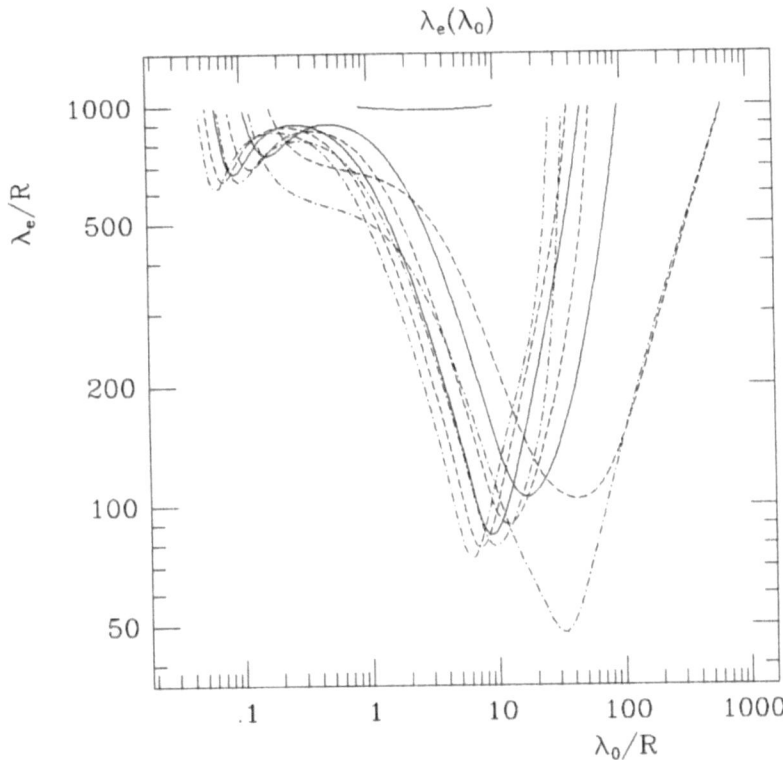

Figure 15. Solutions of the dispersion relation (78) for a beam with velocity 3000 km s^{-1}, density 100 times the density of the ambient medium, and ambient medium sound speed 300 km s^{-1}, bounded by a vortex sheet ($\delta = 0$). The ordinary modes extend to large λ_0, with the $n = 1$ and $n = 2$ modes degenerate at large λ_0 and the $n = 0$ mode much more stable (so that it only appears on this plot near $\lambda_e = 10^3 R$). The reflection modes appear at $\lambda_0 < 100R$. Many higher-order reflection modes have been excluded from this plot, for clarity of presentation.

λ_e^* and λ_0^* result, except that the $n = 0$, $N = 0$ mode is destabilised again. A sufficiently large increase or decrease of ρ_b subject to either a fixed kinetic energy flux or momentum flux may cause v_b to decrease below $c_{Sa} + c_{Sb}$, so that the reflection modes no longer appear.

The pressure perturbations $P^{(1)}(r)$ for the $n = 2$, $N = 1$ modes of Figs 3.4 and 3.4 near maximum instability are illustrated in Figs 3.4 and 3.4. These perturbations are seen to have an appreciable amplitude over a significant region inside and outside the beam, although $P^{(1)}$ tends to zero as $r \to 0$ for $n > 0$. The ordinary mode ($N = 0$) is a

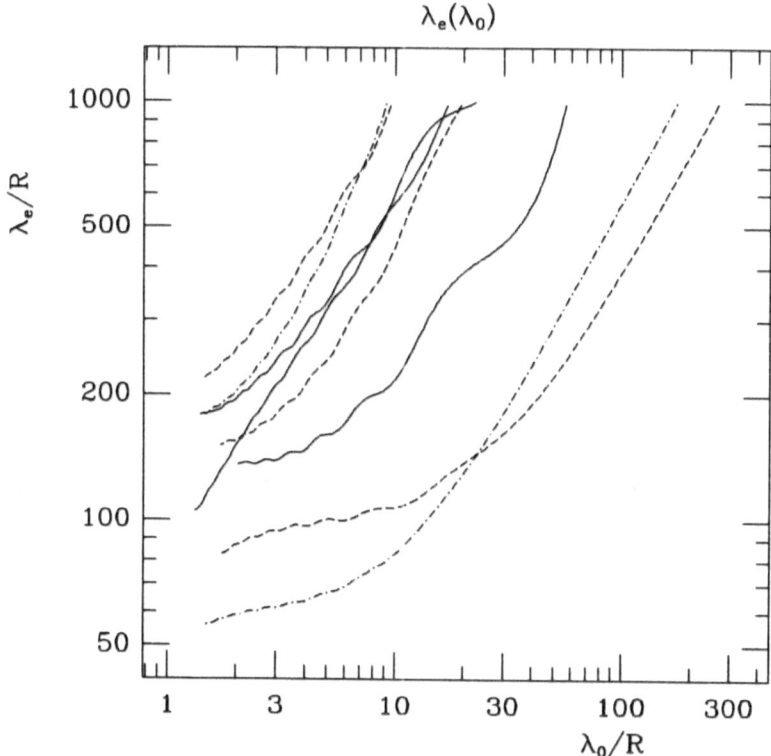

Figure 16. The mode structure for the beam seen in Fig. 3.4, but now bounded by a shear layer with $\delta = 1$. The stability curves show ripples at over the full range of wavelengths, due to resonances with modes in the shear layer, and the ordinary and the reflection modes all have similarly-shaped stability curves without pronounced minima. The growth lengths of these instabilities are large, and the beam is fairly stable.

consequence of the existence of any boundary between a flowing and a static region of a fluid—it arises for planar flows (*e.g.*, Blake 1972; Turland & Scheuer 1976) as well as in other geometries, and is unstable for all wavelengths. Reflection ($N > 0$) modes, on the other hand, require the fluid to contain more than a single planar surface, and are particularly strong where the fluid contains an enclosed region whose walls can vibrate coherently (*e.g.*, the beam–cavity). In that case, instabilities occur as sound waves radiated from one surface are reflected back from the other surface (with some phase delay) to interfere constructively. Thus the beam acts as an acoustic waveguide: standing waves across

$$\lambda_e(\lambda_0)$$

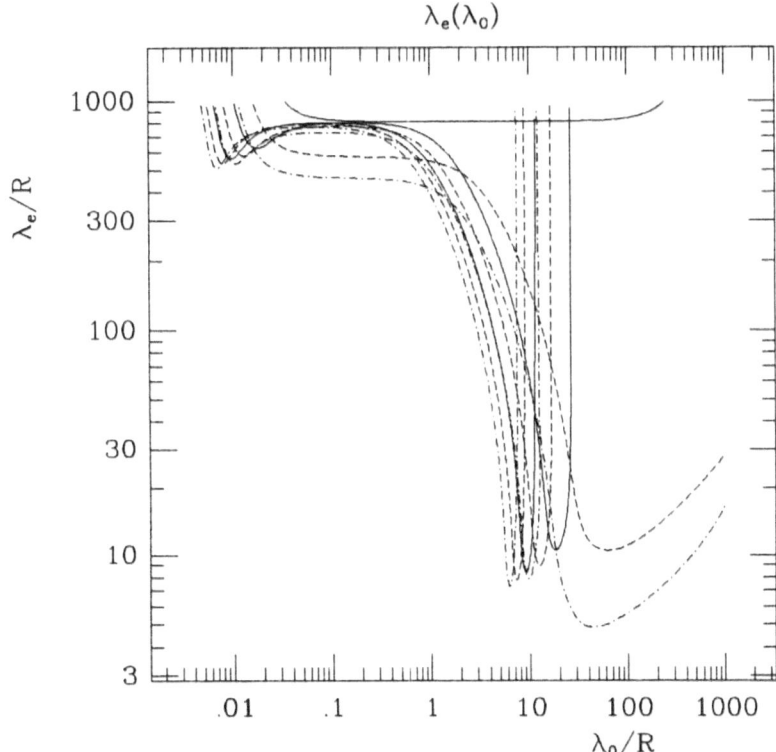

Figure 17. Solutions of the dispersion relation (78) for a beam with velocity 3000 km s^{-1}, density 10^{-2} times the density of the ambient medium, and internal sound speed 300 km s^{-1}, bounded by a vortex sheet ($\delta - 0$). The ordinary modes extend to large λ_0, with the $n = 0$ mode (which appears near $\lambda_e = 10^3 R$) much more stable than the $n = 1$ and $n = 2$ modes. The reflection modes appear at $\lambda_0 < 28R$, and have sharper minima than the the ordinary modes. Many high-order modes have been excluded from this plot.

the waveguide allow co–operative modes of the walls to be excited, and growing perturbations can be produced if the reflection coefficient for a wave striking a wall is greater than unity (over–reflection; Payne & Cohn 1985; Bodo *et al.* 1989). The appearance of reflection modes at regular intervals of λ_0 in Figs 3.4 to 3.4 can be interpreted in terms of this frequency–dependent coupling of internal modes of the beam to radiated magnetosonic waves (Bodo *et al.* 1989). A regular pattern of appearance of such modes as λ_0 is decreased corresponds to the 'fitting

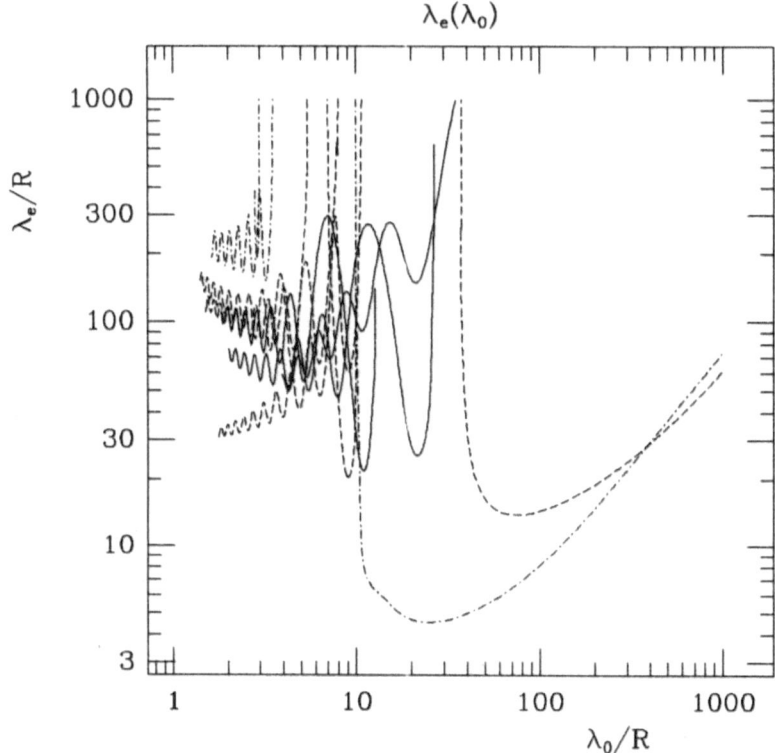

Figure 18. The mode structure for the beam seen in Fig. 3.4, but now bounded by a shear layer with $\delta = 1$. The stability curves show dramatic ripples at short wavelengths for all the modes, caused by strong resonances with waves in the shear layer, and the $n = 1$ and $n = 2$ ordinary modes are absolutely stabilised at wavelengths $\lesssim 35R$ and $12R$, respectively. The zones of instability terminate after the wavelengths have dropped by a factor ≈ 4, when the ordinary modes lie in the same zone of strong resonances as the reflection modes. Note the $n = 2$ ordinary mode is much the most unstable, dominating the stability of the beam for wavelengths from $10 - 400R$. The first few reflection modes also show stabilised regions in this beam, so that the maximum wavelengths λ_0 at which they appear are smaller than would be expected.

in' of increasing numbers of nulls of the pressure perturbation within the beam.

The presence of a shear layer around a beam complicates the instability pattern by allowing resonances within the shear layer itself. Various treatments of this effect have been performed: by Ray (1982), Ferrari *et al.* (1982), Roy Choudhury & Lovelace (1984), and Hardee (1982,

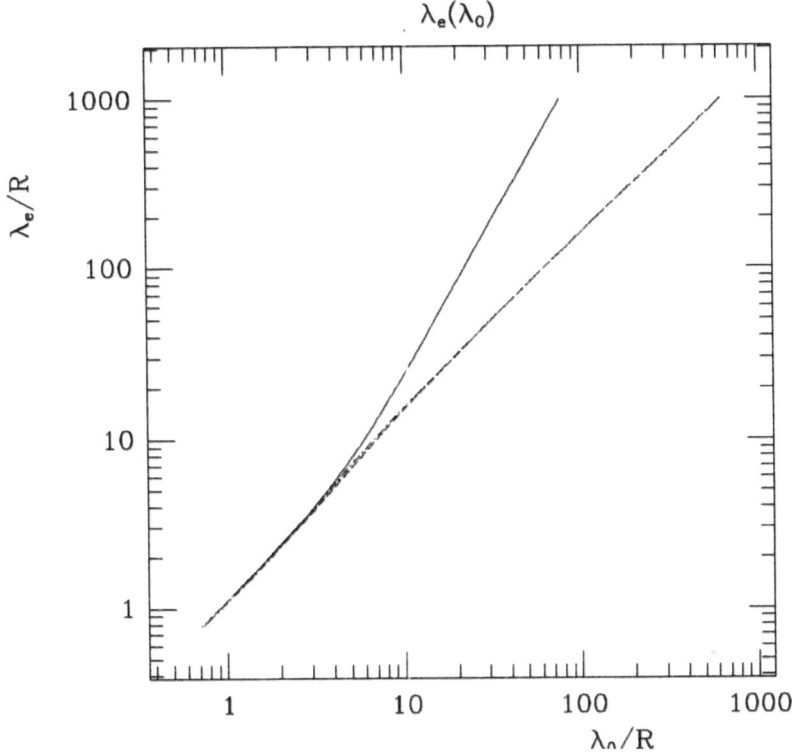

Figure 19. Solutions of the dispersion relation (78) for a beam with velocity 3000 km s^{-1}, density 100 times the density of the ambient medium, and internal sound speed 300 km s^{-1} with a vortex layer boundary. Since $v_b < c_{Sa} + c_{Sb}$ for this beam, no reflection modes will appear. The stability curves are drawn with solid, dashed, and dot-dashed lines for $n = 0$, 1, and 2, respectively. The $n = 0$ mode is more stable than the $n = 1$ and $n = 2$ modes at large wavelength, and all three modes become degenerate and of the form $\lambda_e \propto \lambda_0$ as $\lambda_0 \rightarrow 0$, as would be found for a slab jet.

1983, 1984). Earlier work on the structures of instabilities in shear layers has been done for laboratory jets by Blumen *et al.* (1975) and Drazin & Davey (1977). The solutions for a cylindrical beam bounded by linear shear layers, shown in Figs. 3.4 to 3.4 were obtained by Birkinshaw (1991).

The shear layer resonances modify the appearance of the vortex sheet modes by introducing additional structure into the $\lambda_e(\lambda_0)$ relation. In Fig. 3.4, for example, this can be seen as the additional "ripples"

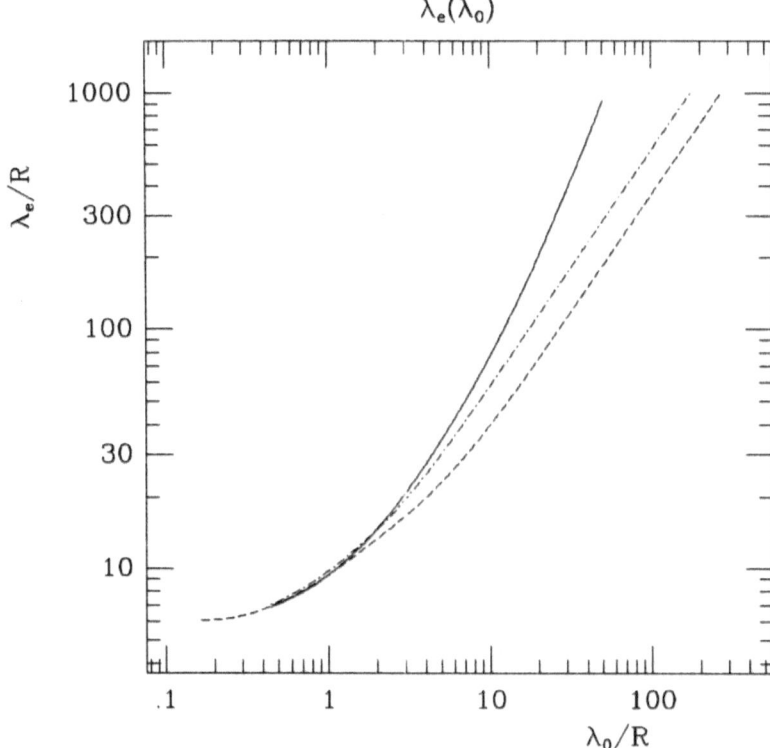

Figure 20. The mode structure for the beam seen in Fig. 3.4, but now bounded by a shear layer with $\delta = 1$. The modes are stabilised at all wavelengths, but remain of the form $\lambda_e \propto \lambda_0$ at the longest wavelengths. As $\lambda_0 \to 0$, they become degenerate again, because the radial wavelengths become much smaller than R. As is usual for subsonic and transsonic beams, the flow is highly unstable to short-wavelength perturbations.

on the curves at small δ. As the dimensionless thickness of the shear layer, δ, is increased, the ordinary modes are slightly stabilized at long wavelengths. For wavelengths near the beam radius, R, and at moderate δ, mode mixing causes slightly *greater* ordinary mode instability than in a vortex layer. The reflection modes are stabilised by the effects of the shear, and the wavelengths at which they appear, λ_0^{crit}, change slightly. Large values of δ cause increasing stabilisation of the flow, and at sufficiently large δ some of the ordinary modes which are unstable for flows bounded by a vortex sheet become stable over a wide range of λ_0.

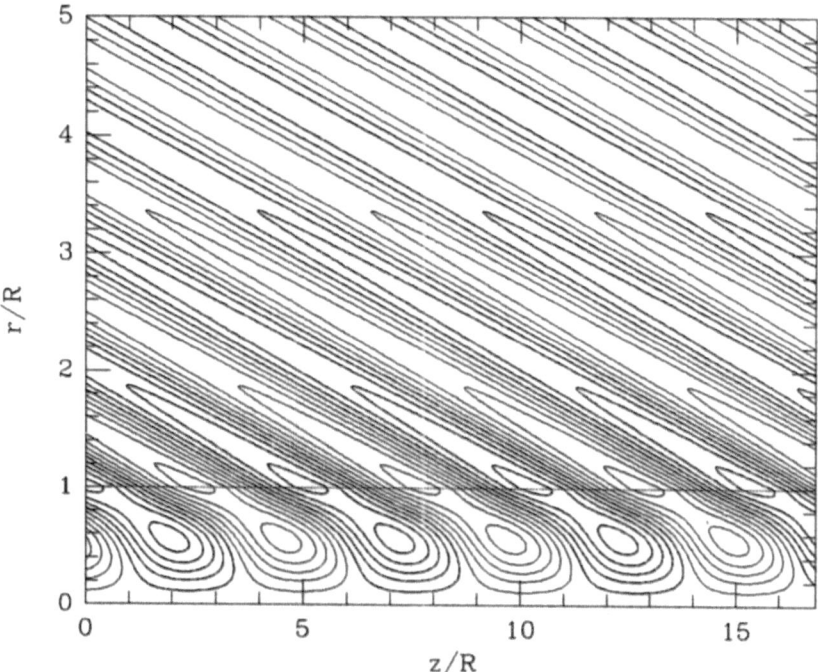

Figure 21. The physical pressure solution, $\mathrm{Re}(P_1(r, \theta, z))$, for the $n = 2$, $N = 1$ mode at its maximum instability of the beam shown in Fig. 3.4 at a particular time, t. The dotted line represents the location of the vortex sheet, and contours of the pressure perturbation are drawn at equal intervals of $0.2\times$ the maximum pressure perturbation from -0.9 to $+0.9\times$ the maximum pressure perturbation in this region of the flow. Thicker lines mark the positive contours. For clarity, the exponential growth in the mode (for which $\lambda_e = 7.8R$) has been suppressed in this diagram. Note that the wave-train is inclined at an an angle $\approx 80°$ to the beam axis in the ambient medium.

At the shortest wavelengths, even the presence of a wide shear layer does not cause complete mode stabilization, and it is precisely these short-wavelength modes that have the shortest growth lengths, and hence which should dominate beam instability.

Magnetic effects have been discussed by Cohn (1983), Ray (1981), Fiedler & Jones (1984), Ferrari *et al.* (1980, 1981) and by Benford (1981) for a cold beam (with no internal dynamics). Bodo *et al.* (1989) have discussed the stability of a rotating beam to the pinching ($n = 0$) mode. Roy Choudhury & Lovelace (1986) and Ray & Ershkovich (1983) have discussed the combined effects of magnetic field and shear on a planar flow. The results of these investigations are that all modes are stable for sub–Alfvénic flows ($v_b < c_{Ab}$) and axial magnetic fields (Ray 1981; Ferrari *et al.* 1981; Fiedler & Jones 1984). $n = 1$ modes stabilise

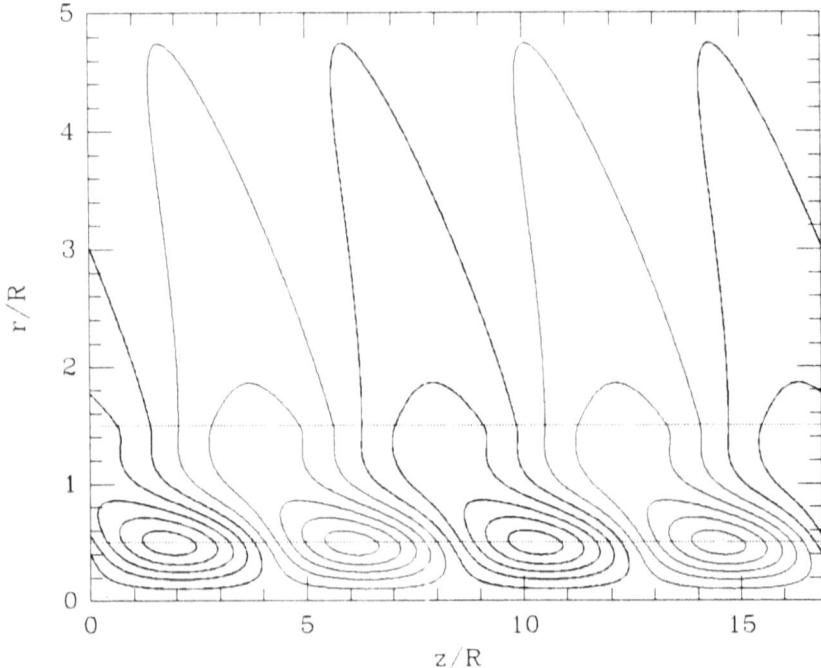

Figure 22. The physical pressure solution, $\mathrm{Re}(P_1(r, \theta, z))$, for the $n = 2$, $N = 1$ mode at its maximum instability of the beam shown in Fig. 3.4 at a particular time, t. Dotted lines represent the inner and outer edges of the shear layer, and contours of the pressure perturbation are drawn at equal intervals of $0.2\times$ the maximum pressure perturbation from -0.9 to $+0.9\times$ the maximum pressure perturbation in this region of the flow. Thicker lines mark the positive contours. For clarity, the exponential growth in the mode (for which $\lambda_e = 12.8R$) has been suppressed in this diagram. Note that the wave-train is inclined at an an angle $\approx 45°$ to the beam axis in the ambient medium, and is more oblique (at an angle $\approx 70°$ to the beam axis) in the shear layer.

at a lower axial field ($c_{\mathrm{Ab}} > 0.5v_{\mathrm{b}}$). By contrast, azimuthal fields are destabilising (as shown for the $n = 1$ modes by Ferrari *et al.* 1981; for the $n = 0$ and 1 modes by Fiedler & Jones 1984; and for $n = 0$ modes on magnetically–confined beams by Cohn 1983). In all cases, the nature of the instability is similar to that of a non–magnetic flow, and the rates of growth of instabilities are not much different from those of non–magnetic flows for axial fields less than the critical value for stabilisation,

$$B_{\mathrm{zb}}^{\mathrm{crit}} \approx (n/\mathrm{m}^{-3})^{1/2} (v_{\mathrm{b}}/c) \quad \mathrm{nT}. \tag{87}$$

None of the above discussions of the stability of magnetic beams identified the presence of slow magnetosonic reflection modes in the dispersion relation, but discuss only the ordinary modes and fast magnetosonic reflection modes. Bodo *et al.* (1989) have described the importance of the slow magnetosonic modes. Only ordinary modes are present at low (subsonic) beam speeds, the slow magnetosonic reflection modes dominate the stability next, and at high Mach numbers it is the fast magnetosonic relection modes that determine the stability of the flow.

The effect of Kelvin–Helmholtz modes on the *survival* of the beam cannot be predicted with certainty from the simple linear instability results—since small–λ_0 modes are the most unstable it is likely that the structure of the beam will adjust to suppress these modes before growth of the longer (λ_0, λ_e) modes (which may produce observable structures). The rapid adjustment of the structure of the beam to small–wavelength modes may invalidate the assumed z–independence of the flow used to derive the dispersion relations. Where longer wavelength modes occur, they should grow on a modified beam structure. The results of simple shear–layer calculations tend to suggest that the modification of the structure of the beam can be ignored in predicting the instability at long wavelengths. If we assume that the shorter–wavelength modes saturate by establishing a thin shear layer, then the shear–layer calculations of Ray (1981), Ferrari *et al.* (1982) and Birkinshaw (1991) suggest that the dominant instabilities will always be those with wavelengths comparable with the width of the shear layer. Since the width of the shear layer that is produced is not known, it is not possible to predict which wavelengths will dominate the stability of a given beam. The best use of the theory in its present state is, therefore, to attempt a consistency check between the wavelengths and growth lengths for modes which are inferred from maps of sources with the values that are predicted using jet parameters determined in other ways.

The linear calculations described above contain no information on the limits to the amplitudes of instability. The method that was used to derive the dispersion relations can be extended to second order, by writing the pressure (for example) as

$$P = P^{(0)}(\mathbf{r}, t) + P^{(1)}(\mathbf{r}, t) + P^{(2)}(\mathbf{r}, t), \qquad (88)$$

and carrying the expansion of $(44 - 47)$ or $(48 - 53)$ to second order to obtain equations linear in $P^{(2)}$ (but quadratic in $P^{(1)}$), and then the behaviour of $P^{(2)}$ might yield limits to the growth of the instability. The analytical difficulties in performing such a calculation are severe, and no direct analytical discussion of the limits to growth has been made. Benford (1981) argues that the growth of the perturbations will cease when the velocity excursions produced by the Kelvin–Helmholtz modes become sonic and develop shocks—at $|\mathbf{v}^{(1)}| = c_S$. This may be regarded as a strong upper limit on the linear phase of growth.

An alternative approach is to use numerical simulations of beams
(*e.g.*, Nepveu 1982a, 1982b; Norman *et al.* 1982, 1983, 1984; Williams
& Gull 1984) to discover the results of the growth of the instability.
Norman *et al.* (1984) have examined their results in the light of analytical instability theory, and conclude that pinch reflection modes (with
$n = 0$ and $N > 0$) do not have a significant effect in disrupting beams
since their non–linear growth saturates with a series of weak conical
shocks which cause alternate divergence and refocussing of the flow
(that is, the growth of the pinch reflection modes terminates by Benford's (1981) sonic criterion). Recent calculations on slab beams (which
can model the $n = 0$ and $n = 1$ modes of cylindrical beams) by Hardee & Norman (1988) and Norman & Hardee (1988) indicate that the
reflection modes saturate with the formation of weak oblique shocks,
whilst the ordinary modes are unsaturated and can cause large-scale
distortions and disruption of the flow. Since the ordinary pinching mode
is only important for trans–sonic flows, this suggests that the ordinary
helical mode may be the most important long-wavelength instability
channel for supersonic beams. The numerical simulations found growth
lengths $\lambda_e^* = (2 - 4)\mathcal{M}_b R$ and wavelengths $\lambda_0^* \approx 5R$ for Mach numbers
$\mathcal{M}_b = 3$ to 6, in fair agreement with linear instability theory.

It is easy to see from Figs. 3.4 to 3.4 that low–density beams are
expected to make a rapid transition to turbulence since their short–
wavelength modes have growth lengths less than R. Higher sound-
speeds in the ambient medium and the beam ameliorate this situation as far as the low–order reflection modes are concerned, but the
higher–order modes will still have small values of λ_e^*. The transition to
turbulence can only be prevented if the beam is well protected from
the influence of modes with wavelengths less than a few R—such pro-
tection might be provided by a hot sheathing layer for which waves
with $\lambda_0 < \delta R$ will propagate poorly. Beams which are more dense
then the ambient medium are rather stable and only slightly affected
by short wavelength modes (until the frequency of the driving perturbation becomes very high—that is, until the wavelength of the driving
mode becomes much less than the beam radius). For simple beam structures, therefore, the single clear prediction of Kelvin–Helmholtz theory
is that light beams will make a fast transition from laminar to turbulent
flow, whilst denser beams can remain laminar longer.

3.5. THE STRUCTURE OF A DISTORTED BEAM

The growth of the modes described above may lead to observable distortions on a beam if the wavelengths and growth lengths are not severely
modified in their transition from the linear to the non–linear phase of

growth, and if the beams are not first disrupted by the influence of the shortest wavelengths of instability. In this Section we apply the results of an analysis of sheared beams to models of 3C 449, NGC 6251, and HH 34 assuming that magnetic fields are not dynamically important and do not suppress the Kelvin-Helmholtz modes.

Since the result of the operation of Kelvin–Helmholtz modes, as observed in the laboratory, is the transition to turbulence of the beam flow, it is logical to interpret the greater or lesser instability of astro-physical beams as an indication of their susceptibility to turbulence. Ferrari et al. (1979) suggested that the radio–bright jets that are com-mon in low–power radio galaxies and high–power quasars are indica-tive of turbulent particle acceleration occurring in beams which are very unstable, and that the 'invisible' jets of the higher–power radio galaxies should be interpreted as beams which are more stable, and hence have not developed strong turbulence and its consequent particle accelera-tion. Since light beams are more unstable than dense beams, it might be suggested that this tendancy to bright radio emission might also be taken as a tracer of beam density, but there is no independent evidence for this hypothesis.

For an identification of the mode structures with the radio mor-phologies to be possible, a number of assumptions are necessary:

— that the observed radio structures trace the actual fluid flow carry-ing energy from the galaxy to the outer parts of the radio source;

— that the wavelengths and growth lengths of the unstable modes are not severely modified in their transition from the linear to the non–linear phase of growth, so that the forms of $P_1(r)$ reflect the observable distortions; and

— that the beams are not disrupted by very short wavelength modes (with $\lambda_0 \ll R$), which are usually the modes that grow most rapid-ly.

Since typical modes of instability involve appreciable pressures over the entire volume of the beam and of the shear layer (Figs 3.4 and 3.4), edge brightening of the beam is not to be expected if the radio bright-ness is related to the pressure perturbation through particle accelera-tion, or if the diffusion of relativistic electrons in the beam is efficient (as might be the case if the magnetic field in the beam is tangled; e.g., Eilek 1982. Note that even the dynamically-unimportant fields that are consistent with the present treatment of a beam as a field-free flow may still dominate the diffusion of relativistic electrons.) Observation-ally, there is little evidence for the presence of limb–brightening, except in circumstances like the first knot in the Cen A radio jet (Clarke *et al.*, 1986a) where it is clear that complicated flow phenomena are occurring. Although the result that distorted beams show no limb brightnening is in accord with the predictions of the Kelvin–Helmholtz model, this is also a feature of other models for beam–bending. It should be stressed that the effects of projection can cause complicated patterns to be seen from simple beam structures (Zaninetti & Van Horn 1988), so that the relative orientation of the beam and the line of sight is of critical importance in interpreting the jet structures. Furthermore, the assumption that the pressure perturbation is simply related to the radio brightness of the beam is certainly unjustified: no convincing discussion of beam appearance with a realistic particle acceleration model has been produced.

3.5.1. *Application to 3C 449*

3C 449 (Sec. 1.2.4) is the archetypal distorted radio jet for which most comparisons between the structure and the Kelvin-Helmholtz instability model (and other models; e.g. Blandford & Icke 1978; Wirth *et al.* 1982) have been made. Hardee (1981) interpreted the large side-to-side oscillations in the radio structure of this two–sided jet in terms of a beam affected by an $n = 1$ Kelvin–Helmholtz mode with $\lambda_0 \approx 16R$ and $\lambda_e \approx 6R$ in a beam with Mach number $\mathcal{M}_b = 10$, and equal densities in the beam and the ambient medium. Assuming that the observed non-linear structure of 3C 449's jet reflects the linear mode structure imposed by the initial growth of Kelvin-Helmholtz modes, we can use the results of Figs 3.4 and 3.4 to discuss the structure of the flow.

In Figs 3.4 and 3.4 we see that the shortest growth length of the $n = 1$ ordinary mode in a $M_b = 10$, equal-density, beam varies from about $10R$ to about $42R$ as the width of a sheathing layer is increased from 0 to R. At the same time, the wavelengths of the modes increase from about $27R$ to about $79R$. To obtain agreement with the structure of the jet in 3C 449, the direction of the jet must be inclined at an angle of about 50 degrees to the plane of the sky and the shear layer in 3C 449 must be thin, since a thick shear layer causes the growth length to be too large relative to the wavelength.

The $n = 1$ reflection modes can also produce helical waves on the beam, since they have the appropriate θ-symmetry to deflect the centre line of the flow, but for $M_b = 10$, all these modes have wavelengths that are too small to produce the observed structure of 3C 449, whether or not the beam is sheathed. Thus if the beam in 3C 449 has $M_b = 10$, and a density equal to that of the external medium, the observed wiggles are produced by the Kelvin-Helmholtz instability only if the edge of the jet is sharp.

This conclusion can be avoided if the observed width of the radio emission in the radio jet of 3C 449 does not reflect the width of the underlying beam—perhaps particle diffusion causes the radio emission to come from a much larger radius than the beam-core or beam-core plus shear layer. In this case, a lower-density or a faster beam may be a better description of 3C 449, and the ratio 16 : 6 of growth length to wavelength needed to fit the radio structure may be achieved by increasing the beam velocity (or density contrast) for a sheathed beam. The view that the beam in 3C 449 should be surrounded by a shear layer is consistent with the very short growth lengths of many reflection modes for the vortex-sheet beam model, so that even if these modes saturate at low amplitudes, their combined effect should be to broaden the edge of the beam. A very low density for the beam relative to the ambient medium may not be appropriate, however: numerical simulations of beam flows (e.g., Norman *et al.* 1984) find that such beams quickly develop a surrounding, lower-density, cocoon so that the beam is in contact with a low-density medium, and it is the cocoon that is in contact with the denser, external, medium.

The good, two-sided, C-type, symmetry of the radio structure of 3C 449 is a major problem for a pure Kelvin-Helmholtz instability model of the structure. Local conditions along the paths of the beams from the core of the galaxy should not produce phase-coherent structures on the two sides of the source. The perturbation driving the Kelvin-Helmholtz modes must therefore be applied to both beams coherently, presumably as perturbations at beam injection by the engine at the centre of the galaxy (UGC 12064). The observation that only a few wavelengths of oscillation appear in the structure of 3C 449 also argues that the perturbations are introduced at large amplitude, or that there is a strong non-linear mode competition between Kelvin-Helmholtz modes excited on the beam.

The apparent purity of the $n = 1$ mode causing the structure in 3C 449 is also a problem for the Kelvin-Helmholtz instability model (Birkinshaw 1984), since modes with growth length λ_e within about λ_e^{*2}/z of the minimum of the growth curve, λ_e^*, are amplified by the instability by within a factor $1/e$ of the maximum growth rate after the beam has travelled distance z. However, in 3C 449, it is clear that all the $n = 1$ modes that are present have wavelengths within $\Delta\lambda_0/\lambda_0 < 0.2$ of the mode that has grown. Since the jet is less than five wavelengths long, the mode appears to have been tuned more precisely than is possible according to Fig. 3.4. This problem is alleviated somewhat if the beam in 3C 449 is surrounded by a thick shear layer—such a layer in a low-density, fast, beam can increase the sharpness of a mode several-fold. This is not, of itself, sufficient to account for the sharpness of

3C 449's structure: it would require a propagation distance greater than $10\lambda_0^*$ for sufficient mode purity to be achieved for the clear structure of 3C 449 to be produced, whereas only two or three waves of the 'helical' structure of 3C 449 are seen on either side of the source. Therefore, if the Kelvin-Helmholtz model is to be adopted, strong mode competition must operate near the peak of the $n = 1$ resonance.

3.5.2. *Application to NGC 6251*

The spectacular jet of NGC 6251 was described in Sec. 1.2.3. It shows a number of wave-like structures. Within the first 40 arcsec, slowly–growing structures with apparent wavelength 5.7 and 9 arcsec can be identified although it is not known whether these correspond to side-ways displacements of the jet ($n = 1$–like modes) or merely brightness enhancements and pinches ($n = 0$–like modes). At intermediate angular scales (120 to 240 arcsec), a rapidly–growing 31–arcsec oscillation which produces knots in the jet can be identified. Finally, on the largest scales (beyond 240 arcsec), the radius of the jet averages ≈ 10 arcsec and increases almost linearly, and the structure is dominated by a 143–arcsec oscillation which produces bulk displacements of the jet, and which has an almost linear growth with distance down the jet. In what follows, I will focus on the last, and best-defined, of the structural distortions of NGC 6251. In this region Perley *et al.* (1984a) estimate the flow to have a Mach number $\mathcal{M}_b \approx 13$, and that the beam is a factor $\gtrsim 3$ times less dense than the ambient medium, if the oscillations are interpreted as helical Kelvin-Helmholtz instabilities.

Hardee (1982, 1984) showed that on linearly expanding beams the Kelvin-Helmholtz instability does not grow exponentially, but that the local stability properties of the beam (that might be calculated using the local parameters of the flow and the local beam radius) may be used as global indicators of the stability of the flow. If the structural distortions of NGC 6251 are caused by the growth of a Kelvin-Helmholtz instability, then it is required that the beam should display a dominant $n = 1$ mode with $\lambda_0^* \sim 15R$. Perley *et al.*'s parameters for the beam are close to those for Fig. 3.4, although the lower beam density and higher speed may cause it to assume some of the (very similar) characteristics of the models in Figs 3.4 and 3.4. Then, just as for 3C 449, the dominant $n = 1$ mode for structures with scales $\gtrsim 10R$ is the ordinary mode, for which $\lambda_0^* \approx 27R$ and $\lambda_e^* \approx 10R$ if $\delta = 0$ and $\lambda_0^* \approx 79R$ and $\lambda_e^* \approx 42R$ if $\delta = 1$. A structure with projected wavelength $\lambda_0 \approx 15R$ cannot appear on this beam unless the apparent radio radius overestimates the radius of the flow.

At the large distance from the centre of NGC 6251 that the 143-arcsec oscillation is seen, the injected profile of the beam in NGC 6251 must have been modified by the growth of high-order modes. If the reflection modes saturate at amplitudes that cause weak shocks (Norman *et al.* 1984), then the modification of the beam profile on the small scale will have been severe, and the beam should have taken up a self-consistent form in which the further growth of small-scale modes is minimised. This argument would suggest that at any distance from the centre of the radio source, the dominant scale of structure that can be seen in the beam will be similar to the distance that the beam has propagated. This is certainly true for NGC 6251, and the question to be asked based on the shear layer analysis is then whether the dominant mode at wavelengths $\approx 15R$ has the correct $(n = 1)$ symmetry to describe the oscillations in the position of the radio jet. According to Fig. 3.4, this is not possible unless the beam flowing through the radio jet is much narrower than the radio structure. The alternative, that the flow parameters estimated by Perley *et al.* (1984) are in error by a large factor, is perhaps more likely: simple results for the growth rates of Kelvin-Helmholtz instabilities were used to derive the flow parameters, and a shear layer could modify these rates substantially. Hardee & Norman (1989), from a slab beam simulation, also found that Perley *et al.*'s flow parameters were inconsistent with the generation of $n = 1$ Kelvin-Helmholtz modes of the correct wavelength, and suggested a smaller Mach number for the flow ($\mathcal{M}_b \sin \theta_{los} \sim 5$, where θ_{los} is the angle between the flow direction and the line of sight).

The one-sidedness of the jet in NGC 6251, and the lack of definition of the helical structure identified by Perley *et al.*, removes two of the arguments against the Kelvin-Helmholtz instability as a cause of the structure in this source, and the discussion above indicates that there is sufficient freedom in the parameters of the model to create agreement with the inferred $n = 1$ mode if the high-order modes are saturated at small amplitude. Further progress on applying the model to NGC 6251 awaits a calculation of the self-consistent structure of a beam that minimises the growth of the Kelvin-Helmholtz instabilities, and more precise data on the physical parameters of the flow and the ambient medium.

3.5.3. *Application to HH 34*

It is widely thought that galactic jets are dense (Section 2), so that their pattern of Kelvin-Holmholtz instabilities should more closely resemble Fig. 3.4 or 3.4 than Fig. 3.4 or 3.4. For almost all dense jets, at large wavelengths the growth rates increase as $\lambda_e \propto \lambda_0$, while at small wavelengths the growth rates are limited by the formation of a shear layer.

The flow pattern seen in HH 34 (Section 1.2.9) is not what we would predict based on these instability calculations. Rather what we are see-

ing resembles a 1-sided ejection of lumps of material, each with its own
bow-shock. This corresponds more with a "bullet" than a "jet" model
for the outflow, and Kelvin-Helmholtz instabilities are not appropriate
for a discussion of the overall flow. Furthermore, the density of the outflowing material is such that radiative losses should be substantial—the
equations of motion adopted for the analysis of the instability, which
include the condition of the conservation of entropy are not appropriate
in this case.

Instabilities will, however, be active in this flow: not only the Kelvin-
Holmholtz instability, but also the Rayleigh-Taylor instability, both at
the sides of the moving blobs of material and at their heads. These
instabilities can be expected to act on very small linear scales, and give
the moving blobs of gas an excited sheath, which may be the site of
the [SII] and other line emission seen in optical images.

3.6. SUMMARY AND FINAL CONSIDERATIONS

The major theoretical difficulty in discussing astrophysical beams is
that they are likely to be strongly turbulent—the Reynolds numbers of
radio jets are $\sim 10^{28}$. An accurate description of jet structures, including bending and the propagation of large–scale waves, requires the use
of a physically–consistent model for turbulence that permits reliable
calculation of the Reynolds stresses.

Work so far has also neglected the effects of current sheets and field
loops, where strong energy releases and anomalous transport coefficients can arise. Other entropy–generating processes have also been
ignored. Chief among these is the transfer of energy from the flow to
radiating particles, and hence the loss of energy through radiation. The
thermal conductivity of the beam fluid may also be relevant if there is
a significant transfer of heat by diffusive conduction.

Since the observed structures in jets are certainly of large amplitude, the theory of Kelvin–Helmholtz modes needs to be extended to
describe their non–linear evolution. Only rudimentary ideas are available at present since the numerical results apply only to the longer–
wavelength modes. These numerical simulations also suggest that the
jet becomes sheathed by a hot, sheared cocoon which might protect
the beam against Kelvin–Helmholtz instabilities. We need to calculate
the structure of the (turbulent) flow pattern and sheathing layer that
should be produced by the operation of short–wavelength instabilities
on the beam.

For large–scale coherent structures to develop on the beam, the per-
turbing force must be coherent for times greater than the period of
the structure. Since the dominant perturbations are most likely to be

those encountered by the beam as it leaves a nozzle and is injected into the ambient medium, either a coherent perturbation is applied there, or mode competition between co–existing modes on the beam operates to set the dominant structure. The steadiness of the beam as it leaves the nozzle is also an issue, since theoretical investigations of the Kelvin–Helmholtz modes rely on the absence of significant variations in the injected flow. If such variations do exist, then it is likely that only full, three–dimensional numerical simulations will be capable of useful investigations of beam stability.

None of the instability calculations so far have taken account of the effects of structure in the external medium, except insofar as beam expansion has been included in Hardee's (1982, 1984) treatments. If the scale of variations in the beam parameters is much larger than the wavelengths and growth lengths of the instabilities, then the theory is changed only by the need to apply the dispersion relation locally. If the gradients of the beam parameters are large, then the effects of Kelvin–Helmholtz waves are unlikely to be important compared with the effects of shocks and other structures produced as the beam adjusts to its environment(e.g., Gouveia Dal Pino *et al.* 1995). Buoyancy effects on the dynamics of the beam in the stratified ambient atmosphere may also be important (Achterberg, 1982).

The present state of the theory of the Kelvin–Helmholtz instability for beams is that laminar, cylindrical, fluid flows are unstable for all beam velocities and densities unless the beam contains a strong, axial magnetic field, and that high–density flows are less unstable than low–density flows of the same beam velocity. Azimuthal magnetic fields, which can help to confine the beam, tend to destabilise the flow. Shear layers can either stabilize or destabilise the flow. The growth lengths of the instability are short, and so that it is likely to produce beam heating and turbulence on small scales (perhaps in a beam–sheath). On larger scales, where observable structural distortions of the flow may result, it has not yet been possible to find unambiguous predictions of the dominant modes because of uncertainties in the beam structure and the lack of information on the non–linear limit of the modes. A number of types of Kelvin–Helmholtz mode occur: the ordinary mode is the simple analogue of the instability as seen on a semi–infinite planar surface, while reflection modes arise from the partial trapping of magnetosonic waves inside the beam and resonances of these waves with the beam/ambient medium interface (over–reflection). The ordinary mode dominates the stability of a beam for low Mach number flows, whilst the reflection modes dominate for fast flows.

The Kelvin–Helmholtz model for the origin of structures in beams has achieved some successes, and such instabilities must influence the development of laminar beam flows, but at present little can be learned about any given flow through the application of the theory. If self–consistent results for the profiles of the jets and better estimates of the flow parameters were available, the present theory would allow strong statements about the stability of laminar flows to be made. Little work has been done so far on the development of coherent structures in turbulent flows.

References

Achterberg, A., 1982. A&A, 114, 233

Arnaud, K.A., Fabian, A.C., Eales, S.A., Jones, C. & Forman, W., 1984. MNRAS, 211, 981

Bailyn, C.D., Orosz, J.A., Girard, T.M., Jogee, S., della Valle, M., Begam, M.C., Fruchter, A.S., González, R., Ianna, P.A., Layden, A.C., Martins, D.H. & Smith, M., 1995. Nature, 374, 701

Bailyn, C.D., Orosz, J.A., McClintock, J.E. & Remillard, R.A., 1995. Nature, 378, 157

Bally, J. & Devine, D., 1994. ApJ, 428, L65

Balsara, D.S. & Norman, M.L., 1992. ApJ, 393, 631

Barsony, M., Scoville, N.Z. & Chandler, C.J., 1993. ApJ, 409, 275

Begelman, M.C., Rees, M.J. & Blandford, R.D., 1979. Nature, 279, 770

Benford, G., 1981. ApJ, 247, 792

Bicknell, G.V., 1984, In *Physics of energy transport in extragalactic radio sources*, eds. Bridle, A.H. & Eilek, J.A. (NRAO: Green Bank, WV) p. 229

Biretta, J.A. & Meisenheimer, K., 1993. In *Jets in Extragalactic Radio Sources*, 159; eds. Röser, H.-J. & Meisenheimer, K.; Springer, Berlin

Birkinshaw, M., 1984. MNRAS, 208, 887

Birkinshaw, M., 1991. MNRAS, 252, 505

Birkinshaw, M., Ho, P.T.P., Reid, M.J. & Zheng, X.W., 1995. In preparation

Birkinshaw, M. & Worrall, D.M., 1991. ApJ, 412, 568

Blake, G.M., 1972. MNRAS, 156, 67

Blandford, R.D. & Eichler, D., 1987. Phys. Rep., 154, 1

Blandford, R.D. & Icke, V., 1978. MNRAS, 185, 527

Blandford, R.D. & Rees, M.J., 1974. MNRAS, 169, 395

Blandford, R.D. & Pringle, J.E., 1976. MNRAS, 176, 443

Blondin, J.M., Fryxell, B.A. & Königl, A., 1990. ApJ, 360, 370

Blumen, W., Drazin, P.G. & Billings, D.F., 1975. J. Fluid Mech., 71, 305

Bodo, G., Rosner, R., Ferrari, A. & Knobloch, E., 1989. ApJ, 341, 631

Burbidge, G.R., 1956. ApJ, 124, 416

Burns, J.O., 1984. In *Physics of energy transport in extragalactic radio sources*, eds. Bridle, A.H. & Eilek, J.A. (NRAO: Green Bank, WV), p. 25

Burns, R.L. & Lovelace, R.V.E., 1982. Ap. J., 262, 87

Carilli, C.L., Perley, R.A. & Dreher, J.W., 1988. ApJ, 334, L73

Carilli, C.L., Bartel,N. & Diamond, P., 1994. AJ, 108, 64

Carilli, C.L., Perley, R.A., Leahy, J.P. & Dreher, J.W., 1991. ApJ, 383, 554

Chandrasekhar, S., 1961, *Hydrodynamic and Hydromagnetic Stability*, (Oxford University Press: Oxford)

Chernin, L.M. & Masson, C.R., 1995. ApJ, 443, 181

Cioffi, D.F. & Blondin, J.M., 1992. ApJ, 392, 458

Cioffi, D.F. & Jones, T.W., 1980. AJ, 85, 368

Clarke, D.A., Burns, J.O. & Feigelson, E.D., 1986. ApJ, 300, L41

Clarke, D.A., Norman, M.L. & Burns, J.O. 1986. ApJ, 311, L63

Clemmow, P.C. & Dougherty, J.P., 1969. *Electrodynamics of Particles and Plasmas*, (Addison-Wesley: Reading, MA)

Cohen, M.H. & Readhead, A.C.S., 1979. ApJ, 233, L101

Cohn, H., 1983. ApJ, 269, 500

Conway, R.G., Garrington, S.T., Perley, R.A. & Biretta, J.A., 1993. A&A, 267, 347

Curtis, H.D., 1918. Lick Obs. Publ., 13, 11

Davis, D.S., Mushotzky, R.F., Mulchaey, J.S., Worrall, D.M., Birkinshaw, M. & Burstein, D., 1995. ApJ, 444, 582

De Young, D.S., 1991. ApJ, 371, 69

De Young, D.S., 1984. In *Physics of energy transport in extragalactic radio sources*, eds. Bridle, A.H. & Eilek, J.A. (NRAO: Green Bank, WV), p. 202

Drazin, P.G. & Davey, H., 1977. J Fl. Mech., 82, 255

Drazin, P.G. & Reid, W.H., 1981. *Hydrodynamic Stability*, (Cambridge University Press: Cambridge)

Dreher, J.W., Carilli, C.L. & Perley, R.A., 1987. ApJ, 316, 611

Drury, L.O'C., 1983. Rep. Prog. Phys., 46, 973

Eilek, J.A., 1982. ApJ, 254, 472

Ershkovich, A.I., 1980. Sp. Sci. Rev., 25, 3

Fanaroff, B.L. & Riley, J.M., 1974. MNRAS, 167, 31P

Ferrari, A., Massaglia, S. & Trussoni, E., 1982. MNRAS, 198, 1065

Ferrari, A., Trussoni, E. & Zaninetti, L., 1978. A&A, 64, 43

Ferrari, A., Trussoni, E. & Zaninetti, L., 1979. A&A, 79, 190

Ferrari, A., Trussoni, E. & Zaninetti, L., 1980. MNRAS, 193, 469

Ferrari, A., Trussoni, E. & Zaninetti, L., 1981.MNRAS, 196, 1051

Fiedler, R.L., 1986. ApJ, 305, 100

Fiedler, R.L. & Jones, T., 1984. ApJ, 283, 532

Ford, H.C., Harms, R.J., Tsvetanov, Z.I., Hartig, G.F., Dressel, L.L., Kriss, G.A., Bohlin, R.C., Davidsen, A.F., Margon, B. & Kochhar, A.J,. 1994. ApJ, 435, L27

Forman, W. & Jones, C., 1982. ARAA, 20, 547

Fridlund, C.V.M. & Liseau, R., 1994. A&A, 292, 631

Gill, A.E., 1965. Phys. Fluids, 8, 1428

Goodman, A.A., Crutcher, R.C., Heiles, C., Myers, P.C. & Troland, T.H., 1989. ApJ, 338, L61

Gouveia Dal Pino, E.M. & Benz, W., 1993. ApJ, 410, 686

Gouveia Dal Pino, E.M., Birkinshaw, M. & Benz, W., 1995. ApJ, submitted

Hardee. P.E., 1979. ApJ, 234, 47

Hardee. P.E., 1981. ApJ, 250, L9

Hardee. P.E., 1982. ApJ, 257, 509

Hardee. P.E., 1983. ApJ, 269, 94

Hardee. P.E., 1984. ApJ, 287, 523

Hardee. P.E., 1986. ApJ, 303, 111

Hardee. P.E., 1987a. ApJ, 313, 607

Hardee. P.E., 1987b. ApJ, 318, 78

Hardee. P.E. & Clarke, D.A., 1992. ApJ, 400, L9

Hardee, P.E. & Norman, M.L., 1988. ApJ, 334, 70

Hardee, P.E. & Norman, M.L., 1989. ApJ, 342, 680

Hardee, P.E. & Norman, M.L., 1990. ApJ, 365, 134

Hardee, P.E., White, R.E., Norman, M.L., Cooper, M.A. & Clarke, D.A., 1992. ApJ, 387, 460

Harms, R.J., Ford, H.C., Tsvetanov, Z.I., Hartig, G.F., Dressel, L.L., Kriss, G.A., Bohlin, R.C., Davidsen, A.F., Margon, B. & Kochhar, A.J,. 1994. ApJ, 435, L35

Harris, D.E. & Stern, C.P., 1987. ApJ, 313, 136

Henriksen, R.N., 1987. ApJ, 314, 33

Hester, J. *et al.*, 1995. NASA press release, 6 June 1995

Hjellming, R.M. & Johnston, K.J., 1988. ApJ, 328, 600
Hjellming, R.M. & Rupen, M.P., 1995. Nature, 375, 464
Ho, C.M. & Huerre, P., 1984. Ann. Rev. Fluid Mech., 16, 365
Hughes, P.A. (ed.), 1991. *Beams and Jets in Astrophysics*, (Cambridge University Press: Cambridge)
Hughes, P.A., Aller, H.D. & Aller, M.F., 1985. ApJ, 298, 301
Jones, T.W. & O'Dell, S.L., 1977. ApJ, 214, 522
Junor, W. & Biretta, J.A., 1995. AJ, 109, 500
Kim, K.-T., Kronberg, P.P., Dewdney, P.E. & Landecker, T.L., 1986. In *Radio Continuum Processes in Clusters of Galaxies*, eds. O'Dea, C.P. & Uson, J.M. (NRAO: Green Bank)a
Kirk, J.G. 1994. In *Plasma Astrophysics*, Saas-Fee 24, eds. Benz, A.O. & Courvoisier, T.J.-L. (Springer-Verlag: Berlin)
Kochhar, R.K. & Trehan, S.K., 1988. MNRAS, 234, 123
Königl, A., 1980. Phys. Fluids, 23, 1083a
Laing, R.A., 1980. MNRAS, 193, 439
Laing, R.A., 1981. Ap. J., 248, 87
Laing, R.A., 1984. In *Physics of energy transport in extragalactic radio sources*, eds. Bridle, A.H. & Eilek, J.A. (NRAO: Green Bank, WV)
Laing, R.A. & Bridle, A.H., 1987. MNRAS, 228, 557
Landau, L.D. & Lifshitz, E.M., 1959. *Fluid Mechanics*; Pergamon Press
Leahy, J.P., 1984. MNRAS, 208, 323
Leahy, J.P., 1991. In *Beams and Jets in Astrophysics*; ed. Hughes, P.A. (Cambridge University Press: Cambridge)
Lerche, I. & Wiita, P.J., 1980. Ap. Sp. Sci., 68, 207
Linfield, R., 1985. ApJ, 295, 463
Loken, C., Burns, J.O., Clarke, D.A. & Norman, M.L., 1992. ApJ, 392, 54
Margon, B., 1984. ARAA, 22, 507
Misner, C.W., Thorne, K.S. & Wheeler, J.A., 1973. Gravitation, (Freeman: San Fransisco)
Morabito, D.D., Preston, R.A. & Jauncey, D.I., 1988. AJ, 95, 1037
Moriarty-Schieven, G.H. & Snell, R.L., 1988. ApJ, 332, 364
Mundt, R. & Fried, J., 1983. ApJ, 274, L83
Nepveu, M., 1982a. A&A, 105, 15
Nepveu, M., 1982b. A&A, 112, 223
Norman, M.L. & Hardee. P.E., 1988. ApJ, 334, 80
Norman, M.L., Smarr, L., Winkler, K.-H.A. & Smith, M.D., 1982. A&A, 113, 285
Norman, M.L., Winkler, K.-H.,A. & Smarr, L.L., 1983. In *Astrophysical Jets*, eds. Ferrari, A. & Pacholczyk, A.G. (Reidel: Dordrecht, Holland), p. 227.
Norman, M.L., Winkler, K-H.A. & Smarr, L., 1984. In *Physics of energy transport in extragalactic radio sources*, eds. Bridle, A.H. & Eilek, J.A. (NRAO: Green Bank, WV), p. 150
O'Dea, C.P. & Owen, F.N., 1986. ApJ, 301, 841
Owen, F.N., Hardee, P.E. & Cornwell, T.J., 1989. ApJ, 340, 698
Payne, D.G. & Cohn, H., 1985. ApJ, 291, 655
Pearson, T.J., Unwin, S.C., Cohen, M.H., Linfield, R.P., Readhead, A.C.S., Seielstad, G.A., Simon, R.S. & Walker, R.C., 1981. Nature, 290, 365
Perley, R.A., Bridle, A.H. & Willis, A.G., 1984a. ApJS, 54, 291
Perley, R.A., Dreher, J.W. & Cowan, J., 1984b. ApJ, 285, L35
Perley, R.A., Willis, A.G. & Scott, J.S., 1979. Nature, 281, 437
Ray, T.P., 1981. MNRAS, 196, 195
Ray, T.P., 1982. MNRAS, 198, 617
Ray, T.P. & Ershkovich, A.I., 1983. MNRAS, 204, 821
Rees, M.J., 1971. Nature, 229, 312 and 510

Reid, M.J., Biretta, J.A., Junor, W., Muxlow, T.W.B. & Spencer, R.E., 1989. ApJ, 336, 112

Reipurth, B., Bally, J., Graham, J.A., Lane, A.P. & Zealey, W.J., 1986. A&A, 164, 51

Roberts, B., 1987. ApJ, 318, 590

Roberts, F.A., Dimotakis, P.E. & Roshko, A., 1982. In *Album of Fluid Motions*, ed, A. Van Dyke, (Parabolic Press: Stanford, California)

Rodriguez, L.F., Canto, J., Torrelles, J.M. & Ho, P.T.P., 1986. ApJ, 301, L25

Roy Choudhury, S. & Lovelace, R.V.E., 1984. ApJ, 283, 331

Roy Choudhury, S. & Lovelace, R.V.E., 1986. ApJ, 302, 188

Rudnick, L. & Edgar, B.K., 1984. ApJ, 279, 74

Sarazin, C., 1986. Rev. Mod. Phys., 58, 1

Scheuer, P.A.G., 1974. MNRAS, 166, 513

Seward, F., Grindlay, J., Seaquist, E. & Gilmore, W., 1980. Nature, 287, 806

Snell, R.L. & Schloerb, F.P., 1985. ApJ, 295, 490

Spitzer, L., 1962. Physics of fully ionized gases, (Interscience: New York)

Stewart, P., 1971. Ap. Sp. Sci., 14, 261

Stocke, J.T., Hartigan, P.M., Strom, S.E., Strom, K.M., Anderson, E.R., Hartmann, L.W. & Kenyon, S.J., 1988. ApJS, 68, 229

Synge, J.L., 1957. *The Relativistic Gas* (North–Holland Publ. Co.: Amsterdam)

Thomson, R.C., Mackay, C.D. & Wright, A.E., 1993. Nature, 365, 133

Trussoni, E., Massaglia, S., Bodo, G. & Ferrari, A., 1988. MNRAS, 234, 539

Turland, B.D. & Scheuer, P.A.G., 1976. MNRAS, 176, 421

Vermeulen, R.C., Schilizzi, R.T., Spencer, R.E., Romney, J.D. & Fejes, I., 1993. A&A, 270, 177

Waggett, P.C., Warner, P.J. & Baldwin, J.E., 1977. MNRAS, 181, 465

Watson, M.G., Willingale, R., Grindlay, J.E. & Seward, F.D., 1983. ApJ, 273, 688

Wiita, P.J. & Norman, M.L., 1992. ApJ, 385, 478

Wiita, P.J., Rosen, A. & Norman, M.L., 1990. ApJ, 350, 545

Williams, A.G., 1991. In *Beams and Jets in Astrophysics*, (Cambridge University Press: Cambridge)

Williams, A.G. & Gull, S.F., 1984. Nature, 310, 33

Williams, A.G. & Gull, S.F., 1985. Nature, 313, 34

Wirth, A., Smarr, L. & Gallagher, J.S., 1982. AJ, 87, 602

Worrall, D.M. & Birkinshaw, M., 1994. ApJ, 427, 134

Zaninetti, L., 1986a. A&A, 156, 194

Zaninetti, L., 1986b. A&A, 160, 135

Zaninetti, L. & Van Horn, H.M., 1988. A&A, 189, 45

Tony Peratt

Advances in Numerical Modeling of Astrophysical and Space Plasmas

Dedicated to the memories of Hannes Alfvén and Oscar Buneman; Founders of the Subject.

Anthony L. Peratt *
Los Alamos National Laboratory
Los Alamos, New Mexico, USA

Abstract. Plasma science is rich in distinguishable scales ranging from the atomic to the galactic to the meta-galactic, i.e., the *mesoscale*. Thus plasma science has an important contribution to make in understanding the connection between microscopic and macroscopic phenomena. Plasma is a system composed of a large number of particles which interact primarily, but not exclusively, through the electromagnetic field. The problem of understanding the linkages and couplings in multi-scale processes is a frontier problem of modern science involving fields as diverse as plasma phenomena in the laboratory to galactic dynamics.

Unlike the first three states of matter, plasma, often called the fourth state of matter, involves the mesoscale and its interdisciplinary founding have drawn upon various subfields of physics including engineering, astronomy, and chemistry. Basic plasma research is now posed to provide, with major developments in instrumentation and large-scale computational resources, fundamental insights into the properties of matter on scales ranging from the atomic to the galactic. In all cases, these are treated as mesoscale systems. Thus, basic plasma research, when applied to the study of astrophysical and space plasmas, recognizes that the behavior of the near-earth plasma environment may depend to some extent on the behavior of the stellar plasma, that may in turn be governed by galactic plasmas. However, unlike laboratory plasmas, astrophysical plasmas will forever be inaccessible to in situ observation. The inability to test concepts and theories of large-scale plasmas leaves only virtual testing as a means to understand the universe. Advances in in computer technology and the capability of performing physics first principles, fully three-dimensional, particle–in–cell simulations, are making virtual testing a viable alternative to verify our predictions about the far universe.

The first part of this paper explores the dynamical and fluid properties of the plasma state, plasma kinetics, and the radiation emitted from plasmas. The second part of this paper outlines the formulation for the particle-in-cell simulation of astrophysical plasmas and advances in simulational techniques and algorithms, as-well-as the advances that may be expected as the computational resource grows to petaflop speed/memory capabilities.

Key words: Numerical Simulation, Astrophysical Plasmas, Space Plasmas, Plasma Universe, Electric Space

---

* Scientific Advisor, Office of Research and Development, United States Department of Energy, Washington D.C.

Astrophysics and Space Science **242**: 93–163, 1997.
©1997 *Kluwer Academic Publishers.*

1. "In-Situ" Observation of Astrophysical and Space Plasmas via Computer Simulation

While it is thinkable that our ability to make in situ measurements can perhaps be extended to the nearest stars, most of the universe beyond a few parsecs will be beyond the reach of our spacecraft forever.

From one's unaided view of the clear night sky, it is tantalizing to believe that the physics of the universe can be unfolded from the observable stars, which may be up to kiloparsecs away, or from the fuzzy "nebula" such as the galaxy M31, nearly a megaparsec away. Our experience in unfolding energetic events in our own solar system suggests otherwise.

The inability to make in situ observations places a severe constraint on our ability to understand the universe, which is to say, astrophysical and space plasmas, even when the full electromagnetic spectrum is available to us. Only in the last $2\frac{1}{2}$ decades, after satellites monitored our near-earth environment and spacecraft discovered and probed the plasma magnetospheres of the planets, could we begin to get a true picture of the highly-energetic processes occurring everywhere in the solar system. These plasma processes included large-scale magnetic-field-aligned currents and electric fields and their role in the transport of energy over large distances. The magnetospheres of the planets are invisible in the visual octave (400-800 nm) and from earth cannot even be positively identified in the X ray and gamma ray regions, which cover 10 times as many octaves and have more than 1,000 times the bandwidth as the visual octave. Only in the low frequency radio region is there a hint of the presence of quasi-static electric fields which accelerate charged particles in the magnetospheres of the planets.

As the properties of plasma immediately beyond the range of spacecraft are thought not to change, it must be expected that plasma sources of energy and the transport of that energy via field-aligned currents exist at even larger scales than that found in the solar system. How then are we to identify these mechanisms in the distant universe?

The advent of particle-in-cell (PIC) simulation of cosmic plasmas on large computer systems ushered in an era whereby in situ observation in distant or inaccessible plasma regions is possible. While the first simulations were simple, with many physics issues limited by constraints in computer speed and memory, it is now possible to study the full three-dimensional, fully-electromagnetic evolution of magnetized plasma over a very large range of sizes. In addition, PIC simulations have matured enough to contain Monte Carlo collisional scattering and energy loss treatments, conducting sheets or surfaces, dielectric regions, space-charge-limited emission from surfaces and regions, and electromagnetic wave propagation. Since a simulation involves the motion of charge or mass particles according to electromagnetic or gravitational forces, all in situ information is available to the simulationist.

If the simulation correctly models the cosmic plasma object under study, replication of observations over the entire electromagnetic spectrum should be expected, to the extent that the model contains sufficient temporal and spatial resolution.

However, before the simulationist can model a particular problem of astrophysical or space plasma interest, fundamental knowledge is required about the nature of the plasma state under study, its characteristic properties and behavior, its evolution, but perhaps most importantly, its mesoscale physics especially in regards to plasmas of larger dimensions which are most probably the source of energy of the observed radiation.

2. The Plasma State

2.1. PLASMA

Plasma consists of electrically charged particles that respond collectively to electromagnetic forces. The charged particles are usually clouds or beams of electrons or ions, or a mixture of electrons and ions, but also can be charged grains or dust particles. Cosmic plasma is also at temperatures comparable to or higher than that in the interior of stars. At these temperatures, only the constituent building blocks of light atoms exist: positively charged bare nuclei and negatively charged free electrons. The name plasma is also properly applied to ionized gases at lower temperatures where a considerable fraction of neutral atoms or molecules are present.

While all matter is subject to gravitational forces, the positively charged nuclei, or ions, and the negatively charged electrons of plasmas react strongly to electromagnetic forces, as formulated by Oliver Heaviside (1850-1925),[1] but now called Maxwell's Equations, after James Clerk Maxwell (1831-1879),[2]

$$\frac{\partial \mathbf{B}}{\partial t} = -\nabla \times \mathbf{E} \tag{1}$$

---

[1] Oliver Heaviside was the first to reduce Maxwell's 20 equations in 20 variables to the two equations 1 and 2 in vector field notation. For some years Eqs. 1- 4 were known as the Hertz-Heaviside Equations, and later A. Einstein called them the Maxwell-Hertz Equations. Today, only Maxwell's name is mentioned [Nahin 1988].

[2] These are "rewritten" in update form ideal for programming. This also emphasizes the causality correctly: curl $\mathbf{E}$ is the cause of changes in $\mathbf{B}$, curl $\mathbf{H}$ is the cause of changes in $\mathbf{D}$.

$$\frac{\partial \mathbf{D}}{\partial t} = \nabla \times \mathbf{H} - \mathbf{j} \tag{2}$$

$$\nabla \cdot \mathbf{D} = \rho \tag{3}$$

$$\nabla \cdot \mathbf{B} = 0 \tag{4}$$

and the equation of motion due to Hendrik Antoon Lorentz (1853-1928),

$$\frac{d}{dt}(m\mathbf{v}) = q(\mathbf{E} + v \times \mathbf{B}); \quad \frac{d\mathbf{r}}{dt} = \mathbf{v}; \tag{5}$$

The quantities $\mathbf{D} = \epsilon\mathbf{E}$ and $\mathbf{B} = \mu\mathbf{H}$ are the constitutive relations between the electric field $\mathbf{E}$ and the displacement $\mathbf{D}$ and the magnetic induction $\mathbf{B}$ and magnetic intensity $\mathbf{H}$, μ and ϵ are the permeability and permittivity of the medium, respectively, and ρ and $\mathbf{j}$ are the charge and current densities, respectively.[3] The mass and charge of the particle obeying the force law (Eq. 5) are m and q, respectively.

Because of their strong interaction with electromagnetism, plasmas display a complexity in structure and motion that far exceeds that found in matter in the gaseous, liquid, or solid states. For this reason, plasmas, especially their electrodynamic properties, are far from understood. Irving Langmuir; (1881-1957), the electrical engineer and Nobel chemist, coined the term plasma in 1923, probably borrowing the term from medical science to describe the collective motions that gave an almost lifelike behavior to the ion and electron regions with which he experimented. Langmuir was also the first to note the separation of plasma into cell-like regions separated by charged particle sheaths. Today, this cellular structure is observed wherever plasmas with different densities, temperatures, magnetic field strengths, chemical constituencies, or matter–antimatter plasmas come in contact.

Plasmas need not be neutral (i.e., balanced in number densities of electrons and ions). Indeed, the study of pure electron plasmas and even positron plasmas, as well as the electric fields that form when electrons and ions separate, are among the most interesting topics in plasma research today. In addition to cellular morphology, plasmas often display a filamentary structure. This structure derives from the fact that plasma, because of its free electrons, is a good conductor of electricity, far exceeding the conducting properties of metals such as copper or gold. Wherever charged particles flow in a neutralizing medium, such as free electrons in a background of ions, the charged particle flow or current produces a ring of magnetic field around the current, pinching the current into filamentary strands of conducting currents.

---

[3] In free space $\epsilon = \epsilon_0 = 8.8542 \sim 10^{-12}$ farad m^{-1} and $\mu = \mu_0 = 4\pi \sim 10^{-7}$ henry m^{-1}.

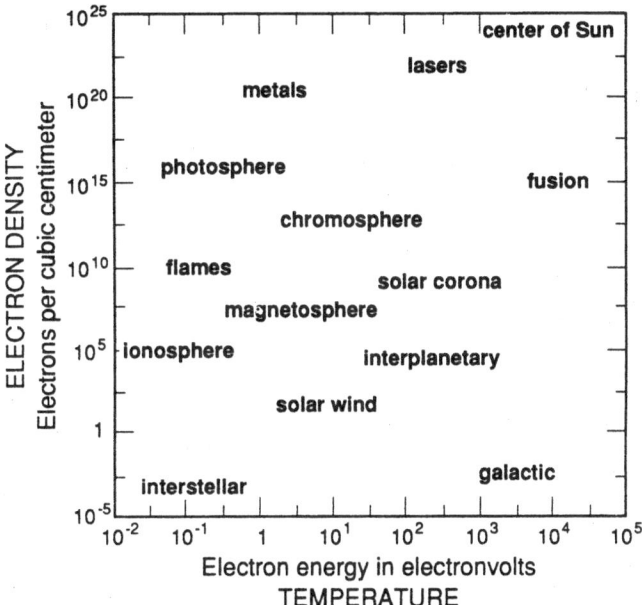

Figure 1. The remarkable range of temperatures and densities of plasmas is illustrated by this chart. In comparison, solids, liquids, and gases exist over a very small range of temperatures and pressures. In "solid" metals, the electrons that carry an electric current exist as a plasma within the more rigid crystal structure.

Matter in the plasma state can range in temperature from hundreds of thousands of electronvolts (1 eV = 11,605 degrees absolute) to just one-hundredth of an electronvolt. In density, plasmas may be tenuous, with just a few electrons present in a million cubic centimeters, or they may be dense, with more than 10^{20} electrons packed per cubic centimeter (Figure 1). Nearly all the matter in the universe exists in the plasma state, occurring predominantly in this form in the Sun and stars and in interstellar space. Auroras, lightning, and welding arcs are also plasmas. Plasmas exist in neon and fluorescent tubes, in the sea of electrons that moves freely within energy bands in the crystalline structure of metallic solids, and in many other objects.

Above all, plasmas are prodigious producers of electromagnetic radiation; from the highest energy gamma rays, to microwaves, to extremely low frequencies which, in our time frame may be considered to be practically dc.

2.2. The Physical Sizes and Characteristics of Plasmas in the Universe

Volume-wise, 99.999% of all the observable matter in the universe exists in the plasma state.

2.2.1. *Plasmas on Earth*

On the earth, plasmas are found with dimensions of microns to meters, that is, sizes spanning six orders of magnitude. The magnetic fields associated with these plasmas range from about 0.5 gauss (the earth's ambient field) to megagauss field strengths. Plasma lifetimes on earth span 12 to 19 orders of magnitude: Laser produced plasmas have properties measurable in picoseconds, pulsed power plasmas have nanosecond to microsecond lifetimes, and magnetically confined fusion oriented plasmas persist for appreciable fractions of a second. Quiescent plasma sources, including fluorescent light sources, continuously produce plasmas whose lifetimes may be measured in hours, weeks, or years, depending on the cleanliness of the ionization system or the integrity of the cathode and anode discharge surfaces.

Lightning is a natural plasma resulting from electrical discharges in the earth's lower troposphere. Such flashes are usually associated with cumulonimbus clouds but also occur in snow and dust storms, tornadoes, active volcanos, nuclear explosions, and ground fracturing. The maximum time duration of a lightning flash is about 2 s in which peak currents as high as 200 kA can occur. The conversion from air molecules to a singly ionized plasma occurs in a few microseconds, with hundreds of megajoules of energy dissipated and plasma temperatures reaching 3 eV. The discharge channel avalanches at about one-tenth the speed of light, and the high current carrying core expands to a diameter of a few centimeters. The total length of the discharge is typically 2-3 km, although cloud-to-cloud discharges can be appreciably longer. Lightning has been observed on Jupiter, Saturn, Uranus, and Venus. The energy released in a single flash on earth, Venus, and Jupiter is typically 6×10^8 J, 2×10^{10} J, and 2.5×10^{12} J, respectively.

Nuclear driven atmospheric plasmas were a notable exception to the generally short-lived energetic plasmas on earth. For example, the 1.4 megaton (5.9×10^{15} J) Starfish detonation, 400 km above Johnston Island, on July 9, 1962, generated plasma from which artificial Van Allen belts of electrons circulating the earth were created. These electrons, bound at about 1.2 earth-radii in a 0.175 G field, produced synchrotron radiation whose decay constant exceeded 100 days.

2.2.2. *Near-Earth Plasmas*

The earth's ionosphere and magnetosphere constitute a cosmic plasma system that is readily available for extensive and detailed in situ observation and even active experimentation. Its usefulness as a source of understanding of cosmic plasmas is enhanced by the fact that it contains a rich variety of plasma populations with densities ranging from more than 10^6 cm^{-3} to less than 10^{-2} cm^{-3}, and temperatures from about 0.1 eV to more than 10 keV.

The earth's magnetosphere is that region of space defined by the interaction of the solar wind with the earth's dipole-like magnetic field. It extends from approximately 100 km above the earth's surface, where the proton neutral atom collision frequency is equal to the proton gyrofrequency, to about ten earth radii ($\sim$ 63,800 km) in the sunward direction and to several hundred earth radii in the anti-sunward direction.

First detected by radio waves and then by radar, the ionosphere is a layered plasma region closest to the surface of the earth whose properties change continuously during a full day. First to be identified was a layer of molecular ionization, called the E layer. This region extends over a height range of 90-140 km and may have a nominal density of 10^5 cm^{-3} during periods of low solar activity. A D region underlies this with a nominal daytime density of 10^3 cm^{-3}. Overlaying the E region is the F layer of ionization, the major layer of the ionosphere, starting at about 140 km. In the height range 100-150 km, strong electric currents are generated by a process analogous to that of a conventional electric generator, or dynamo. The region, in consequence, is often termed the dynamo region and may have densities of 10^6 cm^{-3}. The F layer may extend 1,000 km in altitude where it eventually merges with the plasmas of the magnetopause and solar wind.

The interaction of the supersonic solar wind with the intrinsic dipole magnetic field of the earth forms the magnetosphere whose boundary, called the magnetopause, separates interplanetary and geophysical magnetic fields and plasma environments. Upstream of the magnetopause a collisionless bow shock is formed in the solar wind-magnetosphere interaction process. At the bow shock the solar wind becomes thermalized and subsonic and continues its flow around the magnetosphere as magnetosheath plasma, ultimately rejoining the undisturbed solar wind.

In the anti-solar direction, observations show that the earth's magnetic field is stretched out in an elongated geomagnetic tail to distances of several hundred earth radii. The field lines of the geomagnetic tail intersect the earth at high latitudes ($\approx 60° - 75°$) in both the northern and southern hemisphere (polar horns), near the geomagnetic poles. Topologically, the geomagnetic tail roughly consists of oppositely directed field lines separated by a "neutral" sheet of nearly zero magnetic field. Surrounding the neutral sheet is a plasma of "hot" particles having a temperature of 1-10 keV, density of $\approx$0.01-1 cm^{-3}, and a bulk flow velocity of a few tens to a few hundreds of km s^{-1}.

Deep within the magnetosphere is the plasmasphere, a population of cold ($\leq$ 1 eV) ionospheric ions and electrons corotating with the earth.

2.2.3. *Plasmas in the Solar System*
The space environment around the various planetary satellites and rings in the solar system is filled with plasma such as the solar wind, solar and galactic cosmic rays (high energy charged particles), and particles trapped in the planetary magnetospheres. The first in situ observations of plasma and energetic particle populations in the magnetospheres of Jupiter, Saturn, Uranus, Neptune, and Titan were made by the Voyager 1 and 2 spacecraft from 1979 to 1989. Interplanetary spacecraft have identified magnetospheres around Mercury, Venus, Jupiter, Saturn, Uranus, and Neptune.

Comets also have "magnetospheres." The cometosheath, a region extending about 1.1×10^6 km (for Comet Halley), consists of decelerated plasma of density and temperature $T_e \sim$ 1.5 eV. The peak magnetic field strength found in Comet Halley was 700-800 mG. Confirmation that pinched Birkeland currents also occur in cometary magntospheres was obtained by the detection of X rays from Comet Hyakutake in 1996.

Excluding the Sun, the largest organized structures found in the solar system are the plasma tori around Jupiter and Saturn. The Jupiter-Io plasma torus is primarily filled with sulphur ions at a density of 3×10^3 cm^{-3}. An immense weakly ionized hydrogen plasma torus has been found to encircle Saturn, with an outer diameter 25 times the radius of the planet and an inner diameter of about fifteen Saturn radii.

2.2.4. *Transition Regions in the Solar System*
Examples of transition regions include the boundary layers found in planetary and comet magnetospheres. Transition regions between plasmas of different densities, temperatures, magnetization, and chemical composition offer a rich variety of plasma phenomena in the solar system [Eastman 1990].

2.2.5. *Solar, Stellar, and Interstellar Plasmas*
The nuclear core of the Sun is a plasma at about a temperature of 1.5 keV. Beyond this, our knowledge about the Sun's interior is highly

uncertain. Processes which govern the abundance of elements, nuclear reactions, and the generation mechanism and strength of the interior magnetic fields, are incompletely known.

We do have information about the Sun's surface atmospheres that are delineated as follows: the photosphere, the chromosphere, and the inner corona. These plasma layers are superposed on the Sun like onion skins. The photosphere (T $\sim$ 0.5 eV) is only a very weakly ionized atmosphere, the degree of ionization being $10^{-4} - 10^{-5}$ in the quiet regions and perhaps $10^{-6} - 10^{-7}$ in the vicinity of sunspots. The chromosphere (T $\sim$ 4 eV) extends 5,000 km above the photosphere and is a transition region to the inner corona. The highly ionized inner corona extends some 10^5 km above the photosphere. From a plasma physics point of view, the corona is perhaps the most interesting region of the Sun. The corona is the sight of explosively unstable magnetic-field configurations, X ray emission, and plasma temperatures in the range 70-263 eV.[4] The source of this heating is uncertain.[5]

Solar flares resulting from coronal instabilities raise temperatures to 10-30 keV and produce relativistic streams of electrons and protons. Particles accelerated outward produce radio interference at the earth. Protons accelerated inward collide with ions in the Sun's atmosphere to produce nuclear reactions, whose gamma rays and neutrons have been detected from spacecraft. Solar flares consist of plasma at a temperature of about 1 keV to 10 keV. Although flares represent the most intense energy dissipation of any form of solar activity, releasing energy in the form of gamma rays, X rays, and microwaves, the active sun has many other plasma manifestations.

These include sunspots, photospheric faculae, chromospheric and transition region plages, large coronal loops, and even larger scale coro-

---

[4] One of the Sun's outstanding problems is the temperature of the corona. The temperature rises steadily in the chromosphere, then jumps abruptly in the corona to a level 300 times hotter than the surface. That the Sun is a plasma and not just a hot gaseous object is illustrated by the fact that the temperature increases away from its surface, rather than cooling as dictated by the thermodynamic principle for matter in the nonplasma state.

[5] For decades the preferred explanation has been that energy flows from the Sun's surface to the corona in the form of sound waves generated by convective upswelling motions. However, space-based ultraviolet observations proved that sound waves do not carry energy as high as the corona. One mechanism that may produce coronal heating is electron beams produced in double layers in coronal loops. These are expected to accelerate electrons to energies comparable to those in the corona. Generally, the term acceleration refers to the preferential gain of energy by a population of electrons and ions, while heating is defined as the bulk energization of the ambient plasma. Paraphrasing Kirchoff that "heating is a special kind of acceleration," one may argue, since heating and acceleration are always present in flares and in laboratory relativistic electron beams, that electron beam instabilities may be the source of coronal heating.

nal streamers and occasional coronal mass ejections. In addition, prominences (referred to as filaments when seen in H_α absorption on the disk) frequently form between opposite magnetic polarities in active regions of the sun. These phenomena are dynamic, on time scales ranging from seconds to the complete solar magnetic cycle of 22 years.

The outer corona and solar wind form the heliosphere. At one astronomical unit the solar wind has a plasma density $5 \leq n \leq 60$ cm-3 and a velocity of $200 \leq v_{sw} \leq 800$km s^{-1}. Its temperature can be as high as 50 eV, while B may reach 200 mG. The outer heliosphere has a plasma density $10^{-3} \leq n \leq 10^{-1}$ cm^3, a temperature $10^{-1} \leq T < 10$ eV, and a magnetic field strength ~ 1 mG. The local interstellar medium is characterized by $10^{-2} \leq n \leq 1$, a temperature of order 1 eV, and magnetic field strength $1 \leq B \leq 20\mu$G.

The rotating sun, coupled with its continual radial ejection of plasma, twists its magnetic field (that is referred to as the interplanetary magnetic field or IMF) into a classical Archimedean spiral. Measurements have confirmed that the interplanetary magnetic field is directed toward the sun in certain regions of the solar system and away from the sun in other regions. These regions are separated by a very sharp boundary layer that is interpreted as a current layer. In this situation, the planets find themselves sometimes in a region where the field has a strong northward component and sometimes where it has a strong southward component.

Stellar plasmas have not only the dimension of the star, 0.3×10^6 km to 10^8 km, but also the stellar magnetospheres, a remnant of the interstellar plasma that the star and its satellites condensed out of. The surface temperatures vary from about 0.3 to 3 eV. Estimates of the magnetic fields range from a few gauss to tens of kilogauss or more for magnetic variable stars.

Stellar winds occur in stars of many types, with wind properties probably connected with stellar magnetism.

2.2.6. *Galactic and Extragalactic Plasmas*

Dark clouds within our Galaxy have dimensions of 10^8 km and microgauss strength magnetic fields.

The Galactic plasma has an extent equal to the dimensions of our Galaxy itself; ~ 35 kpc or 10^{21} m. The most salient feature of the Galactic plasma are 10^{-3} G poloidal-toroidal plasma filaments extending nearly 250 light years (60 pc, 1.8×10^{18} m) at the Galactic center. The vast regions of nearly neutral hydrogen (HI regions) found in the Galaxy and other galaxies are weakly ionized plasmas. These regions extend across the entire width of the galaxy and are sometimes found between interacting galaxies. They are detected by the 21 cm radiation they emit.

Galaxies may have bulk plasma densities of 10^{-1} cm^{-3}; groups of galaxies, 3×10^{-2} cm^{-3}; and rich clusters of galaxies, 3×10^{-3} cm^{-3}.

By far the single largest plasmas detected in the Universe are those of double radio galaxies. In size, these sources extend hundreds of kiloparsecs ($10^{21} - 10^{22}$ m) to a few megaparsecs ($10^{22} - 10^{23}$ m). Double radio galaxies are thought to have densities of 10^{-3} cm^{-3} and magnetic fields of the order of 10^{-4} G.

2.3. REGIONS OF APPLICABILITY OF PLASMA PHYSICS

The degree of ionization in interplanetary space and in other cosmic plasmas may vary over a wide range, from fully ionized to degrees of ionization of only a fraction of a percent.[6] Even weakly ionized plasma reacts strongly to electromagnetic fields since the ratio of the electromagnetic force to the gravitational force is 39 orders of magnitude. For example, although the solar photospheric plasma has a degree of ionization as low as 10^{-4}, the major part of the condensible components is still largely ionized. The "neutral" hydrogen (HI) regions around galaxies are also plasmas, although the degree of ionization is only 10^{-4}. Most of our knowledge about electromagnetic waves in plasmas derives from laboratory plasma experiments where the gases used have a low degree of ionization, 10^{-2}-10^{-6}.

Because electromagnetic fields play such an important role in the electrodynamics of plasmas, and because the dynamics of plasmas are often the sources of electromagnetic fields, it is desirable to determine where within the universe a plasma approach is necessary. We first consider the magnetic field. The criterion for neglecting magnetic effects in the treatment of a problem in gas dynamics is that the Lundquist parameter is much less than unity,

$$L_u = \frac{u^{1/2} \sigma B l_c}{\sqrt{\rho_m}} << 1 \qquad (6)$$

where l_c is a characteristic length of the plasma and ρ_m is the mass density. As the conductivity of known plasmas generally varies only over about four orders of magnitude, from 10^2 to 10^6 siemens/m, the value of is largely dependent on the strength of B in the plasma.

The variation of B in plasmas can be 18 orders of magnitude, from microgauss strengths in intergalactic space to perhaps teragauss levels in the magnetospheres of neutron sources. On earth, magnetic field

---

[6] The degree of ionization is defined as $n_p/(n_0 + n_p)$ where n_p is the plasma density and n_0 is the density of neutral particles.

strengths can be found from about 0.5 gauss $(0.5 \times 10^{-4}$ T) to 10^7 gauss $(10^3$ T) in pulsed-power experiments; the outer planets have magnetic fields reaching many gauss, while the magnetic fields of stars are 30-40 kG (3-4 T). Large scale magnetic fields have also been discovered in distant cosmic objects. The center of the Galaxy has milligauss magnetic field strengths stretching 60 pc in length. Similar strengths are inferred from polarization measurements of radiation recorded for double radio galaxies. No rotating object in the universe, that is devoid of a magnetic field, is known.

In cosmic problems involving planetary, interplanetary, interstellar, galactic, and extragalactic phenomena, L_u is usually of the order $10^{15} - 10^{20}$. In planetary ionospheres falls below unity in the E layer. Neglecting lightning, planetary atmospheres and hydrospheres are the only domains in the universe where a nonhydromagnetic treatment of fluid dynamic problems is justified.

2.3.1. *Field-Aligned Currents in Astrophysical Plasmas*
As far as we know, most cosmic low density plasmas also depict a filamentary structure. For example, filamentary structures are found in the following cosmic plasmas, all of which are observed or are likely to be associated with electric currents:

1. In the aurora, filaments parallel to the magnetic field are often observed. These can sometimes have dimensions down to about 100 m.

2. Inverted V events and the in-situ measurements of strong electric fields in the magnetosphere $(10^5 - 10^6$ A, 10^8 m) demonstrate the existence of filamentary structures.

3. In the ionosphere of Venus, "flux ropes", whose filamentary diameters are typically 20 km, are observed.

4. In the sun, prominences $(10^{11}$ A), spicules, coronal streamers, polar plumes, etc., show filamentary structure whose dimensions are of the order $10^7 - 10^8$ m.

5. Cometary tails often have a pronounced filamentary structure.

6. In the interstellar medium and in interstellar clouds there is an abundance of filamentary structures [e.g., the Veil nebula, the Lagoon nebula, the Orion nebula, and the Crab nebula].

7. The center of the Galaxy, where twisting plasma filaments, apparently held together by a magnetic field possessing both azimuthal and poloidal components, extend for nearly 60 pc $(10^{18}$ m).

8. Within the radio bright lobes of double radio galaxies, where filament lengths may exceed 20 kpc (6×10^{20} m).

9. In extended radio sources and synchrotron emitting jets [Gouveia Dal Pino and Olpher 1989]

Regardless of scale, the motion of charged particles produces a self-magnetic field that can act on other collections of particles or plasmas, internally or externally. Plasmas in relative motion are coupled via currents that they drive through each other. Currents are therefore expected in a universe of inhomogeneous astrophysical plasmas of all sizes.

3. Dynamical Characteristics of Plasmas

It is the global dynamics and systematic interactions of astrophysical plasmas that allow energy to be conveyed over great distances. The evolution of cosmic plasma that includes its structuring into cells results in a relative motion, however slow, of plasma clouds whose dimensions may be measured in hundreds or megaparsecs or gigaparsecs. All plasma clouds may be considered a system: they are coupled by electrical currents (charged particles beams) they induce in each other. These beams are the source of energy transfer from large, slow moving plasma to smaller plasma regions that may release the energy abruptly or cause local plasmas to pinch to the condense state.

3.1. POWER GENERATION AND TRANSMISSION

On earth, power is generated by nuclear and nonnuclear fuels, hydro and solar energy, and to a much lesser extent, by geothermal sources and magnetohydrodynamic generators. Always, the location of the supply is not the location of major power usage or dissipation. Transmission lines are used to convey the power generated to the load region. As an example, abundant hydroelectric resources in the Pacific Northwest of the United States produce power ($\sim$1,500 MW) that is then transmitted to Los Angeles, 1,330 km away, via 800 kV high-efficiency dc transmission lines. In optical and infrared emission, only the load region, Los Angeles, is visible from the light and heat it dissipates in power usage. The transmission line is invisible.

This situation is also true in space. With the coming of the space age and the subsequent discovery of magnetospheric-ionospheric electrical circuits, Kirchoff's circuit laws were suddenly catapulted to dimensions eight orders of magnitude larger than that previously investigated in the laboratory and nearly four orders of magnitude greater than that associated with the longest power distribution systems on earth.

On earth, transmission lines consist of metallic conductors or waveguides in which energy is made to flow via the motion of free electrons (currents) in the metal or in displacement currents in a time varying electric field. Often strong currents within the line allow the transmission of power many orders of magnitude stronger than that possible with weak currents. This is because a current associated with the flow of electrons produces a self-magnetic field that helps to confine or pinch the particle flow. Magnetic-insulation is commonly used in pulsed-power technology to transmit large amounts of power from the generator to the load without suffering a breakdown due to leakage currents caused by high electric potentials.

There is a tendency for charged particles to follow magnetic lines of force and this forms the basis of transmission lines in space. In the magnetosphere-ionosphere, a transmission line 7-8 earth radii in length ($R_e = 6,350$ km) can convey tens of terawatts of power, that derives from the solar wind-magnetosphere coupling[7], to the lower atmosphere. The transmission line is the earth's dipole magnetic field lines along which electrons and ions are constrained to flow. The driving potential is solar-wind induced plasma moving across the magnetic field lines at large radii. The result is an electrical circuit in which electric currents cause the formation of auroras at high latitude in the upper atmosphere on earth. This aurora mechanism is observed on Jupiter, Io, Saturn, Uranus, and is thought to have been detected on Neptune and perhaps, Venus.

Only the aurora discharge is visible at optical wavelengths to an observer. The source and transmission line are invisible. Before the coming of space probes, in situ measurement was impossible and exotic explanations were often given of auroras. This is probably true of other non in situ cosmic plasmas today. The existence of a megaampere flux tube of current, connecting the Jovian satellite Io to its mother planet, was verified with the passage of the Voyager spacecraft.

3.2. ELECTRICAL DISCHARGES IN COSMIC PLASMA

An electrical discharge is a sudden release of electric or magnetic stored energy. This generally occurs when the electromagnetic stress exceeds some threshold for breakdown that is usually determined by small scale

---

[7] It is not known how the energy carried by the solar wind is transformed into the energy of the aurora. It has been demonstrated that the southward-directed interplanetary magnetic field is an essential ingredient in causing auroral substorms so that energy transformation appears to occur through interactions between the interplanetary and geomagnetic fields [Akasofu 1981].

properties of the energy transmission medium. As such, discharges are local phenomena and are usually accompanied by violent processes such as rapid heating, ionization, the creation of pinched and filamentary conduction channels, particle acceleration [Melrose 1997], and the generation of prodigious amounts of electromagnetic radiation.

As an example, multi-terawatt pulsed-power generators on earth rely on strong electrical discharges to produce intense particle beams, X rays, and microwaves. Megajoules of energy are electrically stored in capacitor banks, whose volume may encompass 250 m^3. This energy is then transferred to a discharge region, located many meters from the source, via a transmission line. The discharge region, or load, encompasses at most a few cubic centimeters of space, and is the site of high-variability, intense, electromagnetic radiation.

On earth, lightning is another example of the discharge mechanism at work where electrostatic energy is stored in clouds whose volume may be of the order of 3,000 km^3. This energy is released in a few cubic meters of the discharge channel.

The aurora is a discharge caused by the bombardment of atoms in the upper atmosphere by 1-20 keV electrons and 200 keV ions spiraling down the earth's magnetic field lines at high latitudes. Here, the electric field accelerating the charged particles derives from plasma moving across the earth's dipole magnetic field lines many earth radii into the magnetosphere. The potential energy generated by the plasma motion is fed to the upper atmosphere by multi-megaampere Birkeland currents that comprise a transmission line, 50,000 kilometers in length, as they flow into and out of the discharge regions at the polar horns. The generator region may encompass $10^{12} - 10^{13}$ km^3 while the total discharge volume can be $10^9 - 10^{10}$ km^3.

3.2.1. *Flickering of Electromagnetic Radiation*
The flickering of a light in Los Angeles does not mean that the *supply source*, a waterfall or hydroelectric dam in the Pacific Northwest, has abruptly changed dimensions or any any other physical property. The flickering comes from electrical changes at the observed load or *radiative source*, such as the formation of instabilities or virtual anodes or cathodes in charged particle beams that are orders of magnitude smaller than the supply. Bizarre and interesting non-physical interpretations are obtained if the flickering light is interpreted by a distant observer to be both the source and supply. This also holds true for astrophysical plasmas. The flickering and pulsating of the observed electromagnetic radiation from a distant astrophysical source, when interpreted to be local, unattached, and isolated in 'vacuum' space, leads to bizarre 'black hole' type explanations. As discussed earlier, space is not vacu-

um but rather filled with plasma whose properties, volume-wise, differ
little from those in the laboratory or magnetospheres. And plasmas
exhibit large *system* global properties, such as the transfer of energy
over great distances to smaller regions where it may be systematically
or catastrophically released.

3.3. PARTICLE ACCELERATION IN COSMIC PLASMA

3.3.1. *Acceleration of Electric Charges*
The acceleration of a charged particle q in an electromagnetic field is
mathematically described by the Lorentz equation Eq. 5,

$$\mathbf{F} = m\mathbf{a} = q\left(\mathbf{E} + \mathbf{v} \times \mathbf{B}\right) \tag{7}$$

The electric field vector $\mathbf{E}$ can arise from a number of processes that
include the motion of plasma with velocity $\mathbf{v}$ across magnetic fields
lines $\mathbf{B}$, charge separation, and time varying magnetic fields via Eq. 1.

Acceleration of charged particles in laboratory plasmas is achieved
by applying a potential gradient between metallic conductors (cathodes
and anodes); by producing time varying magnetic fields such as in beta-
trons; by radio frequency (RF) fields applied to accelerating cavities
as in linear accelerators (LINACS); and by beat frequency oscillators
or wake-field accelerators that use either the electric field of lasers or
charged particle beams to accelerate particles.

The magnetospheric plasma is essentially collisionless. In such a plasma, electric fields aligned along the magnetic field direction freely accelerate particles. Electrons and ions are accelerated in opposite directions, giving rise to a current along the magnetic field lines.

3.3.2. *Collective Ion Acceleration*
The possibility of producing electric fields by the space-charge effect to
accelerate positive ions to high energies was first discussed by Alfvén
and Wernholm in 1952. They were unsuccessful in their attempt to
experimentally accelerate ions in the collective field of clouds of electrons, probably because of the low intensity of electron beam devices
available then. However, proof of principle came in 1961 when Plyutto
reported the first successful experiment in which ions were collectively
accelerated. By 1975, the collective acceleration of ions had become
a wide-spread area of research. Luce reported collectively accelerating
both light and heavy ions to multi-MeV energies, producing an intense
burst of D-D neutrons and nuclear reactions leading to the identification of several radioisotopes. Luce used a plasma-focus device and
attributed the collective beam to intense current vortex filaments in
the pinched plasma. Subsequently in 1979, Destler, Hoeberling, Kim,

and Bostick collectively accelerated carbon ions to energies in excess of 170 MeV using a 6 MeV electron beam.

Individual ion energies up to several GeV using pulsed-power generators have been suggested in particle-in-cell simulations of collective ion acceleration processes.

Collective acceleration as a mechanism for creating high energy ions in astrophysical plasmas were investigated by Bostick [1986].

3.4. PLASMA PINCHES AND INSTABILITIES

3.4.1. *The Bennett Pinch*
In cosmic plasma the perhaps most important constriction mechanism is the electromagnetic attraction between parallel currents. A manifestation of this mechanism is the pinch effect as first studied by Bennett (1934). Phenomena of this general type also exist on a cosmic scale and lead to a bunching of currents and magnetic fields to filaments. This bunching is usually accompanied by the accumulation of matter, and it may explain the observational fact that cosmic matter exhibits an abundance of filamentary structures.

Consider a fully ionized cylindrical plasma column of radius r, in an axial electric field E_z, that produces an axial current density j_z. Associated with j_z is an azimuthal magnetic field B_ϕ. The current flowing across its own magnetic field exerts a $\mathbf{j} \times \mathbf{B}$, radially inward, pinch force. In the steady-state, the balance of forces is

$$\nabla p = \nabla (p_e + p_i) = \mathbf{j} \times \mathbf{B} \qquad (8)$$

By employing Eq. 2, $\nabla \times \mathbf{B} = \mu_0 \mathbf{j}$, and the perfect gas law $p = NkT$, we arrive at the Bennett relation

$$2Nk \left(T_e + T_i\right) = \frac{\mu_0}{4\pi} I^2 \qquad (9)$$

where N is the number of electrons per unit length along the beam, T_e and T_i are the electron and ion temperatures, I is the total beam current, and k is Boltzmann's constant.

3.4.2. *The Force-Free Configuration*
Sheared magnetic fields are a characteristic of most plasmas. Here, the sheared field is considered a nonpotential field that is caused by shear flows of plasma. A nonpotential field tends to settle into a particular configuration called a "force-free" field, namely

$$(\nabla \times \mathbf{B}) \times \mathbf{B} = 0 \qquad (10)$$

or

$$\mathbf{j} \times \mathbf{B} = 0 \qquad (11)$$

since $\mathbf{j} = \nabla \times \mathbf{B}\mu_0^{-1}$, showing that the electric current tends to flow along $\mathbf{B}$. Substituting 11 into 8 gives $\mathbf{F} = \nabla p = 0$, hence the name "force-free". Force-free fields tend to have a twisted or "sheared" appearance. Examples of force-free fields are chromospheric fibrils and penumbral structures near active sunspots.

The condition 10 can be satisfied in three ways: $\mathbf{B} = \mathbf{0}$ (trivial), (i.e., $\mathbf{j} = \mathbf{0}$), or

$$\nabla \times \mathbf{B} = \alpha \mathbf{B} \qquad (12)$$

where the scalar $\alpha = \alpha(r)$ in general. The essence of a force-free field is simply that electric currents flow parallel to magnetic field lines. Such currents are often called "field-aligned" currents.

The force-free fields with constant α represent the lowest state of magnetic energy that a closed system may attain. This has two important consequences. It proves the stability of force-free fields with constant α, and shows that in a system in which the magnetic forces are dominant and in which there is a mechanism to dissipate the fluid motion, force-free fields with constant α are the natural end configuration. In astrophysical plasmas, the dissipation mechanism may be the acceleration of charged particles to cosmic ray energies.

3.5. Analysis of Beams and Filamentary Plasma

3.5.1. General Plasma Fluid Equations

Fundamental equations for the plasma velocity, magnetic field, plasma density, electric current, plasma pressure, and plasma temperature can be derived from macroscopic averages of currents, fields, charge densities, and mass densities. In this "fluid" treatment, the Maxwell's equations 1-4 are coupled to the moments of the Boltzmann equation for a highly ionized plasma.

The evolution of the distribution function $f(\mathbf{r}, \mathbf{p}, t)$ for particles with charge q and mass m is described by the Boltzmann equation

$$\left[\frac{\partial}{\partial t} + \mathbf{v} \cdot \frac{\partial}{\partial \mathbf{r}} + q \left(\mathbf{E} + \mathbf{v} \times \mathbf{B} \right) \cdot \frac{\partial}{\partial \mathbf{p}} \right] f(r, p, t) = \left(\frac{\partial f}{\partial t} \right)_{collisions} \qquad (13)$$

which is an expression of Liouville's theorem for the incompressible motion of particles in the six-dimensional phase space $(\mathbf{r}, t)$. In the fluid description the particle density n, mean velocity $\bar{v}$, momentum $n\bar{v}$, pressure P, and friction R are defined by

$$n(r, t) \equiv \int d^3p f(r, p, t) \tag{14}$$

$$n(r, t) \bar{v}(r, t) \equiv \int d^3p v f(r, p, t) \tag{15}$$

$$n(r, t) \bar{p}(r, t) \equiv \int d^3p p f(r, p, t) \tag{16}$$

$$P(r, t) \equiv \int d^3p [\bar{p} - p(r, t)] [\bar{v} - v(r, t)] f(r, p, t) \tag{17}$$

$$R(r, t) \equiv \int d^3p [\bar{p} - p(r, t)] \left(\frac{\partial f}{\partial t}\right)_{collisions} \tag{18}$$

where the momentum p and velocity v are related by

$$p = m\gamma v \tag{19}$$

The fields $E(r, t)$ and $B(r, t)$ in Eq. 13 are self-consistently solved from Eqs. 1- 4 with

$$\rho(r, t) \equiv e \int d^3p f(r, p, t) \tag{20}$$

$$j(r, t) \equiv e \int d^3p v f(r, p, t) \tag{21}$$

Taking the moments $\int d^3p$ and $\int d^3p \mathbf{p}$ of the Boltzmann equation yields the two-fluid equations [Alfvén and Fälthammar 1963] for ions and electrons $\alpha = i, e$,

$$\frac{\partial n_\alpha}{\partial t} + \nabla \cdot (n_\alpha \bar{v}_\alpha) = 0 \tag{22}$$

$$n_\alpha \frac{dp_\alpha}{dt} = q_\alpha n_\alpha (E + v_\alpha \times B) - \nabla \cdot P_\alpha + R_\alpha - n_\alpha m_\alpha \nabla \phi_G \tag{23}$$

These are called the continuity and momentum equations, respectively. The continuity equation, as written, is valid if ionization and recombination are not important. Conservation of linear momentum dictates that

$$R_i + R_e = 0 \tag{24}$$

The two fluid equations are the moments, or averages, of the kinetic plasma description and no longer contain the discrete particle phenomena such as double layers from charge separation and synchrotron radiation. Nevertheless, this approach is useful in studying bulk plasma flow and behavior. A single fluid hydromagnetic force equation may be obtained by substituting into Eq. 23 and adding to get,

$$\rho_m \frac{\partial \mathbf{v}_m}{\partial t} = \rho \mathbf{E} + \mathbf{j} \times \mathbf{B} - \nabla p - \rho_m \nabla \phi_G \qquad (25)$$

which relates the forces to mass and acceleration for the following averaged quantities:

$$
\begin{aligned}
\rho_m &= n_e m_e + n_i m_i & \text{mass density} \\
\mathbf{j}_m &= n_e m_e \bar{\mathbf{v}}_e + n_i m_i \bar{\mathbf{v}}_i & \text{mass current} \\
\mathbf{v}_m &= \mathbf{j}_m / \rho_m & \text{averaged velocity} \\
\rho &= n_e q_e + n_i q_i & \text{charge density} \\
\mathbf{j} &= n_e q_e \bar{\mathbf{v}}_e + n_i q_i \bar{\mathbf{v}}_i & \text{current density}
\end{aligned}
\qquad (26)
$$

The first term in Eq. 26 is caused by the electric field, the second term derives from the motion of the current flow across the magnetic field, the third term is due to the pressure gradient [Eq. 26 is valid for an isotropic distribution $\nabla \cdot P \rightarrow \nabla p$, where $p \equiv nkT$], and the fourth term is due to the gravitational potential φ_G. The near absence of excess charge $\rho = e\,(n_i - n_e) \approx 0$, for $q_{i,e} = \pm e$, is a characteristic of the plasma state; however, this does not mean that electrostatic fields [e.g., those deriving from Eq. 3] are unimportant. According to Chen [1985]:

> "In a plasma, it is usually possible to assume $n_i = n_e$ and $\nabla \cdot \mathbf{E} \neq 0$ at the same time. We shall call this the *plasma approximation*. It is a fundamental trait of plasmas, one which is difficult for the novice to understand. *Do not use Poisson's equation to obtain E unless it is unavoidable!*"

Completing the single fluid description is the equation for mass conservation,

$$\frac{\partial \rho_m}{\partial t} + \nabla \cdot (\rho_m \mathbf{v}_m) = 0 \qquad (27)$$

In addition to Eqs. 25 and 27, we find it useful to add the equation for magnetic induction,

$$\frac{\partial B}{\partial t} = \nabla \times (\mathbf{v}_m \times \mathbf{B}) + \frac{1}{\mu \sigma} \nabla^2 B \qquad (28)$$

obtained by taking the curl of Ohm's law

$$\mathbf{j} = \sigma\,(\mathbf{E} + \mathbf{v} \times \mathbf{B})$$

where σ is the electrical conductivity.

3.5.2. *Magnetic Reynolds and Lundquist Numbers*

The significance of Eq. 28 in which $(\mu\sigma)^{-1}$ is the magnetic diffusivity, is that changes in the magnetic field strength are caused by the transport of the magnetic field with the plasma (as represented by the first term on the right-hand-side), together with diffusion of the magnetic field through the plasma (second term on the right-hand-side). In order of magnitude, the ratio of the first to the second term is the *magnetic Reynolds number*

$$R_m = \mu\sigma V_c l_c \tag{29}$$

in terms of a characteristic plasma speed V_c and a characteristic scale length l_c. A related quantity is the Lundquist parameter

$$L_u = \mu\sigma V_A l_c \tag{30}$$

where

$$V_A = \frac{B}{\sqrt{\mu\rho_m}} \tag{31}$$

is the Alfvén speed. It may be written as the ratio

$$L_u = \frac{\tau_d}{\tau_A} \tag{32}$$

of the *magnetic diffusion time*

$$\tau_d = \mu\sigma l_c^2 \tag{33}$$

to the *Alfvén travel time*

$$\tau_A = l_c/V_A \tag{34}$$

3.6. THE GENERALIZED BENNETT RELATION

A generalized Bennett relation follows directly from Eq. 25 and Eqs. 1-4. Consider a current-carrying, magnetic-field-aligned cylindrical plasma of radius a which consists of electrons, ions, and neutral gas having the densities n_e, n_i, and n_n, and the temperatures T_e, T_i, and T_n, respectively. A current of density j_z flows in the plasma along the axis of the cylinder which coincides with the z-axis. As a result of the axial current a toroidal magnetic field B_ϕ is induced. An axial electric field is also present. Thus, there exists the electric and magnetic fields

$$\mathbf{E} = (E_r, E_\phi, E_z)$$
$$\mathbf{B} = (0, B_\phi, B_z)$$

The derivation of the generalized Bennett relation for this plasma is straightforward, but lengthy [Witalis 1981], and the final result is

$$\frac{1}{4}\frac{\partial^2 J_0}{\partial t^2} = W_\perp + \Delta W_{E_z} + \Delta W_{B_z} + \Delta W_k$$
$$- \frac{\mu_0}{8\pi}I^2(a) - \frac{1}{2}G\bar{m}^2 N^2(a) + \frac{1}{2}\pi a^2 \varepsilon_0 \left(E_r^2(a) - E_\phi^2(a)\right) \quad (35)$$

where

$$J_0 = \int_0^{2\pi}\int r^2 \rho_m r dr d\phi = \int_0^a r^2 \rho_m 2\pi r dr \quad (36)$$

is the total moment of inertia with respect to the z axis. (As the mass m of a particle or beam is its resistance to linear acceleration, J_0 is the beam resistance to angular displacement or rotation). The quantities ΔW are defined by

$$\Delta W_{E_z} \equiv W_{E_z} - \frac{1}{2}\varepsilon_0 E_z^2(a)\pi a^2 \quad (37)$$

$$\Delta W_{B_z} \equiv W_{B_z} - \frac{1}{2\mu_0}B_z^2(a)\pi a^2 \quad (38)$$

$$\Delta W_k \equiv W_k - p(a)\pi a^2 \quad (39)$$

where $E_z(a)$, $B_z(a)$, and $p(a)$ denote values at the boundary $r = a$. The individual energies W are defined as follows. The kinetic energy per unit length due to beam motion transverse to the beam axis:

$$W_{\perp\;kin} \equiv \frac{1}{2}\int_0^a \rho_m(r)\left[v_\phi^2 + v_r^2\right]2\pi r dr \quad (40)$$

The self-consistent B_z energy per unit length:

$$W_{B_z} \equiv \frac{1}{2\mu_0}\int_0^a B_z^2(r)2\pi r dr \quad (41)$$

The self-consistent E_z energy per unit length:

$$W_{E_z} \equiv \frac{\varepsilon_0}{2}\int_0^a E_z^2(r)2\pi r dr \quad (42)$$

The thermokinetic energy per unit length:

$$W_k \equiv \int_0^a p\left(r\right) 2\pi r dr \qquad (43)$$

The axial current inside the radius a:

$$I\left(a\right) \equiv \int_0^a j_z\left(r\right) 2\pi r dr \qquad (44)$$

The total number of particles per unit length:

$$N\left(a\right) \equiv \int_0^a n\left(r\right) 2\pi r dr \qquad (45)$$

where $n = n_i + n_e + n_n$ is the total density of ions, electrons, and neutral particles. The mean particle mass is $\bar{m} = n_i m_i + n_e m_e + n_n m_n$. The self-consistent electric field can be determined from the following equation,

$$r^{-1} \frac{d\left(r E_r\right)}{dr} = -\frac{e}{\varepsilon_0}\left(n_e - n_i\right)$$

and is given by

$$E_r\left(r\right) = \frac{-e n_e\left(1 - f_e\right)}{2\varepsilon_0} r, \ 0 \leq r \leq a$$

$$E_r\left(r\right) = \frac{-e n_e\left(1 - f_e\right) a^2}{2\varepsilon_0}, \ r \geq a$$

Neglecting the displacement current, the self consistent magnetic field can be determined from Ampère's law Eq. 2,

$$r^{-1} \frac{d\left(r B_\phi\right)}{dr} = \mu_0 j$$

and is given by

$$B_\phi\left(r\right) = \frac{\mu_0 I}{2\pi} \frac{r}{a^2}, \ 0 \leq r \leq a$$

$$= \frac{\mu_0 I}{2\pi r}, \ r \geq a$$

The positive terms in Eq. 35 are expansional forces while the negative terms represent beam compressional forces. In addition, it is assumed that the axially directed kinetic energy is

$$W_{\|kin} = \frac{1}{2}\gamma m N \beta^2 c^2 \tag{46}$$

Since Eq. 35 contains no axially directed energy, it must be argued that there are conversions or dissipation processes transferring a kinetic beam of energy of magnitude $W_{\|kin}$ into one or several kinds of energy expressed by the positive W elements in Eq. 35:

$$W_{\|kin} = W_{\perp\;kin} + W_{E_z} + W_{B_z} + W_k \tag{47}$$

3.6.1. The Bennett Relation

Balancing the thermokinetic and azimuthal compressional (pinch) energies in Eq. 35,

$$W_k - \frac{\mu_0}{8\pi}I^2 = 0 \tag{48}$$

yields the Bennett relation,

$$\frac{\mu_0}{4\pi}I^2 = 2NkT \tag{49}$$

If there is a uniform temperature $T = T_e + T_i$ and if the current density is uniform across the current channel cross-section, Eqs. 44, 45, and 49 yield a parabolic density distribution

$$n\left(r\right) = \frac{\mu_0 I^2}{4\pi^2 a^2 kT}\left(1 - \frac{r^2}{a^2}\right)$$

3.6.2. Alfvén Limiting Current

Equating the parallel beam kinetic energy to the pinch energy

$$W_{\perp\;kin} - \frac{\mu_0}{8\pi}I^2 = 0 \tag{50}$$

yields the Alfvén limiting current

$$I_A = 4\pi\varepsilon_0 m_e c^3 \beta\gamma/e = 17\beta\gamma \text{ kiloamperes} \tag{51}$$

for an electron beam. This quantity was derived by Alfvén in 1939 in order to determine at what current level in a cosmic ray beam the self-induced pinch field would turn the forward propagating electrons around. It should be noted that this limit is independent of any physical dimensions.

Lawson's (1959) interpretation of Eq. 51 is that the electron trajectories are beam-like when $I < I_A$ and they are plasma-like when

$I > I_A$. In laboratory relativistic electron beam (REB) research, Budker's parameter

$$\nu_{Bud} = \pi a^2 n_b e^2 / mc^2 = Ne^2 / mc^2 \qquad (52)$$

where n_b is the electron density of the beam, finds wide application in beam and plasma accelerators. Yonas (1974) has interpreted the particle trajectories as beam-like for $\nu_{Bud} < \gamma$ and plasma-like for $\nu_{Bud} > \gamma$. The relationship between I_A and ν_{Bud} is

$$\nu_{Bud}/\gamma = I/I_A \qquad (53)$$

The Alfvén limiting current Eq. 51 is a fundamental limit for a uniform beam, charge-neutralized ($f_e = 1$), with no magnetic neutralization ($f_m = 0$), no rotational motion ($v_\phi = 0$), and no externally applied magnetic field ($B_z = 0$). By modifying these restrictions, it is possible, under certain circumstances, to propagate currents in excess of I_A.

3.6.3. *Charge Neutralized Beam Propagation*
Balancing the parallel kinetic, pinch, and radial electric field energies in Eq. 35 gives

$$W_{\perp \, kin} - \frac{\mu_0}{8\pi} I^2 + \frac{1}{2}\pi a^2 \varepsilon_0 E_r^2 \, (a) = 0 \qquad (54)$$

which yields[8]

$$I_{\max} = I_A \beta^2 \left[\beta^2 - (1 - f_e)^2\right]^{-1}, \; 0 \le f_e \le 1 \qquad (55)$$

Depending on the amount of neutralization, the denominator in Eq. 55 can become small, and I_{max} can exceed I_A. However, the unneutralized electron beam cannot even be injected into a drift space unless the space charge limiting current condition is satisfied. For a shearless electron beam this is [Bogdankevich and Rukhadze 1971]

$$I_{sc} = \frac{17 \left(\gamma^{2/3} - 1\right)^{3/2}}{1 + 2\ln(b/a)} \; kiloamperes \qquad (56)$$

where b is the radius of a conducting cylinder surrounding the drift space. Thus, in free space $I_{sc} \to 0$ and an unneutralized electron beam

---

[8] This equation differs from the Alfvén-Lawson limiting current, $I_{\max} = I_A \beta^2 / \left[\beta^2 - 1 + f_e\right]$, because of the differing ways in describing charge neutralization [Witalis 1981].

will not propagate but, instead, builds up a space charge cloud of electrons (a virtual cathode) which repels any further flow of electrons as a beam. Moreover, the space charge limiting current is derived under the assumption of an infinitely large guide field B_z; no amount of magnetic field will improve beam propagation.

3.6.4. Current Neutralized Beam Propagation

For beam propagation in plasma, the electrostatic self-field E_z built up by the beam, efficiently drives a return current through the plasma, thus moderating the compressional term $\mu_0 I^2/8\pi$ in Eq. 35, so that

$$W_{\perp\ kin} - \frac{\mu_0}{8\pi} I^2 \left(1 - f_m\right) + \frac{1}{2}\pi a^2 \varepsilon_0 E_r^2 \left(a\right) = 0 \tag{57}$$

where the magnetic neutralization factor is

$$f_m \equiv |I_{return}/I| \tag{58}$$

From Eq. 57 the maximum current is

$$I_{\max} = I_A \beta^2 \left[\beta^2 \left(1 - f_m\right) - \left(1 - f_e\right)^2\right]^{-1}, \ 0 \le f_e \le 1, \ 0 \le f_m \le 1. \tag{59}$$

Thus, depending on the values of f_e and f_m, the denominator in Eq. 59 can approach zero and the maximum beam current can greatly exceed I_A.

3.6.5. Beam Propagation in Plasma

When a charged particle beam propagates through plasma, the plasma ions can neutralize the beam space charge. When this occurs, $E_r \to 0$ and, as a result, the beam constricts because of its self-consistent pinch field $B\phi$. For beam currents in excess of the Alfvén limiting current, B_ϕ is sufficient to reverse the direction of the beam electron trajectories at the outer layer of the beam. However, depending on the plasma conductivity σ, the induction electric field at the head of the beam [due to $dB/dt \sim dI/dt$ in Eq. 1] will produce a plasma current $I_p = -I_b$. Hence, the pinch field

$$B_\phi \left(r\right) = \frac{\mu_0}{2\pi r} \left[I_b \left(r\right) + I_p\right] \tag{60}$$

can vanish allowing the propagation of beam currents I_b in excess of I_A.

Because of the finite plasma conductivity, the current neutralization will eventually decay in a magnetic diffusion time τ_d given by Eq. 33. During this time a steady state condition exists in which no net self-fields act on the beam particles. While in a steady state, beam propagation is limited only by the classic macro-instabilities such as the sausage instability and the hose (kink) instability.

When the beam undergoes a small displacement, the magnetic field lags behind for times of the order τ_d. This causes a restoring force to push the beam back to its original position, leading to the well-known $m = 1$ (for a $e^{im\phi}$ azimuthal dependency) kink instability.

3.6.6. *Beam Propagation Along an External Magnetic Field*
An axially directed guide field B_z produces an azimuthal current component I_ϕ through Eq. 2. This modification to the conducting current follows by balancing the energies

$$W_{\perp \, kin} - \frac{1}{2\mu_0}B_z^2 \pi a^2 - \frac{\mu_0}{8\pi}I^2 + \frac{1}{2}\varepsilon_0 E_r^2\,(a)\,\pi a^2 = 0 \qquad (61)$$

In the absence of any background confining gas pressure $p(a)$, the maximum current is

$$I_{\max} = I_\phi \beta^2 \left[\beta^2 - (1 - f_e)^2\right]^{-1} \qquad (62)$$

For the case of an axial guide field, the axial current I_z is not limited to I_A and depends only on the strength of the balancing $I_z\,(B_z)$ current (magnetic field). In terms of the magnetic fields, the current flows as a beam when

$$B_z >> \frac{\left[(1 - f_e)^2 - \beta^2\right]^{1/2}}{\beta}\,B_\phi \qquad (63)$$

Note that Eq. 56 still holds; that is, a cylindrical conductor around the electron beam is necessary for the beam to propagate.

Equation 63 finds application in accelerators such as the high-current betatron [Hammer and Rostocker 1970]. In spite of the high degree of axial stabilization of a charged particle beam because of B_z, appreciable *azimuthal* destabilization and filamentation can occur because of the diocotron effect. This can be alleviated by bringing the metallic wall close to the beam.

3.6.7. *Schönherr Whirl Stabilization*
The transverse kinetic energy term $W_{\perp kin}$ in Eq. 35 explains an observation made long ago [Schönherr 1909]. High-current discharges conduct more current if the discharge is subject to an externally impressed rotation v_ϕ. This phenomena can also be expected in astronomical situations if the charged particle beam encounters a nonaxial component

of a magnetic field line that imparts a spin motion to the beam or if a gas enters transversely to an arc discharge-like plasma.

3.6.8. The Carlqvist Relation

An expression having broad applicability to cosmic plasmas, due to Carlqvist [1988], may be obtained from Eq. 35 if the beam is taken to be cylindrical and in a rotationless and steady-state condition:

$$\frac{\mu_0}{8\pi} I^2(a) + \frac{1}{2} G \bar{m}^2 N^2(a) = \Delta W_{B_z} + \Delta W_k \qquad (64)$$

Thus, in a straightforward and elegant way, the gravitational force has been included in the familiar Bennett relation. Through Eq. 64, the Carlqvist Relation, the relative importance of the electromagnetic force and the gravitational force may be determined for any given cosmic plasma situation. This relation will now be applied to the two commonest pinch geometries–the cylindrical pinch and the sheet pinch.

3.6.9. The Cylindrical Pinch

Consider the case of a dark interstellar cloud of hydrogen molecules ($\bar{m} = 3 \times 10^{-27}$ kg and $T = T_i = T_e = T_n = 20$ K). Carlqvist (1988) has given a graphical representation of the solution to Eq. 64 for these values and this is shown in Figure 2 for discrete values of ΔW_{B_z}. Several physically different regions are identified in this figure.

The region in the upper left-hand part of the figure is where the pinching force due to I and the magnetic pressure force due to B_z constitute the dominating forces. Equation (2.52) in this region reduces to

$$\frac{\mu_0}{8\pi} I^2(a) \approx \Delta W_{B_z} \qquad (65)$$

representing a state of almost force-free magnetic field.

Another important region is demarked by negative values of ΔW_B. In this region an outwardly directed kinetic pressure force is mainly balanced by an inwardly directed magnetic pressure force. Hence the total pressure is constant and Eq. 64 is approximately given by

$$NkT + \Delta W_{B_z} \approx 0 \qquad (66)$$

For yet larger negative values of ΔW_B, the magnetic pressure force is neutralized by the gravitational force so that Eq. 64 reduces to

$$\frac{1}{2} G \bar{m}^2 N^2(a) \approx \Delta W_{B_z} \qquad (67)$$

Another delineable region is where $\Delta W_B = 0$, where Eq. 64 reduces to the Bennett relation,

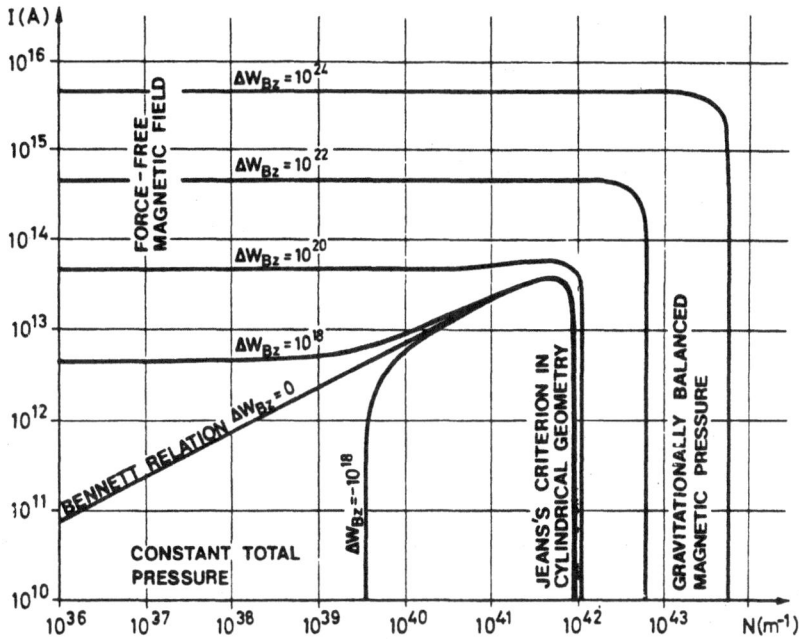

Figure 2. The total current I in a generalized Bennett pinch of cylindrical geometry as a function of the number of particles per unit length N. The temperature of the plasma is $T = 20$ K while the mean particle mass is $\bar{m} = 3 \times 10^{-27}$ kg. It is assumed that the plasma does not rotate ($\omega = 0$) and that the kinetic pressure is much smaller at the border of the pinch than in the inner parts. the parameter of the curves is ΔW_{Bz}, representing the excess magnetic energy per unit length of the pinch due to an axial magnetic field B_z (courtesy of P. Carlqvist).

$$\frac{\mu_0}{8\pi} I^2 (a) \approx \Delta W_k (n, T) \tag{68}$$

Another region of some interest is where the classic Bennett relation line turns over into an almost vertical segment. Here, the pinching force of the current may be neglected, leaving the kinetic pressure force to balance only the gravitational force so that

$$\frac{1}{2} G \bar{m}^2 N^2 (a) \approx NkT \tag{69}$$

This is the Jeans's criterion in a cylindrical geometry.

The size or radius of the cylindrical pinch depends on the balancing forces. For the Bennett pinch Eq. 68 the equilibrium radius is

$$a \approx \frac{I}{2\pi} \sqrt{\frac{\mu_0}{2nkT}} \tag{70}$$

Küppers (1973) has investigated the case of a REB propagating through plasma. Space charge neutralization ($E_r = 0$) is maintained when $n_e +$

$n_b = n_i$. For this case, the replacements $n \rightarrow n_b$ and $T \rightarrow T_b - T_e$ are made to $\Delta W_k (n, T)$, where T_b is the beam temperature and T_e is the background plasma electron temperature. For the space charge neutralized REB,

$$a \approx \frac{I_b}{2\pi} \sqrt{\frac{\mu_0}{2n_b k (T_b - T_e)}} \qquad (71)$$

Note that physically acceptable solutions for the equilibrium radius are obtained only when the beam temperature (in the axial direction for a cold beam) exceeds the plasma electron temperature.

From Eq. 65, the equilibrium radius of a pinch balanced by an internal field B_z is

$$a \approx \frac{\mu_0 I}{2\pi B_z} \qquad (72)$$

4. Radiation Characteristics of the Plasma State

Electromagnetic waves propagated in cosmic space derive from a variety of mechanisms. The major contribution in the optical region of the spectrum is from radiation resulting from bound-bound electron transitions between discrete atomic or molecular states, free-bound transitions during recombination, and free-free transitions in the continuum. In the latter case, when for transitions between levels, radiation classified as bremsstrahlung results from the acceleration of electrons traveling in the vicinity of the atom or ion.

In addition, there are other mechanisms of considerable importance operating in the radio region. In particular, there are noncoherent and coherent mechanisms connected with the existence of sufficiently dense plasmas which are responsible for radiation derived from plasma oscillations, such as the sporadic solar radio emissions. This radiation cannot be attributed to the motion of individual electrons in a vacuum but is due to the collective motion of electrons at the plasma frequency

$$\frac{\omega_{pe}}{2\pi} = \frac{1}{2\pi} \sqrt{\frac{n_e e^2}{m_e \varepsilon_0}} = 9\sqrt{n_e}, \text{ Hz} \qquad (73)$$

for an electron density n_e (m^{-3}). This often occurs in cosmic plasma when electron beams propagate through a neutralizing plasma background.

4.1. SYNCHROTRON RADIATION

When a plasma is subjected to a magnetic field there is yet another mechanism which plays an extremely important role in radio astronomy. The frequency and angular distribution of the radiation from free electrons moving in the presence of a magnetic field undergoes dramatic changes as the electron energy is increased from nonrelativistic to extreme relativistic energies. Essentially three types of spectra are found. Names such as cyclotron emission and magnetobremsstrahlung are used to describe the emission from nonrelativistic and mildly relativistic electron energies, whereas the name synchrotron radiation is traditionally reserved for highly relativistic electrons because it was first observed in 1948 in electron synchrotrons.

Synchrotron radiation is characterized by a generation of frequencies appreciably higher than the cyclotron frequency of electrons (or positrons) in a magnetic field, a continuous spectra whose intensity decreases with frequency beyond a certain critical frequency, highly directed beam energies, and polarized electromagnetic wave vectors.

In astrophysics, nonthermal (nonequilibrium) cosmic radio emission is, in a majority of cases, synchrotron radiation. This is true for general galactic radio emission, radio emission from the envelopes of supernovae, and radio emission from double radio galaxies and quasars (continuum spectra). Synchrotron radiation also appears at times as sporadic radio emission from the sun, as well as from Jupiter. In addition, optical synchrotron radiation is observed in some instances (Crab nebula, the radiogalaxy and "jet" in M87-NGC 4486, M82, and others). This apparently is also related to the continuous optical spectrum sometimes observed in solar flares. Synchrotron radiation in the X ray region can also be expected in several cases, particularly from the Crab nebula.

When cosmic radio or optical emission has the characteristics of synchrotron radiation, a determination of the spectrum makes possible a calculation of the concentration and energy spectrum of the relativistic electrons in the emission sources. Therefore, the question of cosmic synchrotron radiation is closely connected with the physics and origin of cosmic rays and with gamma- and X ray astronomy.

Synchrotron radiation was first brought to the attention of astronomers by H. Alfvén; and N. Herlofson; [1950], a remarkable suggestion at a time when plasma and magnetic fields were thought to have little, if anything, to do in a cosmos filled with "island" universes (galaxies).[9] The recognition that this mechanism of radiation is important in astronomical sources has been one of the most fruitful developments in astro-

---

[9] The "island universes" concept was introduced by the philosopher Kant (1724-1804).

physics. For example, it has made possible the inference that high-energy particles exist in many types of astronomical objects, it has given additional evidence for the existence of extensive magnetic fields, and it has indicated that enormous amounts of energy may indeed be converted, stored, and released in cosmic plasma.

The polarization, spectral, temporal, power, and directivity properties of synchrotron radiation are well known [Peratt 1992].

4.1.1. *Directivity*
The gain, or directivity of the synchrotron radiation zone is given by

$$G\left(\beta_{||}, \beta_{\perp}, \theta\right) = \frac{3}{4}\left(1 - \beta^2\right) F\left(\beta_{||}, \beta_{\perp}, \theta\right) \qquad (74)$$

where

$$F\left(\beta_{||}, \beta_{\perp}, \theta\right) =$$

$$\frac{4g_{||}^2\left[\left(1-\beta_{||}^2\right)\left(1-\cos^2\theta\right)-4\beta_{||}\cos\theta\right]-\left(1-\beta_{||}^2+3\beta_{\perp}^2\right)\beta_{\perp}^2\sin^4\theta}{4\left(g_{||}^2-\beta_{\perp}^2\sin^2\theta\right)^{7/2}} \qquad (75)$$

where

$$\beta^2 = \left(\frac{v_{||}}{c}\right)^2 + \left(\frac{v_{\perp}}{c}\right)^2 = \beta_{||}^2 + \beta_{\perp}^2 \qquad (76)$$

and $v_{||}$ and $v_{\perp}$ are the instantaneous particle velocities along and perpendicular to B_0, respectively.[10]

4.1.2. *Spectrum*
The total emission per unit radian frequency interval is

$$P_{\omega}^T = \frac{\sqrt{3}e^2\omega_b}{8\pi^2\varepsilon_0 c}$$

$$\times \left\{\beta_0\left(1-\beta_0^2\right)\gamma^2\frac{\omega}{\omega_c}\left[2K_{2/3}\left(\frac{\omega}{\omega_c}\right) - \frac{\gamma^2\left(1-\beta_0^2\right)}{\beta_0^2}\int_{\omega/\omega_c}^{\infty}K_{1/3}\left(t\right)dt\right]\right\} \qquad (77)$$

where, by definition,

$$\omega_c = \frac{3}{2}\omega_{\gamma}\gamma^3 = \frac{3}{2}\omega_b\gamma^2$$

$$= 2.64 \times 10^7 B_{0\,gauss}\left(\frac{W_{keV}}{511keV}\right)\,rad/s \qquad (78)$$

---

[10] Noteworthy in $G(\theta)$ is the inversion of gain direction between the cases $\beta_{||} = 0$ as $\beta_{\perp}$ increases and the essentially forward emission in the relativistic case when $\beta_{||}/\beta_{\perp}$ is large.

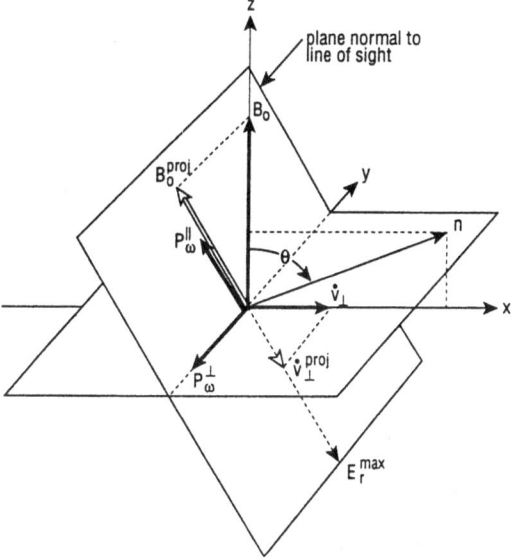

Figure 3. Orientation of the quantities $P_\omega^\perp$ and $P_\omega^{||}$.

4.1.3. *Polarization*

The power radiated into the mth harmonic may be written as the sum of two parts,

$$P_\omega\left(\omega, \beta, \theta\right) = P_\omega^{(1)} + P_\omega^{(2)} \tag{79}$$

where

$$P_\omega^{(1)}\left(\theta\right) = \frac{e^2\omega^2}{8\pi^2\varepsilon_0 c} \sum_{1}^{\infty} \left\{\frac{\cos\theta - \beta_{||}}{\sin\theta} J_m\left(x\right)\right\}^2 \delta\left(y\right) \tag{80}$$

$$P_\omega^{(2)}\left(\theta\right) = \frac{e^2\omega^2}{8\pi^2\varepsilon_0 c} \sum_{1}^{\infty} \left\{\beta_\perp J_m'\left(x\right)\right\}^2 \delta\left(y\right) \tag{81}$$

whose orthogonal field vectors $E_m^{(1)}$ and $E_m^{(2)}$ ($p_\omega \sim E_m^2$) lie parallel and perpendicular, respectively, to the projection of B_0 on the plane normal to the line of sight $\left[B_0^{\perp n} = B_0 - \left(B_0 \cdot n\right) n\right]$ as shown in Figure 3.[11] The easiest way to determine the polarization of the mth harmonic is to take the ratio of the field amplitudes in Eqs.(6.75) and (6.76),

$$R_m = -\frac{E_m^{(1)}}{E_m^{(2)}} = -\left(\frac{\cos\theta - \beta_{||}}{\beta_\perp \sin\theta}\right)\frac{J_m\left(\psi m\right)}{J_m'\left(\psi m\right)} \tag{82}$$

---

[11] While $P_\omega^{(0)}$ has no rotation, $P_\omega^{(1)}$ does rotate.

where $\psi = \beta_\perp \sin\theta / \left(1 - \beta_{\parallel} \cos\theta\right)$. The parameter R_m defines the ellipticity of the propagating wave in the direction θ. When $|R_m| = 1$, the wave is circularly polarized while when $R_m = 0$, the wave is linearly polarized. When $0 < R_m < 1$, the wave is elliptically polarized in the ratio of the field amplitudes oriented along the minor and major axes of the ellipse. The polarization of the extraordinary mode rotates in the same sense as does the radiating electron.

4.2. TRANSITION RADIATION

Even in the absence of a magnetic field, astrophysical plasmas are capable of producing polarized radiation and large scale radiation patterns having diffraction-like patterns. At cellular interfaces delineating astrophysical plasmas of differing constituency, the passage of electrical currents can produce *transition radiation*, first studied by Frank and Ginzburg [1945]. Transition radiation is produced by the propagation of charged particles through the interface between media with differing dielectric constants. It is caused by a collective response of the matter surrounding the particle trajectory to readjust the electromagnetic field of the charged particle.

If for example, electrons pass through the highly conductive interface (e.g., a cell wall or sheet current) separating two relatively tenuous regions, an angular distribution of energy occurs, given by

$$W(\omega, \Theta) = \frac{e^2}{4\pi^3} \sqrt{\frac{\mu_0}{\varepsilon_0}} \beta^2 \frac{\sin^2\Theta}{\left(1 - \beta^2 \cos^2\Theta\right)^2} \tag{83}$$

where Θ is the angle between the electron trajectory and the emitted radiation. For relativistic electrons ($\beta \sim 1$) the emission is sharply peaked in the region of small Θ and is maximum when $\Theta \sim 1/\gamma$. In all cases, the transition radiation is totally polarized. The plane of polarization is given by the electron trajectory and the direction of observation.

If the path of an electron is unobstructed over a distance greater than the formation length

$$\Lambda = \frac{2c}{\omega\left(1 - \beta^2 + \Theta^2\right)} \tag{84}$$

then the angular distribution of Eq. 83 is small. If the free path d is smaller than Λ, the angular distribution is broadened by diffraction and the radiation is strongly suppressed below an angle of

$$\Theta_m \approx \sqrt{\frac{2c}{\omega d}} \tag{85}$$

Integrating Eq. 83 over Θ gives the spectral energy,

$$W(\omega) = \frac{e^2}{3\pi^3} \sqrt{\frac{\mu_0}{\varepsilon_0}} \left\{ \frac{3}{8} \frac{\beta^2 + 1}{\beta^2} \ln \frac{1+\beta}{1-\beta} - \frac{3}{4\beta^2} \right\} \tag{86}$$

If the beam of electrons is bunched, the emitted radiation from different electrons adds coherently when the wavelength λ is comparable to or longer than the beam bunch. Thus the intensity is proportional to the number of electrons squared. Long a topic of theoretical interest, transition radiation has been experimentally verified in the far infrared [Happek et al. 1991].

Thus coherent transition radiation may given information about large scale cellular astrophysical transition regions.

4.3. THERMAL RADIATION

The spectral distribution of the radiation of a black body in thermodynamic equilibrium, for a single polarization, is given by the Planck formula

$$B_0(\omega, T) = \frac{\hbar\omega^3}{8\pi^3 c^2} \frac{1}{e^{\hbar\omega/kT} - 1} \tag{87}$$

Non-black body radiation from plasma can be thermalized by filaments or even carbonaceous needles in space. The radiation intensity is

$$I_{\omega M} = S_\omega \left(1 - e^{-M\alpha_\omega L} \right) \tag{88}$$

Note that the radiation intensity increased for larger M because of each additional filament current source. The absorption coefficient at $\theta = \pi/2$ is [Trubnikov 1958, Peratt 1992]

$$\alpha_\omega = \frac{\omega_p^2}{\omega_b c} \sum \Phi_m (\omega/\omega_b \mu) \tag{89}$$

The quantities $\Phi_m = \Phi_m(\omega/\omega_b, \mu)$ are defined by

$$\Phi_m = \sqrt{2\pi} \frac{\mu^{5/2}}{(\omega/\omega_h)^4} m^2 \sqrt{m^2 - (\omega/\omega_b)^2} e^{-\mu[m(\omega/\omega_b)-1]} A \left[m/(\omega/\omega_b) \right] \tag{90}$$

where $\mu = m_0 c^2/kT$. The quantities $A_m = A_m(\gamma)$ are given by

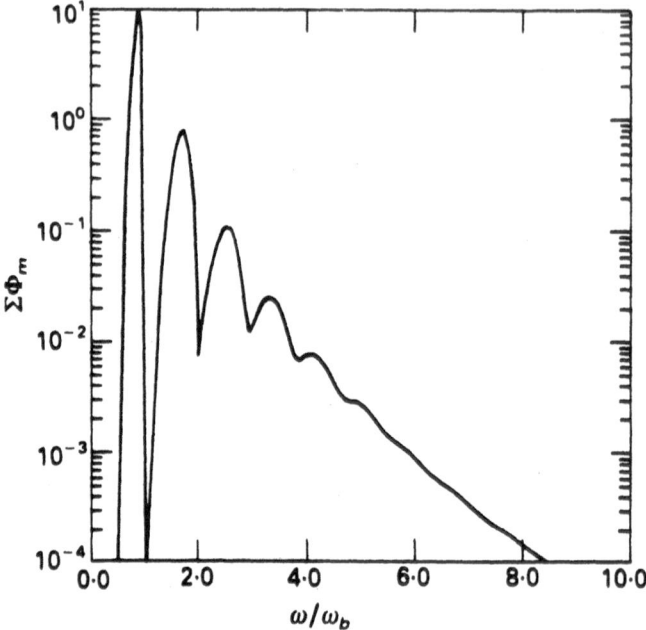

Figure 4. Calculated spectrum of radiation emitted by a plasma with electron temperature of 30 keV. Self-absorption effects are included.

$$A_m^{(X,O)}(\gamma) = \begin{cases} \frac{(m\beta)^{2m}}{(2m+1)!}\left[1; \frac{\beta^2}{2m+3}\right] & m\beta \ll 1 \\ \frac{e^{2m/\gamma}}{\sqrt{16\pi m^3 \gamma}}\left(\frac{\gamma-1}{\gamma+1}\right)^m \left[1; \frac{\gamma(\gamma^2-1)}{2m}\right] & \gamma^3 \ll m \\ \frac{1}{4\sqrt{3}\pi m \gamma^2}\left[\int_{2m/3\gamma^3}^{\infty} K_{5/3}(t)\,dt \pm K_{2/3}(2m/3\gamma^3)\right] & \gamma, m \gg 1 \end{cases}$$

$$(91)$$

The optical depth for M Birkeland currents is $\tau_\omega = \alpha_\omega M L$ or

$$\tau_\omega = \left(\frac{\omega_p^2 L M}{\omega_b c}\right) \sum_m \Phi_m \qquad (92)$$

The spectral characteristics of the emission are contained in the function Φ. Figure 4 shows a plot of $\Sigma\Phi_m$ for the first one hundred harmonics as a function of ω/ω_b for $T = 30$ keV. This value is typical of the thermal temperatures in a plasma filament but is appreciably less than the energies of particles in a relativistic beam. Only the extraordinary wave is considered; the contributions from the ordinary wave are usually small.

The broadening of the individual lines is due to the relativistic change of mass. A given line contributes only to frequencies $\omega \leq m\omega_b$ with the highest energy electrons being responsible for the emission at the lowest frequency. The smearing of the successive harmonics produces an almost monotonically decreasing spectrum at higher frequency. For $T = 30$ keV, $m \sim 5$ is the harmonic above which smearing prevails. To a fair approximation, the total intensity leaving the filaments is

$$I\left(\theta = \pi/2\right) \cong \int_0^{\omega^*} B_0\left(\omega, T\right) d\omega = \frac{\omega_b^3 kT}{24\pi^3 c^2} \left(m^*\right)^3 \qquad (93)$$

where $m^* = \omega^*/\omega_b$ is the harmonic number beyond which the emission effectively ceases to be black-body. An empirical relation for m^* for mildly relativistic plasma has been derived by Trubnikov [1958]; and modified to the case of M filaments,

$$\left(m^*\right)^6 = 0.57 \left(\frac{20\omega_p^2}{3\omega_b c}\right) LMT \qquad (94)$$

Equation 94 is valid under the approximation $mc^2 >> kT$.

5. Formulation for the Particle-in-Cell Simulation of Astrophysical Plasmas

The modern study of the numerical simulation of astrophysical and space plasma can be separated into two categories:

1. Magnetohydrodynamics, or the fluid-like behavior of plasma in the presence of magnetic fields.

2. Particle kinetics, or the discrete electron-ion behavior in self-consistent and external electromagnetic fields.

The first category, Magnetohydrodynamics, was invented by Alfvén to study instability mechanisms associated with the solar atmosphere. Alfvén felt that the magnetohydrodynamics approximation had to be treated with caution and instead favored the particle approach to problems in plasma physics.[12] Examples that require a particle treatment include cases where electric fields parallel to a magnetic field exist (a common occurance in astrophysical plasmas), double layers, critical ionization velocity mechanisms, transition radiation, and synchrotron

---

[12] For example, the concept of a 'frozen-in magnetic field' has no physical meaning in the presence of an axial electric field.

radiation. Likewise, Buneman suggested that the magnetohydrodynam-ic synthesis could be replaced entirely by the particle-in-cell methodol-ogy, once computer resources attained a capability for handling multi-millions of particles and cells in tractable timescales. According to Buneman [1976]:

> "We may conclude from these preliminary results that a feasible full 3D, EM simulation can yield a lot of good physics. In partic-ular, it can reproduce hydromagnetic phenomena (such as Alfvén waves) as well as typical plasma phenomena. In fact, particle sim-ulation may eventually turn out to be the easiest method of doing hydromagnetics on the computer."

Dawson [1993] echoes this sentiment:

> "Magnetohydrodynamic models have existed since the beginning of plasma modeling. Over the years more complete fluid models have been developed that include resistivity, viscosity, heat conduction and other nonideal effects.
>
> Simple fluid models leave out the physics that is responsible for much of the important behavior of plasmas. they leave out kinetic effects, which contribute to damping and to nonlinear saturation of unstable modes and can drive instabilities. Such effects are prob-ably at the heart of determining properties of plasma and heat transport across magnetic fields."

In fact, the numerical magnetohydrodynamic method has seen great use, especially in problems where the plasma is dense enough that collisions are frequent enough that the plasma is in local thermody-namic equilibrium. The approach has had great success in the mod-eling of inertially confined plasmas with sophisticated codes such as LASNEX.[13] Even though collisions are infrequent in a space plasma and it is far from equilibrium, MHD is used to model the interaction of the solar wind with the earth's magnetosphere and the dynamics of the solar corona [Brackbill 1987] as-well-as magnetic merging in astro-physical plasmas [Biskamp 1997].

---

[13] LASNEX is a two-dimensional, azimuthally symmetric Lagrangian code that is often used to model laser-matter interactions and plasma expansions (Zimmerman and Kruer, 1975). The equation of motion for the single fluid, which may consist of a composite of different materials, includes contributions to the pressure from ions, thermal and suprathermal electrons, radiation, magnetic fields, and ponderomotive effects. The nonlocal thermodynamic equilibrium atomic physics model includes a solution of the time–dependent ionization and radiation equations in an average atom approximation. Radiation is also treated by separate photon groups, with emission and transport calculated from a set of coupled flux-limited multigroup diffusion equations.

5.1. THE BASIC LAWS OF PLASMA PHYSICS

The numerical simulation by particles of plasma physics began in the 1950s by Dawson at Princeton and Buneman at Stanford, where various plasma phenomena were identified and studied. It should be mentioned that, in the beginning, it was not at all apparent that the technique developed to study pure electron beam propagation in microwave devices could be applied to the plasma state of matter. Unlike the cold electron beam with charges of all one sign, plasmas often consist of thermal distributions with essentially equal density of charges of opposite sign and greatly different masses. In studying cold electron beams, a few dozen particles sufficed to reproduce the essence of the experiment. However, in laboratory plasmas one has scale lengths greater then the Debye length ($L >> \lambda_D$) and the number of particles in a Debye cube $N_D \equiv n\lambda_D^3$ is much greater than one ($N_D >> 1$). For example, the earth's ionosphere has $N_D \approx 10^4$ and the literal simulation of it over its scale length appears infeasible. However the general character of plasmas can often be found by studying the *collective behavior of collisionless plasmas at wavelengths longer than the Debye length,* $\lambda \geq \lambda_D$. It was found that another characterization of a plasma is that (1) the thermal kinetic energy is much greater than the microscopic potential energy, and (2) the ratio of collision to plasma frequencies is much less than one. Both requirement can be met with rather low values of N_D [Birdsall and Langdon 1985]. Conditions 1 and 2 may be met for finite sized particles called clouds. Clouds occur naturally in simulations which use a spatial grid for interpolation, as well as in simulations which employ spectral methods where the particle profile (usually gaussian) is specified in **k** space.

The term "particle-in-cell" derives from Frank Harlow and his group's work at Los Alamos in the 1950s in investigating the fluid nature of matter at high densities and extreme temperatures. Modern descriptions of the particle-in-cell technique as related to plasma physics are found in the two texts *Computer Simulation Using Particles* [Hockney and Eastwood 1981] and *Plasma Physics via Computer Simulation* [Birdsall and Langdon 1985].

We begin our approach by stating the laws of plasma physics in more or less the form which it has been found convenient to program: the equation of motion for the particles with the Lorentz force Eq. 5, and the Maxwell laws for the electric and magnetic fields Eqs. 1- 4.

5.2. Multidimensional Particle-in-Cell Simulation

5.2.1. *Sampling Constraints in Multidimensional Particle Codes*

The particle-in-cell technique for the analysis of complex phenomena in science has evolved from 1D through $1\frac{1}{2}$D, $1\frac{2}{2}$D, 2D, $2\frac{1}{2}$D, to 3D particle simulations. While at first one has to face certain limitations of an analytic nature, ultimately the limits are set by data management problems the resolution of which depends critically on the available hardware.

A trivial reason for the increasing difficulty of higher dimensional particle simulations is their demand for substantially greater particle numbers. With each added dimension the number of sampling particles has to be multiplied by a certain factor.

This also applies to "half-dimensions." It is customary to denote the inclusion of extra velocity components by referring to them as "half-dimensions." Typically, a pure 1D simulation simulates the plasma as rigid sheet particles, all parallel to the $y-z$ plane, say, and moving in the x direction. It ignores y and z motions of the planes. A $1\frac{1}{2}$D simulation keeps a record of possible y motions, uniform within each plane. Then the x-ward Lorentz force in the presence of a z-directed magnetic field can be taken into account, as well as the y-wards Lorentz force due to x motions. In a $1\frac{2}{2}$D simulation, dz/dt would be recorded as well. In a $2\frac{1}{2}$D simulation, $x, y, dx/dt$, and dz/dt are tracked (but not z); the particles are rigid straight rods whose motion along their axis is taken into account. One-and-a-half dimensional and $1\frac{2}{2}$D simulations have recently found application in space plasma work, namely for simulating the critical ionization phenomenon.

It is reassuring that relatively few samples can often give very good statistics. In many applications the velocity distributions stay close to maxwellian and a modest factor (typically four) in the sample number may suffice to deal with an added half-dimension. One exploits the favorable feature of statistics when initializing thermal velocity distributions: each velocity component is made up as the sum of four random numbers (each uniformly random in a certain interval). The resulting distribution (the "four dice curve," or cubic spline) is almost indistinguishable from the Gaussian.

However, when incrementing by a full dimension, sampling requirements jump dramatically. It is easily checked that the statistical potential energy fluctuations in a granular plasma compete with thermal energies when the particles are spaced on the order of a Debye length apart. Such a plasma would be essentially collision dominated. One is mostly interested in collective effects, since fluid codes are adequate for collision-dominated phenomena. Obviously, one needs several particles per Debye length; again, a modest number suffices. Now, most of the interesting phenomena to be resolved by simulation are on the scale of many Debye lengths (hundreds or thousands). Therefore, the addition of each full dimension calls for an increase of the number of particles by, typically, two orders of magnitude. In 1990 a 1D (electrostatic) simulation could barely be squeezed onto a personal computer, a 2D simulation called for a minicomputer or workstation, and a 3D simulation needed a supercomputer[14]. By the mid 1990's, rudimentary 3D simulations became possible on laptop computers.

5.2.2. *Discretization in Time and Space*

One-dimensional, $1\frac{1}{2}$D, and $1\frac{2}{2}$D simulations can be done without discretizing in space. The electrical interaction of sheets is independent of distance and one only needs to order the sheets to calculate their accelerations. Even if the sheets are of finite thickness or if they are "soft" (i.e., they have a smooth density profile across), only a few operations per sheet are needed to move them one time step. This is an effort of order N where N is the number of sheets.

However, in all simulations, time must be discretized. By studying the simulation of a simple 1D problem, namely, electrostatic oscillations in a cold plasma, and by Fourier transforming one's numerical procedure in time, one finds that while a time step $\delta t = \omega_p^{-1}$ yields the plasma frequency to 5% accuracy, for $\delta t > 2\omega_p^{-1}$ one runs into an instability.

To get over this severe limitation of the speed of simulation in cases where the phenomena of interest are much slower than electrostatic oscillations (typically ion responses), one can either use implicit methods, or one can make one's ions lighter than real ions. Much has been learned from simulations with ion-electron mass ratios as low as 16:1.

The big analytical problems in simulation arise when one advances to two dimensions. Interactions between rods of charge depend on distance, and the many remote rods are as important as the few near ones. For N rods, one has to calculate N^2 interactions and N itself might be typically two orders of magnitude larger than for a 1D simulation.

In order to get back to an effort of order N per step in the particle advance, one tabulates the field over a spatial grid and calculates the self-consistent field from a grid record of the charge and current-densities that each particle contributes.

---

[14] The approximate relationship of supercomputer performance and performance of those in other categories can be shown proportionately. If the performance of contemporary supercomputers is assigned a value of 100, the values in proportion to supercomputers are: minicomputers 0.1 to 5, workstation 0.1 to 1.0, and personal computers 0.001 to 0.1.

The permissible coarseness of the grid mesh becomes a critical issue and the problem of integrating the finite difference version (now in both space and time) of the field equation is far from trivial. Fortunately, both these subjects have been advanced to a state of relative completion and are exhaustively covered in two texts.

Very briefly and broadly, one can state that the grid should be fine enough to resolve a Debye length, and that smoothing or filtering of high spatial frequencies should be practiced in order to minimize "aliasing." This is the stroboscopic phenomenon of high frequencies parading as low frequencies (long wavelengths). Many physical instabilities set in preferentially at long wavelengths and can thus be excited numerically through aliasing. Grid effects can often be studied and checked in 1D where grids are optional. Smoothing can be achieved by particle shaping (i.e., spreading point particles into soft balls with a smooth bell-shaped internal density profile). Likewise, splines and finite-element techniques help.

Regarding the field update from the charge-current record, fast non-iterative methods for solving Poisson's equation over an L-by-M mesh have been developed. These include cyclic reduction in rows and columns, or Fourier transforming in one of the two dimensions (say, that of M). This is an effort to the order $LM \log_2 M$. Two-dimensional simulations go back historically to Hartree who initiated the simulation of the pure electron plasma which circulates in the magnetron. Hartree; also pioneered the time-centered update of the particles from Lorentz's equation Eq. 5

$$\frac{d\mathbf{v}}{dt} \pm \frac{e\mathbf{B}}{m} \times \mathbf{v} = \pm \frac{e\mathbf{E}}{m} \tag{95}$$

using $(v^{new} + vold)/2$ in the second (Lorentz) term and solving the linear equation for v^{new} explicitly. No limitation of $\omega_b \delta t = (eB/M)\, \delta t$ arises from this method except that for large values of $\omega_b \delta t$ the phases of the gyromotion are misrepresented. For small $\omega_b \delta t$ one gets the same results as with cycloid fitting, i.e., joining solutions of the type

$$\mathbf{v}^\perp = \mathbf{E} \times \mathbf{B}/B^2 + gyration\ at\ frequency\ \omega_b \tag{96}$$

for the components of the velocity transverse to $\mathbf{B}$. As regards this particle update, there is no significant increase in effort when advancing from $1\frac{1}{2}$ to 2D and 3D.

A further time-step limitation is encountered when one wants to integrate the full electromagnetic equations over the grid. Because Maxwell's equations (not including Poissons's) are hyperbolic (i.e., they contain a natural $\partial/\partial t$ or "update" term), they can be solved in an effort which is of the order of magnitude of the number of grid points,

LM in the 2D example discussed earlier. Essentially, one solves a wave equation. However, this process becomes unstable unless one observes the Courant speed limit $\delta t < \delta x/c$ in 1D, $\delta t < \delta x/c\sqrt{2}$ in 2D, and $\delta t < \delta x/c\sqrt{3}$ in 3D for square and cubic meshes of side δx. In many applications, scales chosen from other considerations are such that c is a large number and this restriction of δt results in a severe slowdown.

5.2.3. *Spectral Methods and Interpolation*

The Courant condition can be overcome by doing the entire field update in the transform domain. The Maxwell-Hertz-Heaviside laws for the electric and magnetic fields Eqs. 1- 4 can be conveniently combined into one equation for the complex field vector $\mathbf{F} = \mathbf{D} + i\mathbf{H}/c$. When Fourier transforming, this equation becomes

$$\frac{d\mathbf{F}}{dt} - c\mathbf{k} \times \mathbf{F} = -\mathbf{j} \tag{97}$$

for the spatial harmonic which goes like $(\exp i\mathbf{k}\cdot\mathbf{r})$. This field equation is surprisingly similar to that for the particle velocities Eq. 95 and has the corresponding solution for the transverse part of $\mathbf{F}$:

$$\mathbf{F}^{\perp} = \mathbf{j} \times \mathbf{k}/k^2 \ (magnetostatic\ field)+$$
$$circularly\ polarized\ wave\ rotating\ at\ angular\ frequency\ ck \tag{98}$$

The time intervals at which one joins successive solutions of this form are dictated by the rate at which $\mathbf{j}$ changes, not by the magnitude of $c\mathbf{k}$.[15]

To Eq. 95 we should add an initial condition, namely, Poisson's

$$i\mathbf{k}\cdot\mathbf{F}_k = \rho_k \tag{99}$$

Fourier transforming all field-like quantities has many advantages. For instance, the longitudinal part of $\mathbf{F}$ (which is just $\mathbf{D}$) can be obtained from Poisson's equation as $\mathbf{k}\rho/k^2$. Of course, transforming in two dimensions rather than only one (as in the fastest Poisson solvers) makes for an effort of the order $LM\,(\log_2 L + \log_2 M)$. On the other hand, the ready availability of well-programmed FFTs and the additional benefits of spectral methods make up for this increase in effort.

In the transform domain one can perform the filtering, the particle shaping, an optimization for the spline fitting process, and the truncation of the interaction to be discussed in the section on boundary conditions. One does not have to use any spatial finite difference calculus for the field equations. However, a grid is still necessary since we

---

[15] The wavevector $k = \pi/\delta x, \pi\sqrt{(2)}/\delta x$, or $\pi\sqrt{(3)}/\delta x$, according to the number of dimensions.

have only *discrete* numerical Fourier transforms between **r** space and **k** space.

This leaves the problem of interpolation in the mesh. By using high-order interpolation, one can greatly reduce aliasing and improve accuracy. Quadratic and cubic splines have been used, but this soon becomes expensive.[16] Linear interpolation is most commonly used. Interpolation is also needed when the particles contribute their charge and current to the $\rho, \mathbf{j}$ arrays.

Linear interpolation is then, in 2D, equivalent to "area weighting". For 3D, we have cut down the data look-up (or deposit) effort for linear interpolation by using a tetrahedral mesh. Each particle references only the four nearest mesh-point data. The tetrahedra result from drawing the space diagonals into a cubic mesh and introducing cubic center data. Interpolation of currents must be done twice in each step of each practice, once at its old position and once at its new position, since the current is that due to the movement between the two.

5.3. TECHNIQUES FOR SOLUTION

The crucial equations, Eqs. 1 and 2, are in the "update" form, ideally suited to computers which are themselves devices whose function it is to update the state of their memory continually, albeit not continuously. If the time interval δt between updates is so chosen that during this interval changes of **E** and **B**, as seen by any particle, can be ignored in Eq. 1 and changes of **j** can be ignored in Eq. 2, each equation can be solved exactly for the entire interval no matter how long this interval is: The Lorentz equation Eq. 5 then produces cycloidal motion in a plane perpendicular to **B**, composed of a drift and a gyration Eq. 96. This may be accompanied by free fall parallel to **B**, generated by a parallel electric field component. Given the initial velocity and position, or given the position and displacement during the preceding time interval, the displacement during any subsequent interval, and the new position, can be computed precisely.

5.3.1. *Leap-Frogging Particles Against Fields*
The average value of the fields **E** and **B** for Eq. 95 or the current **j** to be used in Eq. 97 is taken to be the actual value at the middle of the time interval. Figure 5 shows how the updating from average values proceeds at equal intervals along a time axis. This involves the following:

---

[16] In a 3D, EM code, cubic splines would require each particle to look up 384 data to interpolate the **E** and the **B** that acts on it!

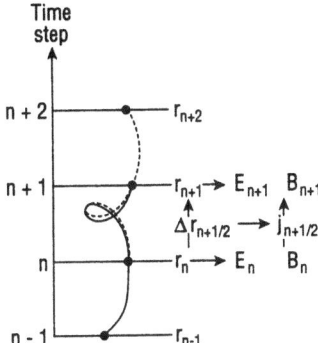

Figure 5. Leap-frogging particles and fields: At time n, cycloids with drifts and gyrations due to $\mathbf{E}_n$ and $\mathbf{B}_n$ are fitted through r_n and r_{n-1}, to be continued through r_{n+1}. Displacements Δr_{n+1} determine currents $\mathbf{j}_{n+1/2}$ from which the transverse fields $\mathbf{E}_n$, $\mathbf{B}_n$ are advanced to $\mathbf{E}_{n+1}$, $\mathbf{B}_{n+1}$. The longitudinal $\mathbf{E}_{n+1}$ is obtained from the r_{n+1}. Then, at time $n+1$, new cycloids (dotted) with drifts and gyrations due to $\mathbf{E}_{n+1}$ and $\mathbf{B}_{n+1}$ are fitted through r_{n+1} and r_n, to be continued through r_{n+2}, etc.

1. Construct the cycloidal orbit of a particle from t_{n-1} through t_n to t_{n+1} using the known mean values of $\mathbf{E}_n$ and $\mathbf{B}_n$ at t_n (the middle of the interval) and the known particle positions at t_{n-1} and t_n.

2. This gives the displacement of the particles from t_n to t_{n+1}, and their final positions at t_{n+1}.

3. The harmonic $\mathbf{j}_k$ of the mean current flowing during the last interval is obtained by summing all the displacements δr with phase factors $\exp i\mathbf{k} \cdot \mathbf{r}$ given by their mean positions, times $q/\delta t$.

4. The transverse fields $\mathbf{F}_k$ are now advanced by $\mathbf{j}_k$ Eq. 97 from t_n to t_{n+1}.

5. The longitudinal fields $\mathbf{F}_k$ are obtained from ρ_k at t_{n+1} by summing with the new positions $\mathbf{r}$ at t_{n+1}.

6. The process is repeated from t_n through t_{n+1} and then to t_{n+2}.

Each time interval is covered by two cycloids for each particle, one with drift and gyrofrequency as given by the fields at the beginning of the interval, the other as given by the fields at the end of the interval. Both cycloids pass through the same points at the termini Figure 5.

5.3.2. *Particle Advance Algorithm*
The time interval δt must be smaller than $\sim \omega_p^{-1}$ for proper resolution of electrostatic plasma oscillations, and $\omega_b \delta t \leq 1$ to account for

synchrotron radiation or for ∇B drifts. The interpretation and implementation of this is as an electric acceleration followed by a magnetic rotation and another acceleration. The updating of the particle positions and velocities is done using a time-centered second-order scheme, valid for relativistic particle velocities,

$$\frac{v_{new} - v_{old}}{\delta t} = \frac{q}{\gamma m_0} \left[E + \left(\frac{v_{new} + v_{old}}{2} \right) \times B \right] \tag{100}$$

with the scaling[17]

$$E \leftarrow \frac{q \delta t}{2m_0} E, \quad B \leftarrow \frac{q \delta t}{2m_0} B, \tag{101}$$

Eq. 100 is solved by the following sequence[18],

$$\gamma_1 = \left(1 - v^2/c^2 \right)^{-1/2}$$
$$u_1 = \gamma_1 v_{old}$$
$$u_2 = u_1 + E \quad (first \ half \ of \ electric \ acceleration)$$
$$\gamma_2 = \left(1 + u_2^2/c^2 \right)^{+1/2}$$
$$u_3 = u_2 + \left(\frac{2}{\gamma_2^2 + B^2} \right) (\gamma_2 u_2 + u_2 \times B) \times B$$
$$u_4 = u_3 + E \quad (second \ half \ of \ electric \ acceleration)$$

$$v_{new} = u_4 \left[1 + (u_4/c)^2 \right]^{+1/2} \tag{102}$$
$$x_{new} = x_{old} + \delta t v_{new}$$

This process is equivalent to rotating the deviation from the gyrocenter drift through the angle $2 \arctan(qB\delta t/2m)$ –the "cycloid fitting"– combined with uninhibited electric acceleration along the magnetic field. It is second-order and time reversible.

Since this algorithm now properly accounts for the effects of relativity, particles are automatically restrained from exceeding the speed of light and need not be artificially braked at c. This limit on the distance traveled by a particle during a single time step plays an important role in particle data management.

---

[17] $\mathbf{E}$ describes half the electric acceleration and the magnitude of $\mathbf{B}$ is half the magnetic rotation angle during the time step.
[18] The sequence is mathematically concise when $\gamma = 1$. The quantity u is velocity in the sense of momentum per unit restmass. The relativistic γ is obtained from it as the square root of $1 + (u/c)^2$. The equation for u_3 is executed by first dividing B by γ^2 and then using "1" in place of γ^2. This accounts for the m rather than m_0 in the angle.

5.3.3. *Field Advance Algorithm*

The advance of the fields through one time step of arbitrary length (subject to $\mathbf{j}_k$ = constant) is mathematically just like that of the particles. According to Eq. 98, $\mathbf{F}_k$ consists of a constant component plus a rotating component. The constant part represents the magnetostatic field generated by the currents; the rotating part represents a circularly polarized electromagnetic wave. Again, the advance through any time interval δt is straightforward. The longitudinal (purely electric) component of the field is updated from the record of ρ at the end of the time step using Eq. 99.

We note, so far, we have not invoked finite difference calculus either in the space or time domain and, typically, the advance of the fields from Maxwell's equations is not restricted by any "Courant condition." However, δt is constrained by the fact that $\mathbf{E}$ and $\mathbf{B}$ should not change across the range of the orbit excursions during δt.

Equation 97 is used to trace the evolution of the transverse field only. The longitudinal, electrostatic field is constructed "from scratch" at the new time, using the charge density records:

$$F_k^{e-s} = ik\rho_k/k^2 \tag{103}$$

The longitudinal field, then, need not be held over through the particle move phase: it can be generated directly by Fourier transforming the charges accumulated during that phase. The transverse field is calculated as follows. A particular solution is constructed from $\mathbf{j}_k$ using

$$F_k^{m-s} = k \times j_k/k^2 \tag{104}$$

To $\mathbf{F}_k^{m-s}$ one has to add the rotating "electromagnetic" component

$$F_k^{e-m}\,(new) = F_k^{e-m}\,(old)\cos k\delta t - (k/k) \times F_k^{e-m}\,(old)\sin k\delta t \tag{105}$$

The new fields are then reconstructed from the updated pieces according to

$$F_k\,(new) = F_k^{m-s} + F_k^{e-m}\,(new) + F_k^{e-s} \tag{106}$$

and this is kept on record for the next field update.

The field seen by a particle must then be obtained by summation over the entire available spectrum

$$
\begin{aligned}
F\,(r) = D\,(r) + iH\,(r)\,/c &= (2\pi)^{-3} \int_k F_k e^{-ik\cdot r} dk \\
&= (2\pi)^{-3} \sum_{k_x}\sum_{k_y}\sum_{k_z} F_k e^{-ik\cdot r}
\end{aligned}
\tag{107}
$$

To calculate $\mathbf{F}$, we must introduce a grid over which field values are generated from the spectrum by FFTs, and we must interpolate the local field from the grid record. Líkewise, charge and current harmonics must be built up by interpolation into a grid and subsequent FFTs.

Having avoided spatial grids and spatial finite-difference calculus so far, the introduction of a grid to obtain the electromagnetic fields from the spectrum F_k leads to difficulties associated with grids: inaccuracies and stroboscopic effects. These problems are reduced using higher-order interpolation methods [Buneman et al. 1980].

5.4. Issues in Simulating Cosmic Phenomena

5.4.1. *Boundary Conditions*
A major problem in space applications is to simulate free-space conditions outside the computer domain. Complex Fourier methods [with $\exp(i\mathbf{k}\cdot\mathbf{r})$] imply periodic repeats of the computed domain in all dimensions. If the simulation is to represent phenomena in a rather larger plasma, such repeats are acceptable, but for an isolated plasma of limited extent they become unrealistic. This problem can be overcome by keeping a generous empty buffer zone around the domain containing particles and truncating the interaction between charges beyond a certain radius so that the nonphysical repeats introduced by the Fourier method cannot influence the central plasma. This was first applied to gravitational simulations.

The most elusive boundary problem for space plasmas is the radiation condition. To decide what part of the field in the charge-current-free space outside the plasma is outgoing and what is incoming presents no problem in 1D and the incoming part can be suppressed.

In 2D the decision is more difficult. It requires information not only in the source-free boundary layer at any time but also over its past history. It almost seems as if, in principle, the entire past history is needed for the decision. However, Lindman found that a fairly short history (such as three past time steps) of the boundary suffices for an algorithm which will suppress all but 1% of the incoming radiation at all but the shallowest angles of incidence. However, just carrying an absorption layer in an outer envelope seems quite successful. This method simply multiplies the electric and magnetic field by a factor which smoothly approaches zero away from the plasma.

5.4.2. *Relativity*
A reason for keeping the mass coupled with the velocities in the update steps is that under relativistic conditions one really updates momenta rather than velocities. Note, however, that in the $qB\delta t/2m$ terms one

needs $m^{-1} = (m_0^2 + p^2/c^2)^{-1}$ where $p = mv$. During the rotation, this magnitude of the momentum does not change, but in the electric acceleration it does. After the full update of momenta, one must again divide by m in order to get $\mathbf{v} = \delta\mathbf{r}/\delta t$. In practice, v rather than the momentum is stored for each particle which means that at the beginning of the update one must calculate $m = m_0(1 - \beta^2)^{-1/2}$. Thus, there are three separate calls to a reciprocal square root in the relativistic advance of each particle. As system supplied square roots are time consuming and more accurate than needed for particle pushing, a Padé-type rational first approximation, followed by a Newton iteration, is used instead.

5.4.3. *Compression of Time Scales*
The number of steps required to simulate a significant epoch in the evolution of a real plasma configuration would be many million, typically, if t has to be of the order ω_p^{-1} or ω_b^{-1}. In order to bring this down to the more acceptable range of several thousand steps, one must compress the time scales. Compressing time scales can be achieved by (1) decreasing the ions' rest mass in relation to the electrons' rest mass, and (2) increasing temperatures so that typical particle velocities get closer to the velocity of light.

For an ion (proton) to electron simulation mass ratio of 16, ion gyrofrequencies $\omega_{bi} = eB/m_i$ will be high by a factor of $1836/16 = 115$, ion plasma frequencies $\omega_{pi} = \sqrt{n_i Z^2 e^2/m_i \varepsilon_0}$, ion thermal velocities $v_{Ti} = \sqrt{kT_i/m_i}$, the Alfvén velocity $v_A = \sqrt{B^2/\mu_0 n_i m_i}$, and the relative velocity in Biot-Savart attraction $v = I_z\sqrt{\mu_0 L/2\pi \sum m_i}$ will be high by factor of $\sqrt{1836/16} = 10.7$.

The exaggeration of temperatures provides one of several motivations for incorporating relativity into our codes. Note, incidentally, that even a 10-kV electron, a temperature typical of many space plasmas, already moves at $1/5\ c$.

The exaggeration of "temperatures" of beam or current electrons can also be achieved by exaggerating the external electric field E_z responsible for accelerating the particles. This technique greatly reduces the number of time steps required to study a phenomenon such as Birkeland current formation and interaction in cosmic plasma. Since the current density is proportional to the electric field (i.e., $j_z = I_z/A = n_e e v_z \sim (n_e e^2/m_e)\ E_z t$), both the time required for the pinch condition Eq. 9 to be satisfied, and the relative velocity between parallel currents, are linearly related to E_z.

Of course, when economy necessitates time compression, the time-scales must be "unfolded" upon simulation completion.

5.4.4. *Collisions*

Just as in real plasmas, there are encounters between particles and these give rise to collisional effects which influence the physics of the model. Since computer models are limited to some 10^6 particles whereas a laboratory plasma may have $10^{18} - 10^{20}$ particles and a galaxy, 10^{65} particles, each particle in the model is a "superparticle" representing many plasma electrons or ions. Thus the forces between model particles are much larger than in a real plasma and the collisional effects are much greater. Fortunately there is a way to reduce the model collisions to rates comparable with real plasmas. This involves the finite-size particle method [Birdsall and Langdon 1985].

Here we use a gaussian profile for particles. The shaping is done in k-space. This is achieved by first building up ρ_k and $\mathbf{j}_k$ for $|k| < k_{\max}$ (truncation of harmonics at maximum k), as if each computer particle were a point and then applying a gaussian filter in k-space [Buneman et al. 1980]. The particle shape is then of the form $\exp\left(-r^2 k_p^2/2\right)$, where the particle profile factor k_p is left as a users option: many simulations have used a profile which keeps the spectrum flat up to a fairly large k and then makes a rapid but smooth slope-off to zero at some desired k_{max}.

A limit to the maximum acceptable radius of the finite-sized particles is set by the collective properties of the plasma. If the effective radius of a gaussian particle is increased much beyond the Debye length, it takes over the role of the Debye length, causing collective effects to be altered.

In simulating a physical system, plasma or gravitating, it is usually sufficient to determine if the system models a collisionless one over the simulation time span. Experimental determination of the effective collisional frequency ν_c in 2D models closely follows the empirical law [Hockney and Eastwood 1981]

$$\frac{\nu_c}{\omega_p/2\pi} = N_D^{-1}\left[1 + \left(\frac{w}{\lambda_D}\right)^2\right]^{-1}$$

where w is the width of the particle and $N_D = n\lambda_D^2$ is the number of particles in a Debye square. In 3D simulations, the reduction of ν_c is achieved, without increasing N_D, by "softening the blow" of collisions– making the particles into fuzzy balls. Values of $\nu_c/\omega_p \approx 10^{-3}$ for gaussian profile particles for N_D of order unity have been calculated. This value is consistent with most plasmas, in laboratories or space.

So far we have only considered collisions between particle species that are charged. However, in weakly ionized plasmas where the number of uncharged particles may be hundreds or thousands of times more prevalent, it is often the collision between the massive ions and the massive neutral atoms that cause a redistribution of energy, and concomitant effects, such as plasma heating. Collisions in weakly ionized plasmas have been successfully treated by melding PIC algorithms with MCC–Monte Carlo Collision–algorithms.

PIC codes involve deterministic classical mechanics which generally move all particles simultaneously using the same time step. The only part left to chance is usually limited to choosing initial velocities and positions and injected velocities. The objective for highly-ionized space plasma is usually seeking collective effects due to self and applied fields. On the other hand, MCC codes are basically probabilistic in nature, seeking mostly collision effects in relatively weak fields. For example, let a given charged particle be known by its kinetic energy W_{kin} and its velocity relative to some target particles. This information produces a collision frequency $\nu_{coll} = n_{target}\sigma W_{kin}v_{relative}$ and a probability that a collision will occur. This information is then used to describe electron collisions with neutrals (elastic scattering, excitation, and ionization) and ion collisions with neutrals (scattering and charge exchange).

The method is to use only the time step of the PIC field solver and mover, δt, and then to collide as many particles as is probable P in that δt separately. The actual fraction of particles in collision is $P = 1 - \exp(-\nu_{coll}\delta t)$. Note that we have slipped into treating our computer particles as single electrons, not as superparticles; the implication is that with a sufficient number of collisions, the resultant scatter in energy and velocity will resemble that of the single particles.

The end result of current efforts at including collisions in PIC codes due to Monte Carlo methods is the change in velocities of the particles. Thus, the only change from a collisionless run is that the particle velocities are varied in a time step. The last task at the end of a time step is to determine the new (scattered) velocity, and new particle velocities if ionization occurs (if ionization and/or recombination processes have been included in the MCC model). Each process is handled separately. Elastic collisions change the velocity angles of the scattered electrons; charge exchanges decrease ion energy and change velocity angles; ionizations do these and create an ion-electron pair, with new velocities. The effect on the neutral gas is not calculated because the lifetimes of the excited atoms are generally less than a time step.

When recombination rates are high, and if the source of energy to the plasma is terminated, gravitational effects must soon be included in the particle kinematics.

5.5. GRAVITATION

The transition of plasma into stars involves the formation of dusty plasma, the sedimentation of the dust into grains, the formation of stellesimals, and then the collapse into a stellar state [Alfvén and Carlqvist 1978]. While the above process appears amenable to particle simulation, a crude approximation of proceeding directly from charged particles (actually a cloud of charged particles) to mass particles is made. The transition of charge particles to mass particles involves the force constant, that is, the ratio of the coulomb electrostatic force between two charges q separated a distance r,

$$F_q(r) = \frac{q^2}{4\pi\varepsilon_0 r^2} \tag{108}$$

to the gravitational force between two masses m separated a distance r,

$$F_G(r) = -\frac{Gm^2}{r^2} \tag{109}$$

In the particle algorithm this change is effected by the following:

1. Changing all particles to a single species.

2. Limiting the axial extent of the simulation to be of the order of less than the extent or the radial dimension (i.e., about the size of the expected double layer dimension).

3. Setting the axial velocities to zero.

4. Setting the charge-to-mass ratio equal to the negative of the square-root of the gravitational constant ($\times 4\pi\epsilon_0$).

This last change produces attractive mass particles via the transformation $\varphi_G(r) = \varphi_q(r)$ in the force equation $F = -\nabla\varphi$, where

$$\varphi_q(r) = -\frac{q^2}{4\pi\varepsilon_0 r} \tag{110}$$

and

$$\varphi_G(r) = -\frac{Gm^2}{r} \tag{111}$$

are the electrostatic and gravitation potentials, respectively.

5.6. SCALING LAWS

The scaling of plasma physics on cosmical and laboratory scales generally involves estimates of the diffusion in plasma, inertia forces acting on the currents, the Coriolis force, the gravitational force, the centrifugal force, and the $\mathbf{j} \times \mathbf{B}$ electromagnetic force.

Specification of plasma density, geometry, temperature, magnetic field strength, acceleration field, and dimension set the initial conditions for simulation. The parameters that delineate the physical characteristics of a current-carrying plasma are the electron drift velocity

$$\beta_z = \frac{v_z}{c} \tag{112}$$

the plasma thermal velocity

$$\beta_{th} = \frac{v_{th}}{c} = \frac{(\lambda_D/\Delta)(\omega_p \delta t)}{c\delta t/\Delta} \tag{113}$$

and the thermal/magnetic pressure ratio

$$\beta_p = \frac{n_e k T_e + n_i k T_i}{B^2/2\mu_0} = \frac{(\lambda_D/\Delta)^2 (\omega_p \delta t)^2 \, 4 \, (1 + T_i/T_e)}{(c\delta t/\Delta)^2 (\omega_{c0}/\omega_p)^2} \tag{114}$$

The parameter δt is the simulation time step, Δ is the cell size, and c is the speed of light. All dimensions are normalized to Δ and all times are normalized to δt. The simulation spatial and temporal dimensions can be changed via the transformation

$$\frac{c\delta t}{\Delta} = \frac{c\delta t'}{\Delta'} = 1 \tag{115}$$

where $\Delta' = \alpha \Delta$ and $\delta t' = \alpha \delta t$, for the size/time multiplication factor α. The values of $n, T, B,$ and E remain the same regardless of whether the simulations are scaled to Δ and δt or to Δ' and $\delta t'$.

One immediate consequence of the rescaling is that, while the dimensionless simulation parameters remain untouched, the resolution is reduced, that is,

$$\omega \delta t = \omega' \delta t' \tag{116}$$

where $\omega' = \omega/\alpha$ rad s^{-1} is the highest frequency resolvable.

To convert simulation results to dimensional form, it is sufficient to fix the value of one physical quantity (e.g., B_ϕ).

5.7. THREE DIMENSIONALITY

As in laboratory plasmas, astrophysical and space plasmas are three dimensional in nature even if their source is symmetrically two-dimensional. Lindberg [1978] found that full physics models and approximations were of secondary importance to a three dimensional spatial description in benchmarking observations of a charged particle beam flowing along a curved magnetic field. The problem is now known as the *Reverse Deflection Problem.*

The Reverse Deflection Problem is a relatively simple experimental setup consisting of a plasma gun source with external coil magnets arranged to produce a curved magnetic field. The plasma gun is energized and charged particles are emitted into an initially linear magnetic 'guide' field. However, as the beam encountered the bend in the field, all *a priori* theoretical predictions and two dimensional simulations of the beam kinetics were incorrect (Figure 6). Furthermore, the initially cylindrical beam contracted into a flat slab.

Once this behavior was observed, Lindberg demonstrated that the reverse deflection could be qualitatively understood on the basis on classical and electric circuit theory if the backward drift of high energy electrons were accounted for [Alfvén 1981]. When the plasma enters the curved field, it induces a transverse electric field $\mathbf{E} = -\mathbf{v} \times \mathbf{B}$. This region then becomes a generator driving the current in an upward, transverse, third dimension, that then becomes polarized because of its low transverse conductivity (Figure 7).

6. Further Developments in Plasma Simulation

6.1. PARALLELISM

As pointed out, data management problems dominate the subject of 3D plasma simulation using particles-in-cell. In the novel computer architectures, with their high degree of parallelism, data transport becomes an even more important issue. Computing efficiency depends critically on (topological or physical) data proximity in the basic procedure of a problem. "Local" algorithms, such as finite-difference equations, have preference over "global" algorithms, such as Fourier transforms (Note that the calculation of each single Fourier harmonic requires the entire data-base). With this in mind, new 3D plasma codes have been constructed. In these the particles are advanced just as in Eqs.(8.6)-(8.8), but Maxwell's equations are integrated locally over a cubic mesh in the form:

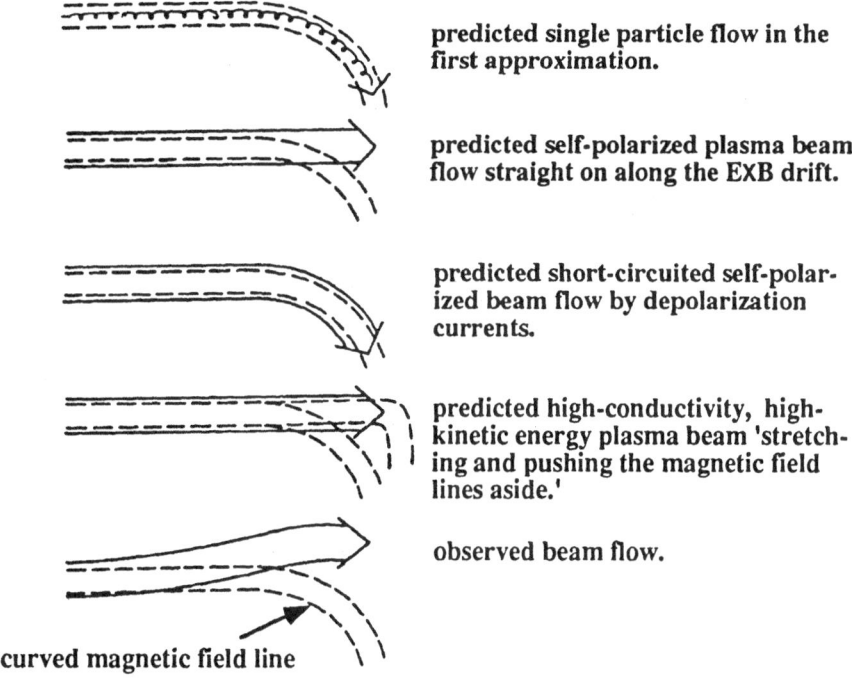

predicted single particle flow in the first approximation.

predicted self-polarized plasma beam flow straight on along the EXB drift.

predicted short-circuited self-polarized beam flow by depolarization currents.

predicted high-conductivity, high-kinetic energy plasma beam 'stretching and pushing the magnetic field lines aside.'

observed beam flow.

curved magnetic field line

Figure 6. Classical predictions and the real behavior of a plasma beam, initially moving parallel to a guide magnetic field, when entering a curved field region. The last frame shows, contrary to all predictions, a reverse deflection of the beam out of the guide field. Only when the experiment is simulated in three dimensions is the correct solution, including beam flattening, obtained. Figure courtesy of H. Alfvén.

1. change of **B**-flux through a cell-face = - circulation of **E** around it.

2. change of **D**-flux through a cell-face = circulation of **H** around it - charge flow through it.

The **E**− or **D**− data mesh is staggered relative to the **B**− or **H**− data mesh both in space and time.

This method has the advantage that needs to be satisfied only at the beginning of a run (where it becomes a triviality of initialization): it is automatically carried forward in time by consistent determination of the charge flow between cells. Thus Poisson's equation does not have to be solved. Poisson's equation is "global": The solution anywhere depends on the data everywhere.

The algorithms for a simplified version of TRISTAN, a fully three-dimensional, fully electromagnetic, and relativistic PIC code, are found in *Physics of the Plasma Universe* [Peratt 1992].

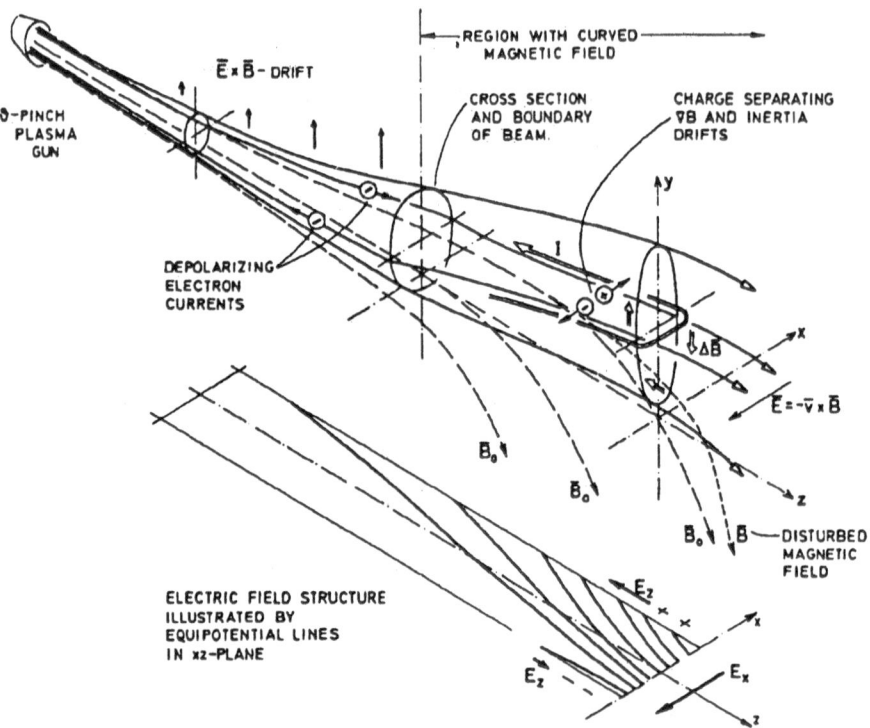

Figure 7. Kinetic and Electromagnetic components in the *Reverse Deflection Experiment.* Figure courtesy of H. Alfvén.

6.2. ISIS

ISIS is a multidimensional electromagnetic particle code, including three-dimensional capability, developed to treat particle-beam propagation in vacuum or plasma for realistic problems in pulsed-power transmission lines, high-power microwave generators, particle accelerators, and high-energy-density plasma applications. A description of the ISIS algorithms are given elsewhere [Jones and Peter 1985].

A major difference between ISIS and the older Fourier Transform versions of SPLASH or TRISTAN is that the ISIS uses finite-difference equations that are integrated over a spatial mesh having great generality in shape and coordinate system; e.g., the meshing may be nonlinear with many more zones inserted where known complex plasma motion occurs. Another major difference between SPLASH and ISIS is that in the former all quantities are scaled to the Debye length λ_D whereas in the latter all quantities are scaled to the Electromagnetic Skin Depth λ_E. In addition, affordable simulation with SPLASH/TRISTAN is usually accomplished by using 'light ions.' A hybrid version of ISIS overcomes this problem.

All space-time relationships in ISIS are in cgs units[19]. Variables are scaled by a plasma frequency, ω_p^0, or equivalently, by a plasma density n_p^0, which is arbitrary. Code units are denoted by a tilde and are related to real units via the expressions:

$$\tilde{t} = t\omega_p^0, \; time$$

$$\tilde{x} = x\omega_p^0/c, \; length$$

$$\tilde{v} = v/c, \; velocity$$

$$\tilde{\gamma} = \gamma mc^2, \; energy$$

$$\gamma\tilde{v} = \gamma vmc, \; momentum$$

$$\tilde{E} = E\left(4\pi n_p^0 mc^2\right)^{-1} = eE/mc\omega_p^0 = (\lambda_E/511)\,E_{kV/cm}, \; electric\,field$$

$$\tilde{B} = B\left(4\pi n_p^0 mc^2\right)^{-1} = eB/mc\omega_p^0 = \omega_c^0/\omega_p^0, \; magnetic\,field$$

The value of ω_p^0 may be defined in either of the following two ways:

1. Choose $\omega_p^0 = $ constant such the electromagnetic skin depth $\lambda_E = c/\omega_p^0$ is equal to an easily recognizable unit.

2. Once one parameter is fixed in a simulation, all other parameters can be scaled to it. For example, allow ω_p^0 to remain arbitrary by setting the code density of one species in the simulation to unity. Then lengths, times, electromagnetic fields, currents, etc., are scale to any arbitrary value of n_p^0 for that species. The density value of the other species are set by the ration of their density specified in the calculation to n_p^0.

Selection of the values of cell sizes ΔX, ΔY, ΔZ, and step size ΔT (where $X, Y, Z \equiv x, y, z/\lambda_E$, and $T \equiv \omega_p^0 t$) are very important to insure that the simulations has sufficient spatial and time resolution to solve the physics of the problem. In practice the stability constraints of ΔT are more demanding than those imposed by resolution and will determine its value. Spatially, at least six cells are required to resolve

---

[19] the text of this paper is in MKS units

a wavelength in any direction. A more appropriate number is twelve cells.

6.2.1. *Electrons as a massless fluid*

A hybrid ISIS model exists where the ions are treated kinetically via the PIC methods described above while the electrons are treated as a massless fluid. Hybrid simulation methods are useful in modeling low frequency plasma phenomena. The basic idea is to treat the electrons as a fluid, usually neglecting the electron mass and to treat the ions by the particle-in-cell method. This method allows following the dynamics of the plasma on the ion time scale, i.e., the ion gyroperiod in a magnetized plasma. The particle-in-cell method essentially solves the collisionless Boltzmann equation for the ions, giving a more complete description on the phase space evolution than a hydrodynamic model. Some approximations have been made that attempt to model the electron-ion collisional interaction, and ion-ion collisions have been treated through particle-pairing methods. As a result, hybrid methods, although suitable for following the time scales of interest, have not been optimally applied to high density collisional plasmas.

In general, the momentum equation for the electron fluid is

$$
\begin{aligned}
m_e \left(\frac{\partial v_e}{\partial t} + v_e \cdot \nabla v_e \right) &= -e \left(E + v_e \times B \right) \\
&- \frac{\nabla (n_e k T_e)}{n_e} - m_e \sum_i \nu_{ie} \left(v_e - v_i \right)
\end{aligned}
\tag{117}
$$

where v_e and v_i are the electron and ion fluid velocities and ν_{ie} is the electron-ion collision rate. In taking the limit of this equation as m_e vanishes, it is assumed that ν_{ie} becomes infinite so that the product $\nu_{ie} m_e$ remains finite. Furthermore, we will assume quasineutrality, which allows us to replace n_e with $\sum_i Z_i n_i$, the ion density.

Taking the limit as m_e goes to zero and using the quasi-neutral assumption, we obtain

$$
E = \frac{-j_e \times B}{\rho_e} - \frac{\nabla (\rho_e k T_e)}{\rho_e e} - \frac{m_e}{e} \sum_i \nu_{ie} \left(v_e - v_i \right)
\tag{118}
$$

where the ion charge density $\rho_i = -\rho_e = e n_e$ and j_e is the electron current density. In the usual hybrid approximation, this equation is solved for E under the assumption of a zero displacement current. The magnetic field is then advanced in time by Faraday's law Eq. 1.

The first term on the right hand side of Eq. 118 is used separately to advance the magnetic field in a subcycling procedure. The electron current density in this equation is determined from Ampère's law neglecting the displacement current and is given by

$$\mathbf{j}_e = \nabla \times \mathbf{B} - \mathbf{j}_i$$

In vacuum regions, the electric field is advanced with Ampère's law Eq. 2, including the displacement current. These equations are subcycled to satisfy the vacuum Courant condition. After the subcycling is completed, the last two terms of Equation 118 are added to the electric field and the corresponding magnetic field from Eq. 1 is also included. Using these electric and magnetic fields, the ions are treated in the usual particle-in-cell manner.

The second term of Equation 118 is the gradient of the electron pressure and is the usual hydrodynamic force term. The last term of Equation 118 is due to the collisional drag.

The electron temperature T_e in Eq. 118 can be entered via a number of different ways. For example, the simplest model would have only isothermal electrons. However, in general, the electron temperature would derive from an electron energy equation that includes electron-ion temperature coupling, electron-radiation coupling, heat transport, advection, and other electron energy source/sink mechanisms.

7. Advances in Numerical Modeling Techniques

7.1. MULTI-LEVEL CONCURRENT SIMULATION

The numerical modeling of complex physical phenomena requires the usage of a number of models of varying complexity. The near-Earth ionosphere-magnetosphere coupling problems is a prime example. Parts of the system require only simple electrical circuit analysis, while parts require a full first-principles treatment in fully three dimensional space. Not every step of the modeling process requires the most sophisticated models. In the past, it has been the modeler's responsibility to make intelligent choices as to which model to use and then the modeler would rely on experience and insight to interface the results of one code as inputs to the next code. This painstaking process has evolved today into what is called Multi-Level Concurrent Simulation (MLCS) [Jones 1995].

As a paradigm, MLCS is a way of facilitating the design by interfacing the various components of the process. Recent advances in distributed computing naturally lend themselves to this structure, using various computational tools from a variety of platforms to concurrently model various levels of the problem. This concept involves various pieces of computer hardware networked to access program libraries, empirical design curves and powerful first principles physics models through a common interface.

In spite of the fact that modelers of complex plasma phenomena have developed expertise in areas such as implicit numerical techniques, hybrid models, adaptive grid methods and multiple time scale perturbation methods, modelers are still often faced with inadequate tools because of disparate time and spatial scales. Advances in computing hardware promised by the Accelerated Scientific Computing Initiative (ASCI)[20] offer the hope that more first principles approaches can be applied to complex problems. However, the demand always exceeds the resources. Furthermore, many models are not stable enough to be used for the extremely long time scales needed to model some systems from first principles, even if the computing resources are available.[21] Thus, robustness in new modeling techniques must match the new capabilities in hardware performance.

7.2. THE HIERARCHY OF SCALING

The need for multiple time and spatial scale modeling in science is ubiquitous. Often modeling of physical systems is essentially impossible because first principles physical models are too resource intensive to be used on the space and time scales needed to answer the pertinent question at hand. Even if massively parallel computers can, in the near future, reach the teraflop regime as promised, present day algorithms and models may not be numerically stable enough to be useful. Systematic methods of treating multiple scales would significantly impact the modeling of complex plasma processes of laboratory or astrophysical dimensions and numerous other scientific disciplines. A NASA space weather initiative has need for accurate plasma models that reflect both large scale and small scale phenomena.

The coupling of disparate spatial and/or temporal scales is at the heart of many scientific modeling efforts and indeed forms the core research in modeling the effects of microphysics on macrophysics scales, including the effects of turbulence on hydrodynamic modeling, hybrid plasma simulation models for magnetic fusion and space plasmas, and subgrid models in electromagnetic modeling. Advanced modeling also has a need for physics of radiation transport, hydrodynamics, plasma physics, and atomic physics in integrated modeling tools. Most plasma

---

[20] ASCI is a U.S. Department of Energy program to advance the state of numerical computation by accelerating the current multi-hundred-megaflop, large platform, computer speed to the 100 teraflop (10^{14} floating operations per second), by the year 2002.

[21] In some circumstances involving multitasking applications on a Cray-YMP, where 6 processors are used simultaneously, 350 Mflops speeds can be reached.

physics problems exhibit an extended range of time and spatial scales. In particular space plasmas have multiple scales involving electron and ion motion, sheaths, and other phenomena that may involve the growth of instabilities, the generation of large electric fields, and the subsequent acceleration of the charged particles to high energies.

The difficulty in using existing tools concurrently is that there often exists a disparity in scale between the various models. Distributed computing is only part of the solution to this problem. Execution times for different models could be tailored to specific hardware on the network. For example, simple systems models may run concurrently on a personal computers while sophisticated physics models are running on the ASCI machines, filling in the unknown parts of the system database. Obviously one is not always lucky enough to have the answers come out at the right time to make this process efficient [Jones 1995].

Optimization of fidelity in physical modeling is paramount in the development of high-performance computing techniques. Multi-level concurrent computing now appears possible because of recent advances in numerical differencing techniques, and the availability of distributed computing resources. However, to make Multi-level concurrent simulation viable, development is required with regards to reduced space and time scales modeling, robust algorithms, distributed computing, and modular code construction.

7.3. REDUCED SPACE AND TIME SCALES

The basic problem is how to represent the fundamental, relatively small-scale, physical processes, which often take place at relatively small or microscopic length scales, in models which describe the relatively large or macro scale response of systems of interest with larger dimensions. Multiple time/spatial scale perturbation theory and other techniques may be used to formulate the effects of micro scale physics on longer space and time scales.

An example is the recent development of "collision field" methods to incorporate Coulomb collisions into a hybrid particle-in-cell code in order to model semicollisional plasmas. The effect of short range collisions is mediated through a field to study the nonfluid behavior of plasma in which the ion mean free path is comparable to the dimensions of the system being studied [Jones et.al. 1996]. This technique has been applied to ICF and magnetospheric problems [Jones et. al. 1995; Miller et. al. 1995].

7.4. ROBUST ALGORITHMS

An example of robust algorithms is the work on coupling lumped circuit models to 3D Maxwell equation solvers. In this example, the disparate time scales make the problem numerically stiff. A complex form of time differencing involving multiple time scale analysis to integrate over the fast time scale to provide difference equations for the slower time scale proved to be a remarkably robust algorithm [Thomas et. al. 1994].

7.5. DISTRIBUTED COMPUTING

The subject of Multi-Level Concurrent Simulation (MLCS) is closely related to distributed computing, object oriented computing, and aspects of distributed computing that are gaining use such as object request brokers. The purpose of distributed computing is to couple models together. However a problem exists in distributed computing in that there are no accepted, platform independent methods presently for this type of approach to computing, and also the networks are rather slow as well. This problem may be solved in the marketplace in the next few years. A problem regarding the necessity of high data rate transfers apparently will be solved with introduction of optical networks leading to 38–42 GHz 'broadcast' transmitters at central locations with transceiver/antenna cards located on site.

The reason why the multi-processing client-server paradigm and object oriented programming fit well together is that the two deal with very different aspects of software. Whereas object oriented programming deals with abstractions of state and behavior, multiple processing deals with abstractions of schedule and sequence. These abstractions have very little overlap, so the two paradigms work well together.

7.6. MODULAR CODE DESIGN

The advantage of linking models together over multiple scales suggests a modular code development approach that adapts itself well to a hierarchy of space and astrophysical plasmas.

One can then envision modularizing the 3D full physics codes by using the MLCS approach to link independently developed modules for magnetohydrodynamics, transition regions, double layers, particle dynamics, and electromagnetic radiation.

8. Advances in Numerical Modeling Platforms

The necessity for ever faster and larger memory and storage capabilities in the computing environment is illustrated by the following example involving the PIC simulation of an inertially confined plasma. Consider as the driver a neodymium glass laser beam at wavelength $\lambda_0 = 0.35\mu$m in the form of a pulse lasting several hundred picoseconds interacting with a cryogenic deuterium–tritium pellet with a total interaction volume of 1 mm^3. At this wavelength, the scale size of the simulation is approximately $3 \times 10^{10}\lambda_0^3$. For a mesh size of $\lambda_0/20$, this problem requires 2×10^{14} zones to simulate the laser-plasma evolution.

The field memory associated with this problem is ten or more words per zone[22]. The particle memory associated with this problem is six words per particle. Plasma spatial resolution requires about 100 particles per zone while 3 to 25 μs may be required to advance one particle per timestep in three dimensions. Particles dominate both memory and time. The particles require 600 words per zone and up to 2.5 ms per zone per timestep. Thus, this problem may require a total of 2×10^{16} particles, 1.2×10^{17} words (one cpu), and some 5×10^{11} seconds per timestep, i.e., sixteen thousand years.

Thus, the simulation of this problem appears intractable and when this analysis is applied to galactic dimensioned plasmas (of order 10^{65} particles), PIC simulation appears impossible.

However, postulate a computer capability of one central processor unit (cpu) per 60 megawords. This corresponds to ten million particles or a maximum of 250 seconds per timestep. For a one picosecond simulation, the timestep at the laser frequency $f = c/\lambda_0 = 8.96 \times 10^{14}$ Hz is approximately 1 ps/8.96×10^{14}Hz, or 4×10^6 cpu seconds (46 days).

Current three-dimensional PIC simulations generally employ 10-50 million particles, but as many as 250 million particles have been used. Expenditures of 350 cpu hours on an 8–processor, 32 Megaword, YMP machine to simulate a 3 million particle, 1 million cell plasma problem are not uncommon. However, to double the resolution of a PIC simulation requires a performance increase of 2^4, a factor 2 in each dimension plus a 1/2 timescale due to the Courant condition.

For a problem involving only a few hundred timesteps, a 3D PIC simulation on a Cray J90 dimensioned as $128 \times 128 \times 200$, with 206 megawords of memory might require 1.3×10^3 seconds (1.5 days) cpu time. If the dimensionality is increased by a factor $16 \times 16 \times 5$, the memory is increased to 264 gigawords and the time to 164×10^6 seconds (1900 days). For a dimensionality increase of $16 \times 16 \times 25$, the memory is 1.32 terawords and the time is 822×10^6 seconds (9500 days).

The increase in performance demand is also true of MHD codes. A 2D MHD code with 45×100 zones can require 9 hours cpu time on a YMP for problem completion. The addition of a third dimension

---

[22] This requirement depends on whether 32 bit words are used on a workstation or 64 bit (8 byte) words are used on a supercomputer.

Table I. Cray J90 hardware parameters.

Item	Parameter
Technology (CPU)	CMOS
Clock period	10 ns
Number of CPUs	32
CPUs per processor module	4
Peak performance	200 MFLOPS per CPU
Technology (memory)	70 ns CMOS DRAM
Memory size	1 - 8 Gbytes, (128 - 1024 Mwords)
Total memory bandwidth	51.2 Gbytes/s
Number of IOSs	16
I/O bandwidth	1.6 Gbytes/s

$45 \times 100 \times 100$ costs 100 times more and the inclusion of full radiation analysis takes 50 times as much time as the magnetohydrodynamics, so that 2000 equivalent YMP days are required. For convenience, typical operational parameters of a Cray J90 machine are tabulated in Table I.

However, these problems can be solved in about a day on a teraflop machine and in minutes on a petaflop machine.[23] Figure 8 illustrates the history of the performance speed of supercomputers and microprocessors versus year of operation.

In the United States, five governmental agencies are sponsoring the Petaflops initiative, the advancement of high performance computing capabilities into the petaflop (10^{15} flop) range. For example, the Department of Energy has established an Accelerated Strategic Computing Initiative (ASCI) whose purpose is to advance the availability of tightly coupled massively parallel processor and coupled shared memory processor clusters; and incorporating high-speed interconnects in a seamless, distributed computing environment. To achieve this goal, ASCI is supporting three memory options for the development of teraflop machines, respectively designated 'red', 'blue Pacific', and 'blue mountain'.

8.1. ASCI Red

The ASCI Red platform effort will bridge the gap between giga-scale and tera-scale computing to accommodate the five-order-of-magnitude increase in performance required by "full-physics", "full-system" sim-

---

[23] 1 Teraflop is equivalent to 5,455 YMP equivalents while a Petaflop is equivalent to 5 and one-half million YMP equivalents.

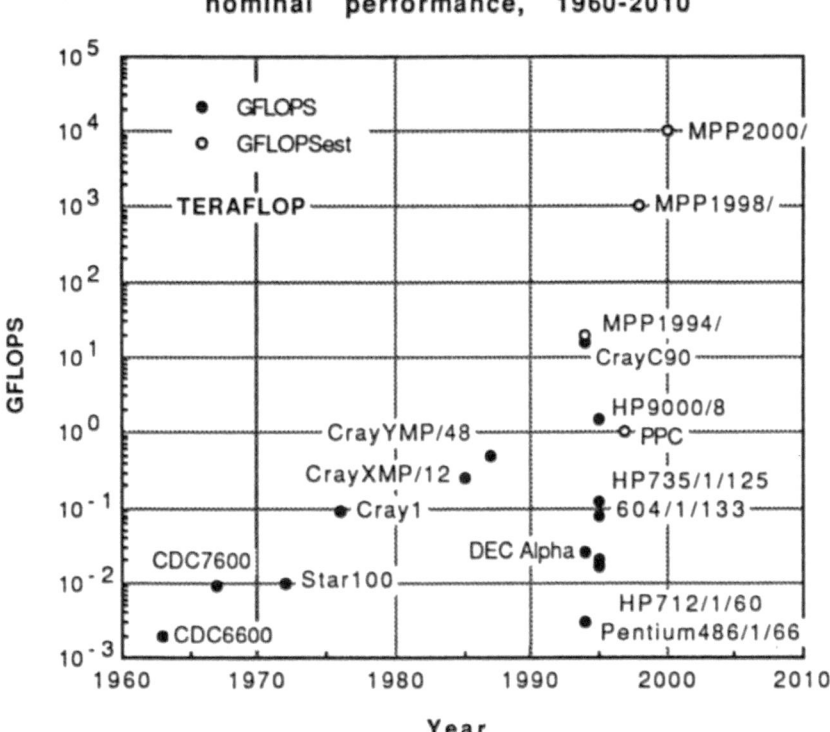

Figure 8. Performance speed of supercomputers and microprocessors versus year of operation.

ulation. The first deliverable in this effort is the 1.8 TFLOPS ASCI Red Supercomputer, a distributed memory system utilizing 9000 Intel P6 microprocessors and is nominally rated at 1.8 teraflops (1.8×10^{12}) flops with 512 gigabytes of memory (Figure 9). The machine will be located at Sandia National Laboratories.

8.2. *ASCI Blue Pacific*

A second platform called ASCI 'blue' is a shared memory processor system that consists of 512 nodes, each composed of eight-processor SMP units, for a total of 4096 IBM-Motorola PowerPC 604 RISC microprocessors. This machine has a rated capacity of 3×10^{12} flops and 2.5×10^{12} bytes of main memory. This machine is located at Lawrence Livermore National Laboratory.

Table II. ASCI Red system hardware parameters.

Item	Parameter
Compute Nodes	4,536
Service Nodes	32
Disk I/O Nodes	32
System Nodes (Boot and Node Station)	2
Network Nodes (Ethernet, ATM)	6
System RAM	594 MBytes
Topology	$38 \times 32 \times 2$
Node to Node bandwidth - Bi-directional	800 MBytes/sec
Bi-directional -Cross section Bandwidth	51.6 GBytes/sec
Total number of Pentium Pro Processors	9,216
Processor to Memory Bandwidth	533 Mbytes/sec
Compute Node Peak Performance	400 MFLOPS
System Peak Performance	1.8 TFLOPS
RAID I/O Bandwidth (per subsystem)	1.0 Gbytes/sec
RAID Storage (per subsystem)	1 TByte

Table III. ASCI Blue Pacific system hardware parameters.

Item	Parameter
Compute Nodes	512
Number of Microprocessors	4096
Microprocessors	PowerPC 604 (to be upgraded to to 630 chip)
RAM	64 MByte
Memory bandwidth	256 bit
Internal disk storage	2.2 Gbyte
Bus speed	160 MByte/sec
System Peak Performance	3 TFLOPS

8.3. *ASCI Blue Mountain*

A third machine, *ASCI Blue Mountain,* installed at Los Alamos National Laboraory is to have a nominal performance of three teraflops using 256 Silicion Graphics/Cray MIPS R10000(TM) microprocessors. A total of 3,072 processors will be in operation by 1999 providing a combined computational capability of 4 teraflops. The speed/memory delopment plan is outlined in Figures 9 and 10.

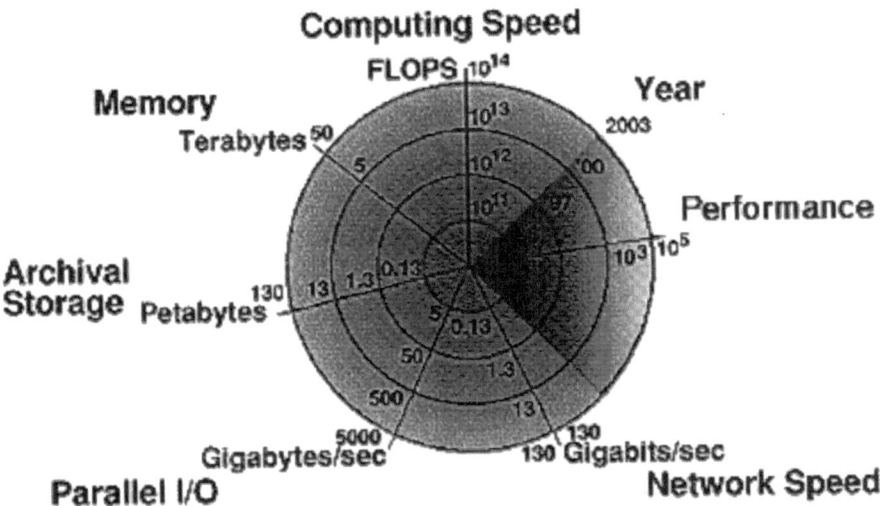

Figure 9. Estimated computing speed, memory, storage, transfer rates, and network speed versus year, 1996-2003.

9. Conclusions

Contrary to popular and scientific opinion of just a few decades ago, space is not an 'empty' void. It is actually filled with high energy particles, magnetic fields, and highly conducting plasma. The ability for plasmas to produce electric fields, either by instabilities brought about by plasma motion or the movement of magnetic fields, has popularized the term 'Electric Space' in recognition of the electric fields systematically discovered and measured in the solar system.[24] Today it is recognized that 99.999% of all observable matter in the universe is in the plasma state.[25]

---

[24] The metaphor 'Electric Space' was coined by Dr. Carolyn T. Brown, Assistant Associate Librarian for Area Studies (management of non-English collections and their scholarly use) at the Library of Congress and is the theme of the Space Science Institute traveling exhibition on the plasma universe supported by the National Oceanic and Atmospheric Administration and the National Science Foundation.

[25] The importance of electromagnetic forces on cosmic plasma cannot be overstated; even in neutral hydrogen regions ($\sim 10^{-4}$ parts ionized), the electromagnetic force to gravitational force ratio is 10^7.

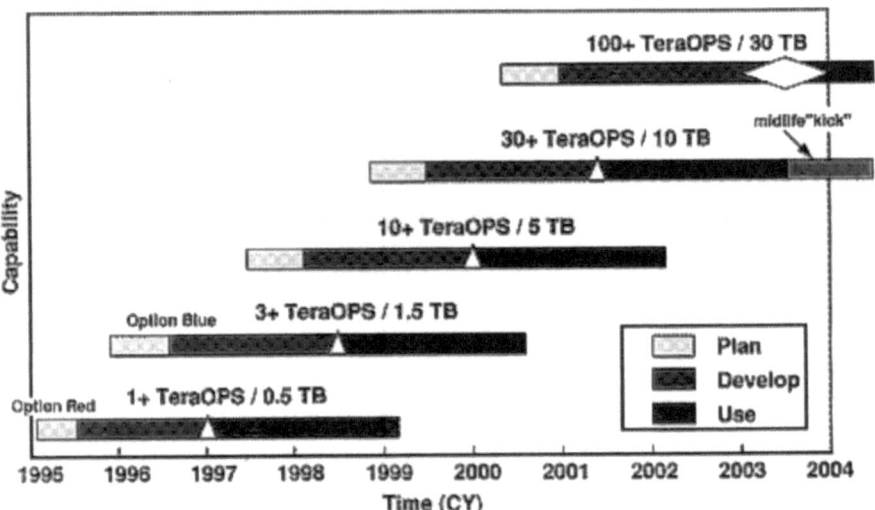

Figure 10. ASCI development and useage plan. Speed/memory in TeraOPS/Terabytes vs Calendar Year.

Plasma science is rich in distinguishable scales ranging from the atomic to the galactic to the meta-galactic. Plasmas of differing size or constituency also tend to be systematically coupled by electromagnetic forces. The problem of understanding the linkages and couplings in multi-scale processes is a frontier problem of modern science involving fields as diverse as high-energy-density plasmas in the laboratory to galactic dynamics.

Unlike the first three states of matter, plasma, the fourth state of matter, involves the mesoscale. Thus, in the study of astrophysical and space plasmas, the behavior of the near-earth plasma environment may depend to some extent on the solar and stellar plasmas that are part of the galactic plasmas. However, unlike laboratory plasmas, astrophysical plasmas will forever be inaccessible to in situ observation. The inability to test concepts and theories of large-scale plasmas leaves only virtual testing as a means to understand the universe. Advances in in computer technology and the capability of performing physics first principles, fully three-dimensional, particle–in–cell simulations, are making virtual testing a viable alternative to verify our predictions about the far universe when the simulations are verified by benchmarking to high-energy-density plasmas produced in the laboratory by intense laser beam and pulsed-power facilities.

The advocacy of first principles numerical modeling of the plasma state requires the elimination of constructs or 'inventions' that were demanded by lesser capability machines used to support one and two-dimensional simulations. Thus, in synergism with availability of multi-teraflop computer platforms, particle-in-cell simulations devoid of the magnetohydrodynamic fluid approximations and in fully three-dimensional space are expected to find increasing application in modeling capabilities. The first principles approach will require a reexamination of the formulation used to describe the plasma state. The move from pure MHD to PIC simulation also makes possible the benchmarking of the radiation observed from either laboratory or astrophysical plasmas over the full electromagnetic spectrum, to the extent that the advanced platform will support the necessary space/temporal resolution.

To fully utilize the availability of sub-petaflop or petaflop computer speed/memory architectures, a new simulation methodology such as parallelism for multiprocessors and multi-level concurrent simulation techniques with its concomitant demands of the advancement in multiple time and spatial scaling, robust algorithms, distributed computing, and modular code design is required. The end benefit is sobering: simulations that today would require cpu times of decades will be completed in days and ultimately, minutes.

10. Acknowledgments

This work was performed under the auspices of the U.S. Department of Energy. The author thanks R. J. Barker, C. K. Birdsall, C.-G. Fälthammar, A. B. Langdon, and M. E. Jones for useful discussions.

References

Alfvén, H., Carlqvist, P.: 1978, *Astrophys. Space Sci.***Vol. no. 55**, 484
Alfvén, H., Herlofson, N.: 1950, *Phys. Rev.***Vol. no. 78**, 616
Alfvén, H. and Fälthammar, C.-G. : 1963, *Cosmical Electrodynamics*, Oxford University Press, New York
Akasofu, A.-I.: 1981, "Energy coupling between the solar wind and the magnetosphere", *Space Sci.Rev.***Vol. no. 28**, p.21
Bennett, W. H.: 1934, "Magnetically self-focusing streams", *Phys. Rev.***Vol. no. 45**, 890
Birdsall, C. K., Langdon, A. B. : 1985, *Plasma Physics via Computer Simulation*, McGraw-Hill, New York

Biskamp, D.: 1997, "Magnetic Reconnection in Plasmas", *Astrophys. Space Sci.,This issue*

Bogdankevich, L. S., Rukhadze, A. A.: 1971, "Sov. Phys. Usp", *Sov. Phys. Usp.***Vol. no. 14**, 163

Bostick, W. H.: 1986, "What laboratory produced plasma structures can contribute to the understanding of cosmic structures both large and small", *IEEE Trans. Plasma Sci.***Vol. no. 14**, 703

Brackbill, J.: 1987, "Fundamentals of Numerical Magnetohydrodynamics, International School for Space Simulation, La londe les Maures, France, 1987", *Los Alamos National Laboratory Report, LA-UR-87-2052.*

Buneman, O.: 1976, "The advance from 2D electrostatic to 3D electromagnetic particle simulations", *Computer Phys. Comm.***Vol. no.12**, pp. 21–31

Buneman, O., Barnes, C. W., Green, J. C., Nielsen, D. E.: 1980, "Principles and capabilities of 3D, EM particle simulations", *J. Comp. Phys.***Vol. no.38**, 1

Buneman, O.: 1986, "Multidimensional particle codes: their capabilities and limitations for modeling space and laboratory plasma", *IEEE Trans. Plasma Sci.***Vol.14**, 661

Carlqvist, P.: 1988, "Cosmic electric currents and the generalized Bennett Relation", *Astrophys. Space Sci.***Vol. no. 144**, 73

Chen, F. F. : 1984, *Introduction to Plasma Physics and Controlled Fusion*, Plenum Press, New York

Dawson, J. M., Decyk, V., Sydora, R., Liewer, P.: 1993, "High-performance computing and plasma physics", *Phys. Today.*, March

Eastman, T.: 1990, "Transition regions in solar system and astrophysical plasmas", *IEEE Trans. Plasma Sci.***Vol. no. 18**, p18.

Frank, I., Ginsburg, V.: 1945, "Radiation of a uniformly moving electron due it its transition from one medium into another", *Journal of Phys.***Vol. no. IX**, pp. 353–362

Gouveia Dal Pino, E. M., Opher, R.: 1989, "The origin of filaments in extended radio sources", *Astrophys. J.***Vol. no. 342**, pp. 686–699

Lindberg, L.: 1970, *Astrophys. Space Sci.***Vol. no. 55**, 203

Hammer, D. A., Rostocker, N.: 1970, *Phys. Fluids***Vol. no. 13**, 1831

Happek, U., Sievers, A. J., Blum, E. B.: 1992, "Observation of coherent transition radiation", *Phys. Rev. Lett*

Hockney, R. W., Eastwood, J. W.: 1981, *Computer Simulation Using Particles*, McGraw-Hill, New York

Jones, M. E., Peter, W. K.: 1985, *IEEE Trans. Nucl. Sci.***Vol. no. NS-32**, p. 1794.

Jones M. E. , D. S. Lemons, R. J. Mason, V. A. Thomas, & D. Winske: 1996, "A Grid-Based Coulomb Collision Model for PIC Codes", *J. Comput. Phys.***Vol. no. 123**, in press.

Jones, M. E., D. Winske, S. R. Goldman, R. A. Kopp, V. G. Rogatchev, S. A. Bel'kov, P. D. Gasparyan, G. V. Dolgoleva, N. V. Zhidkov, N. V. Ivanov, Yu. K Kochubej, G. F. Nasyrov, V. A. Pavlovskii, V. V. Smirnov, and Yu. A. Romanov: 1996, "An Adiabatic Fluid Electron Particle-in-Cell Code for Simulating Ion-Driven Parametric Instabilities", *Phys. Plasmas***Vol. no. 3**, pp. 1096-1108.

Jones, M. E.: 1995, "Multi-Level Concurrent Simulation: A White Paper", *unpublished*

Küppers, G., Salat, A., Wimmel, H. K.: 1973, "Macroscopic equilibria of relativistic electron beams in plasmas", *Plasma Phys.***Vol. no. 15**, 441

Melrose, D. B.: 1997, "Particle Acceleration and Nonthermal Radiation in Space Plasmas", *Astrophys. Space Sci.,This issue*

Miller, R. H., Combi, M. R.: 1995, *Geophys. Res. Lett.***Vol. no. 21**, 1735

Nahin, P. J. : 1988, *Oliver Heaviside: Sage in Solitude*, IEEE Press, New York

Peratt, A. L. : 1992, *Physics of the Plasma Universe*, Springer-Verlag, New York

Thomas, V.: 1995, "Multi-Level Concurrent Simulation", *Los Alamos National Laboratory Report LA-UR-95-3454*

Trubnikov, B. A.: 1958, "Plasma radiation in a magnetic field", *Sov. Phys. 'Doklady'* **Vol. no. 3**, 136

Vu, H. X.: 1996, "An Adiabatic Fluid Electron Particle-in-Cell Code for Simulating Ion-Driven Parametric Instabilities", *J. Comput. Phys.* **Vol. no. 123**, in press.

Witalis, E.: 1981, *Phys. Rev. A* **Vol. no. 24**, 2758

Yonas, G., Poukey, J. W., Prestwich, K. R., Freeman, J. R., Toepfer, A. J., Clauser, J. J.: 1993, *Nucl. Fusion* **Vol. no. 14**, 731

Zimmerman, G. B., Kruer, W. L.: 1975, *Comments Plasma Phys.* **Vol. no. 2**, p. 51.

Dieter Biskamp

Magnetic Reconnection in Plasmas

Dieter Biskamp
Max-Planck-Institut für Plasmaphysik,
85748 Garching, Germany

Abstract. A review of the present status of the theory of magnetic reconnection is given. In strongly collisional plasmas reconnection proceeds via resistive current sheets, i.e. quasi-stationary macroscopic Sweet-Parker sheets at intermediate values of the magnetic Reynolds number R_m, or mirco-current sheets in MHD turbulence, which develops at high R_m. In hot, dilute plasmas the reconnection dynamics is dominated by nondissipative effects, mainly the Hall term and electron inertia. Reconnection rates are found to depend only on the ion mass, being independent of the electron inertia and the residual dissipation coefficients. Small-scale whistler turbulence is readily excited giving rise to an anomalous electron viscosity. Hence reconnection may be much more rapid than predicted by conventional resistive theory.

Key words: Magnetic Fields, Reconnection, MHD Turbulence, Cosmic Plasmas

1. Introduction

The term magnetic reconnection derives from the simple picture of two magnetic field lines being cut and reconnected in a topologically different way. Naive as this might appear at first sight, it addresses a fundamental problem in understanding the dynamics of magnetized plasmas. Since in electrically conducting fluids the property of magnetic flux conservation gives field lines - defined as thin flux tubes - a concrete identity, which can only be destroyed by the effect of finite resistivity or some equivalent effect, magnetic reconnection processes should be slow in most astrophysical plasmas, where such non-ideal effects are extremely weak. A wide range of observations, however, indicate that large amounts of magnetic flux may be reconnected on very short time scales, which seem to be completely unrelated to the weak non-ideal processes responsible for reconnection to occur. Since rapid large-scale magnetic processes in plasmas in general require reconnection, the primary objective of reconnection theory is to identify the basic mechanisms giving rise to such fast decoupling of plasma and magnetic field. Having identified the relevant reconnection process, one may then try to understand the partition of the large-scale magnetic energy released due to fast reconnection into the different channels, i.e. heating, bulk flows and energetic particles. Since energetic particles are omni-present in astrophysical plasmas carrying a finite amount of the total plasma energy, the efficiency of particle acceleration in reconnec-

Astrophysics and Space Science **242**: 165–207, 1997.

tion processes is an important issue. A third area concerns the onset
conditions for large-scale energy release by reconnection. How can free
energy be stored without continuous leakage, which would only result
in to some low-level dynamics, and what effect triggers the sudden
large-scale release as for instance in a flare.

Most of the theoretical research on reconnection performed over
the last more than thirty years deals with the first issue, i.e. to find
simple configurations, usually in the framework of resistive magnetohydrodynamics (MHD), allowing efficient reconnection. After it had
become clear that reconnection in a macroscopic current sheet as proposed by Sweet (1958) and Parker (1963) is far too slow to explain for
instance the time scale of the flash phase of a major flare, Petschek's
slow shock model (proposed at a meeting on solar flares 1964), where
reconnection occurs in a micro-current sheet and thus allows reconnection rates essentially independent of the value of the resistivity, was
instantly accepted as a major break-through. In fact for the following
twenty years reconnection theory was focussed, at least in the western
hemisphere, on Petschek's model, until its major inconsistency became
apparent. The progress of the last ten years in the theory of reconnection, much as in many other branches of physics, was primarily due to
numerical simulations, which allowed to study nonstationary, even turbulent configurations, to broaden the framework from resistive MHD to
general two-fluid theory and even kinetic theory, and also to touch on
3-D effects, though resolution in the latter case is still only marginally
adequate.

In this review, I give an overview of the present status of our per-
ception of fast reconnection mechanisms, limiting consideration to the
framework of fluid theory. In section 2, I introduce the two-fluid equations, from which the MHD approximation is easily derived. Section
3 deals with the traditional, rather well understood case of quasi stationary resistive reconnection, which takes place in macroscopic current
sheets. In section 4, I then introduce several paradigmatical reconnecting systems, the tearing mode and the resistive kink mode, in particular their nonlinear behavior, the process of flux bundle coalescence, and
the generation and dynamics of plasmoids. At sufficiently high magnetic Reynolds number a magnetized plasma becomes turbulent similar to
the case of a neutral fluid. Small-scale turbulence makes reconnection
much more efficient than in a stationary configuration. Section 5, therefore, gives an introduction to MHD turbulence. In section 6, collisionless
reconnection processes are discussed, which dominate the dynamics in
high temperature, dilute plasmas. In magnetized plasmas the Hall term
is the dominant effect in Ohm's law, which allows a decoupling of the
electrons from the ions on small scales $l < c/\omega_{pi}$. At these scales the ions

only form an immobile neutralizing background, while the dynamics is carried by the electrons, which is described by the equations of electron magnetohydrodynamics (EMHD). Reconnection rates are found to be independent of the electron dynamics, and hence depend only on the ion inertia. In the collisionless limit reconnection occurs via whistler turbulence, giving rise to anomalous electron viscosity. In section 7, I briefly discuss the other two areas, mentioned above, particle acceleration during reconnection and the problem of fast reconnection onset.

2. Two-fluid theory and the MDH approximation

The fluid description of plasmas (see, e.g. Braginskii, 1965) is a very convenient and reliable approximation of the bulk behavior, even for quasi-collisionless conditions. We are primarily interested in the long-wavelength regime $\lambda \gg \lambda_D$ (λ_D = the Debye length), where the charge density is negligible compared to the densities of ions and electrons, which is called the quasi-neutral limit. Hence the densities are equal $n_e = n_i = n$ (considering only one singly-charged ion species), which follows the continuity equation

$$\partial_t n = -\nabla \cdot \mathbf{v}_i n = -\nabla \cdot \mathbf{v}_e n \tag{1}$$

The equations of motions are

$$m_i n(\partial_t \mathbf{v}_i + \mathbf{v}_i \cdot \nabla \mathbf{v}_i) = -\nabla p_i + en(\mathbf{E} + \frac{\mathbf{v}_i}{c} \times \mathbf{B}) - \nabla \cdot \pi_i - ne\eta \mathbf{j}, \tag{2}$$

$$m_e n(\partial_t \mathbf{v}_e + \mathbf{v}_e \cdot \nabla \mathbf{v}_e) = -\nabla p_e - en(\mathbf{E} + \frac{\mathbf{v}_e}{c} \times \mathbf{B}) - \nabla \cdot \pi_e + ne\eta \mathbf{j}, \tag{3}$$

where $p_{i,e}$ are the scalar pressures, i.e. the isotropic part of the total pressure tensors, $\pi_{i,e}$ the anisotropic parts called the stress tensors. The friction term in eqs. (2), (3) is written in terms of a scalar resistivity η and the current density $\mathbf{j} = ne(\mathbf{v}_i - \mathbf{v}_e)$, which is coupled to the magnetic field by Amperes law

$$\nabla \times \mathbf{B} = \frac{4\pi}{c} \mathbf{j} . \tag{4}$$

In the collisional-dominated case the stress tensor is given in terms of the velocity derivatives and the viscosity coefficients. For quasi-collisionless conditions no simple expression can be given for the $\pi_{i,e}$, but we assume that they are small and that the most important contribution is the residual viscosity effect, which we write in terms of simple scalar viscosities $\mu_{i,e}$, $\nabla \cdot \pi_{i,e} = n\mu_{i,e}\nabla^2 \mathbf{v}_{i,e}$. Since heat conduction can often be neglected, the pressures follow the adiabatic law,

$$\partial_t p_{i,e} + \mathbf{v}_{i,e} \cdot \nabla p_{i,e} = -\gamma p_{i,e}\nabla \cdot \mathbf{v}_{i,e} . \qquad (5)$$

Addition of eqs. (2) and (3) gives an equation for the mass flow $\mathbf{v} = (m_i\mathbf{v}_i + m_e\mathbf{v}_e)/(m_i + m_e) \simeq \mathbf{v}_i$, from which the electric field is eliminated:

$$m_i n(\partial_t \mathbf{v} + \mathbf{v} \cdot \nabla \mathbf{v}) = -\nabla p + \frac{1}{c}\mathbf{j} \times \mathbf{B} - \nabla \cdot \pi_i \qquad (6)$$

with $p = p_i + p_e$ and $\pi_i + \pi_e \simeq \pi_i$. The electric field appears in Faraday's law

$$\partial_t \mathbf{B} = -c\nabla \times \mathbf{E} . \qquad (7)$$

To express $\mathbf{E}$ in terms of the other dynamic quantities the electron equation of motion (3) is written in the form of a generalized Ohm's law

$$\mathbf{E} + \frac{\mathbf{v}}{c} \times \mathbf{B} = \mathbf{R}, \qquad (8)$$

where $\mathbf{R}$ comprises the remaining terms. Introducing the normalizations $x/L, v/v_A, t/t_A$ with L a typical spatial scale, $v_A = B_0/\sqrt{4\pi n m_i} = $ Alfvén velocity of a typical magnetic field B_0, $t_A = L/v_A$, $\mathbf{R}$ in eq. (8) becomes

$$\mathbf{R} = \eta\mathbf{j} + \mu_e\nabla^2\mathbf{v}_e + d_i[\mathbf{j} \times \mathbf{B} - \beta\nabla p_e - d_e^2(\partial_t\mathbf{v}_e + \mathbf{v}_e \cdot \nabla\mathbf{v}_e)] , \qquad (9)$$

where $d_i = c/\omega_{pi}L, d_e = c/\omega_{pe}L$ are the normalized ion and electron inertia scales, $\beta = 4\pi n T_e/B_0^2$ is a measure of the plasma pressure compared to the magnetic pressure, η is the normalized resistivity $c^2\eta/4\pi L v_A \equiv S^{-1}$, $S = $ Lundquist number, and μ_e the normalized electron viscosity μ_e/Lv_A.

For the large-scale dynamics all terms in $\mathbf{R}$ are usually very small, e.g. $d_i \sim 10^{-7}, \eta \sim 10^{-12}$ in the solar corona, such that to lowest order $\mathbf{R} = 0$. Hence eq. (7) becomes

$$\partial_t \mathbf{B} = \nabla \times (\mathbf{v} \times \mathbf{B}) . \qquad (10)$$

Equations (1), (5), (6), (10) are called (ideal) magnetohydrodynamics (MHD), which describe the behavior of one fluid with mass density $\rho = m_i n$, mass flow $\mathbf{v}$ and pressure p together with the magnetic field. The most important property distinguishing a highly conducting plasma from a neutral fluid is the conservation of magnetic flux across a surface

$F(t)$ bounded by a curve $l(t)$ moving with the fluid

$$\frac{d}{dt} \int \mathbf{B} \cdot d\mathbf{F} = \int_F \partial_t \mathbf{B} \cdot d\mathbf{F} + \int_{\partial_t F} \mathbf{B} \cdot d\mathbf{F} \tag{11}$$

$$= \int_F \nabla \times (\mathbf{v} \times \mathbf{B}) \cdot d\mathbf{F} + \int_l \mathbf{B} \cdot (\mathbf{v} \times d\mathbf{l}) = 0 \tag{12}$$

using eq. (10) and Stokes' theorem. Sweeping the boundary curve l along $\mathbf{B}$ defines a flux tube and in the limit of vanishing tube diameter a field line. Hence in ideal MHD field lines preserve their individuality indefinitely, however strongly convoluted they may become following the fluid motion.

Observations, however, clearly indicate that most dynamic processes in plasmas involve a change of field line topology, i.e. field lines must be cut and reconnected. Since such processes, for instance solar flares or magnetospheric substorms, are very rapid, occurring on time scales close to the Alfvén time, while reconnection relies on the small flux conservation breaking effects in $\mathbf{R}$ on the right side of eq. (8), the main problem is to describe *fast* reconnection.

Conventionally, it has been assumed that reconnection requires some dissipative process. Hence the nondissipative terms in the expression (9), those proportional to d_i, have generally been ignored. Of the two dissipative effects resistivity is usually considered as the more important one, such that $\mathbf{R} \simeq \eta \mathbf{j}$. In fact most previous reconnection studies are confined to the framework of resistive MHD. In the following sections 3-5 of this review consideration is therefore restricted to this framework, while in section 6 I will discuss the conditions, when the resistive theory is no longer adequate and the collisionless effects become dominant.

In addition reconnecting motions are usually assumed to be incompressible, since they are slow compared with the phase velocities of the compressional waves, essentially v_A in a low-β plasma. This approximation simplifies the basic equations considerably. For $\nabla \cdot \mathbf{v} = 0$ the assumption of homogeneous density $n = n_0$ is consistent with the continuity equation. Taking the curl of eq. (6) eliminates the pressure. Using the same normalizations the incompressible resistive MHD equations for the vorticity $\boldsymbol{\omega}$ and $\mathbf{B}$ now become

$$\partial_t \boldsymbol{\omega} - \nabla \times (\mathbf{v} \times \mathbf{B}) - \nabla \times (\mathbf{j} \times \mathbf{B}) = \mu_i \nabla^2 \boldsymbol{\omega}, \tag{13}$$

$$\partial_t \mathbf{B} - \nabla \times (\mathbf{v} \times \mathbf{B}) = \eta \nabla^2 \mathbf{B}, \tag{14}$$

$$\nabla \cdot \mathbf{v} = \nabla \cdot \mathbf{B} = 0, \quad \boldsymbol{\omega} = \nabla \times \mathbf{v}, \quad \mathbf{j} = \nabla \times \mathbf{B}.$$

An important dimensionless parameter characterizing the dynamical state of a resistive fluid is the magnetic Reynolds number

$$R_m = \frac{Lv}{\eta} = S\frac{v}{v_A}, \tag{15}$$

where v is a typical average fluid velocity. Roughly speaking R_m is the ratio of the convective to the diffusive terms in eq. (14).

Since reconnection theory is often limited to 2D systems, I also give the equations for this case. With z as ignorable coordinate it is convenient to introduce the magnetic flux function ψ (essentially the vector potential component A_z) and the stream function ϕ by writing

$$\mathbf{B} = \hat{\mathbf{z}} \times \nabla\psi + \hat{\mathbf{z}}B_z, \quad \mathbf{v} = \hat{\mathbf{z}} \times \nabla\phi + \hat{\mathbf{z}}v_z,$$

such that eqs. (13), (14) become

$$\partial_t\omega + \mathbf{v}_i \cdot \nabla\omega - \mathbf{B} \cdot \nabla j = \mu_i \nabla^2\omega, \tag{16}$$

$$\partial_t\psi + \mathbf{v} \cdot \nabla\psi = \eta\nabla^2\psi, \tag{17}$$

$$\omega = \omega_z = \nabla^2\phi, \quad j = j_z = \nabla^2\psi.$$

Knowing ψ and ϕ the remaining quantities B_z, v_z, p can be computed a posteriori.

3. Quasi-stationary resistive reconnection

For $R_m \gg 1$ fast reconnection can only arise when the gradient scales occurring in the reconnection process are much shorter than the global scale L. Hence the process is localized around regions, where the convective term in eq. (14) is small, i.e. at neutral points of the magnetic configuration. For topological reasons only X-type neutral points can give rise to reconnection. Figure 1 illustrates the behavior in the vicinity of such a point, plasma flowing into the region from above and below and leaving sidewise. Figure 1 also shows that the configuration is likely to be flattened into a sheet.

In low-β systems, where only a small fraction of the total magnetic energy is available as free energy, in general only a small component of $\mathbf{B}$, the poloidal field $\mathbf{B}_\perp$, is reconnected, while the main field (assumed in z direction) remains unchanged. Hence the X-point appears only in the poloidal projection.

3.1. BASIC PROPERTIES OF CURRENT SHEETS

A dynamical current sheet, usually called a Sweet-Parker sheet (Sweet, 1958; Parker, 1963), is a quasi one-dimensional stationary configuration

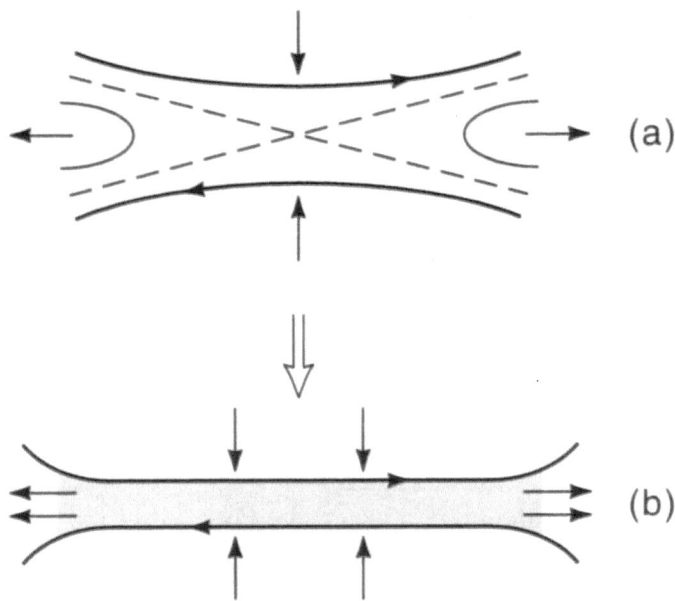

Figure 1. Reconnection flows of an X-point (a) leading to current sheet formation (b).

as illustrated in Fig.1b, where the plasma inflow balances resistive diffusion. Assuming incompressibility a current sheet is characterized by six quantities, the magnetic field immediately outside the sheet called the upstream field B_0 (the downstream field at the sheet edges is small), the upstream flow u_0 into the sheet perpendicular to the field, the downstream flow v_0 along the sheet, the width δ and the length Δ, and the resistivity η. These quantities are connected by three relations derived from the continuity equation, Ohm's law and the equation of motion, assuming stationarity. Integrating $\nabla \cdot n\mathbf{v} = n_0 \nabla \cdot \mathbf{v} = 0$ over the sheet one obtains

$$u_0 \Delta = v_0 \delta \tag{18}$$

ignoring profile effects. The z component of Ohm's law (8) with $\mathbf{R} = \eta\mathbf{j}$ and $E_z = E = $ const. gives

$$E = u_0 B_0 = \eta j \simeq \eta \frac{B_0}{\delta}. \tag{19}$$

Since in general $u_0 \ll B_0$, the inertia term is negligible in the force balance across the sheet, hence

$$B_0^2/2 = p_m - p_0, \tag{20}$$

where p_0 is the upstream pressure and p_m the pressure maximum in the sheet. Along the sheet, where the magnetic component normal to the sheet and hence the magnetic force vanish, the pressure gradient accelerates the fluid. Integration between sheet center and edge yields $v_0^2/2 = p_m - p_0$, assuming that at the edge the pressure has dropped to its upstream value. We thus obtain the important relation, that the outflow velocity equals the upstream Alfvén velocity

$$v_0 = B_0 = v_A .\tag{21}$$

Equations (18) and (19) can be used to express two of the remaining five quantities by the other three, for instance

$$\frac{u_0}{v_A} = \left(\frac{\eta}{B_0 \Delta}\right)^{1/2} \equiv S_0^{-1/2},\tag{22}$$

$$\frac{\delta}{\Delta} = S_0^{-1/2}.\tag{23}$$

Equation (22) shows that the reconnection rate is $\partial_t \psi = E = u_0 B_0 \sim \eta^{1/2}$ for macroscopic sheet length $\Delta \sim L$, which is far too slow to explain the time scales observed for instance in a solar flare. To achieve sufficiently fast reconnection current sheets of much smaller length are required, as for instance assumed in Petschek's theory (erroneously, see section 3.6), or as develop in MHD turbulence (Biskamp and Welter, 1989a).

It should be emphasized that current sheets are formed under quite general conditions. Figure 2 gives an example of a dynamic, initially smooth magnetic configuration ψ with four X-points (Fig. 2a), where current sheets develop shrinking in width (Fig. 2b,c), until resistivity becomes important setting up a quasi-stationary Sweet-Parker sheet. The location of current sheets in a general 3D sheared magnetic config-uration depends on the structure of the flows as excited for instance by an instability. The presence of a strong axial field $B_z \gg B_\perp$ provides a justification of the incompressibility assumption, which might appear doubtful in the sheet, where the velocity becomes large reaching the Alfvén speed of the reconnected field component $B_\perp$. If the latter is small compared with the axial component, the total Alfvén velocity is still much greater than the flow speed, which guarantees that the perpendicular flow is incompressible.

I should point out that a Sweet-Parker current sheet is also a vor-ticity sheet, as is clear from the corresponding flow pattern. The vor-ticity distribution has a quadrupole structure (instead of the monopole structure of the current density), because ω vanishes on the symmetry

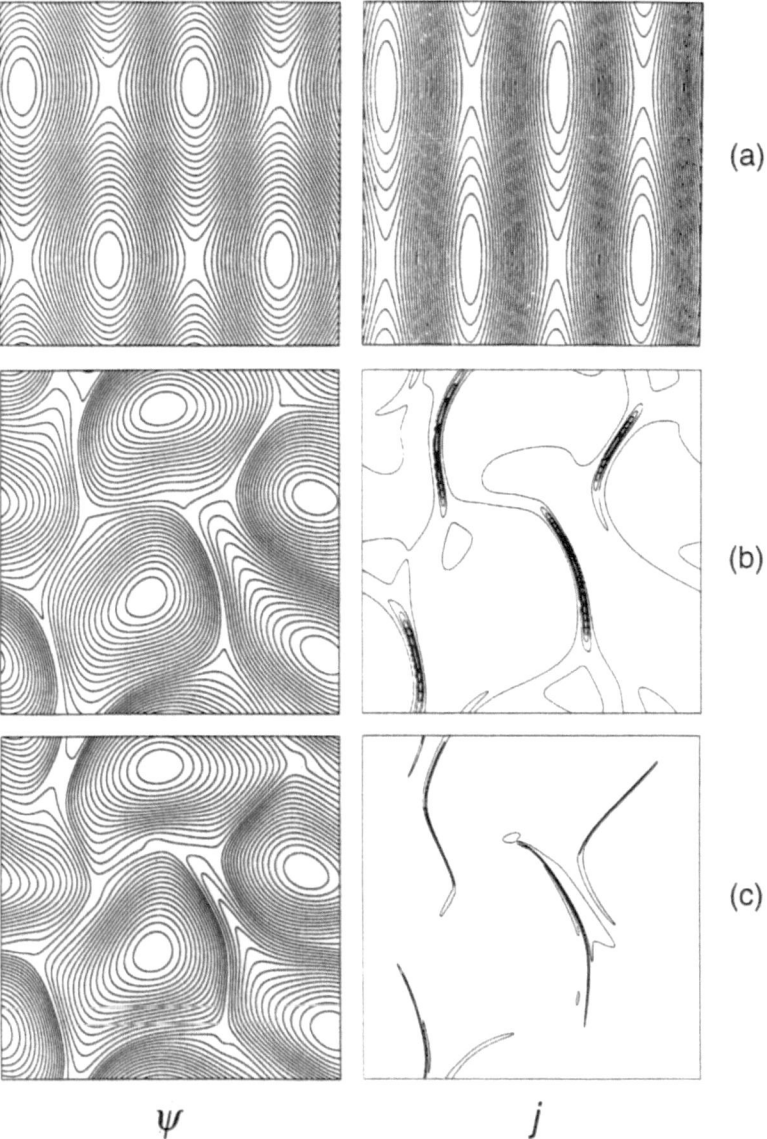

ψ j

Figure 2. Numerical simulation showing the development of current sheets from a smooth dynamic configuration.

lines of the sheet. Since $u_0 \sim \eta^{1/2} B_0$, the magnitude of the vorticity is, however, small $\omega \sim \eta^{1/2} j$, hence for $\mu \sim \eta$ viscous dissipation $\varepsilon_\mu = \mu \int \omega^2 dV$ is smaller than Ohmic dissipation $\varepsilon_\eta = \eta \int j^2 dV$.

It can be shown by a simple power expansion (Cowley, 1975), that in a Sweet-Parker sheet the magnetic field structure in the vicinity of the neutral point in the sheet center differs basically from a normal

X-point. In particular the separatrix branches do not intersect at a
finite angle but osculate as shown in Fig. 3. This result underlines the
inherent tendency toward current sheet formation at an X-point.

While the main part of a current sheet configuration is smooth,
the outflow or edge regions show a rather complicated quasi-singular
behavior. In a seminal paper Syrovatskii (Syrovatskii, 1971) presents
a simple theory of current sheets, which appear as branch cuts of a
complex potential function. He derives an expression for the sheet current density integrated across the sheet width $J(y)$, where $y = $ the
coordinate along the sheet:

$$J(y) = 2 \left(\frac{I}{2\pi} + \frac{\Delta^2}{2} - y^2 \right) \Big/ \sqrt{\Delta^2 - y^2}. \qquad (24)$$

Here I is the total sheet current and Δ the length of the sheet. Expression (24) has the interesting feature that $J(y)$ reverses sign being posi-
tive, i.e. in the direction of the total current in the central part $|y| < y_0$,
and negative in the outer parts $|y| > y_0$, with $y_0^2 = (I/2\pi) + \Delta^2/2$. At
the sheet edges $|y| = \Delta$ the current density becomes singular.

Resistive MHD simulations essentially confirm Syrovatskii's prediction (24). The current density in the sheet is found to reverse sign
and to exhibit a quasi-singular structure in the edge regions of the
sheet (Biskamp, 1986). The current density changes sign at the position, where the velocity reaches its maximum along the sheet, $v = v_A$.
The fluid is subsequently decelerated in the negative current density part and finally completely blocked, turned around and accelerated
again along smaller secondary sheets, formed on both sides of the quasi-
singular edges of the primary sheet. For still smaller η tertiary sheets
become visible, leading to an infinite sequence of higher order sheets
in the limit $\eta \to 0$. However, such regular scaling behavior rests on
the assumptions of spatial symmetry and stationarity. For more general nonstationary conditions a turbulent state consisting of an irregular
ensemble of micro-current sheets is generated (Biskamp and Welter,
1989a), as discussed in section 5.

3.2. SCALING LAWS FOR DRIVEN RESISTIVE RECONNECTION

The concept of driven or forced reconnection plays an important role
in reconnection theory. While originally the term referred only to open,
externally forced systems in contrast to closed systems, where reconnection occurs spontaneously as an internal process, the concept can be
applied much more generally. Assuming that reconnection is localized
in space, one may restrict consideration to a subregion L around this
location instead of the entire system of size Λ, $L \ll \Lambda$. On the other

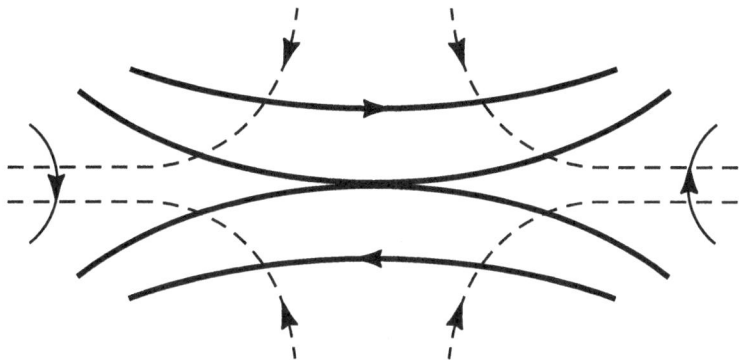

Figure 3. Osculating separatrix branches at the neutral point of a Sweet-Parker sheet.

hand, L should be large compared with the scales of the reconnecting structures, for instance $L \gg \Delta$ in the case of a single current sheet of length Δ, so that these are not affected by the artificial boundaries of the subsystem L. This procedure allows to simplify the geometry and also to assume stationarity even for a nonstationary global system. Since the coupling to the latter occurs through the boundary conditions imposed on the subsystem, which vary on the global time scale Λ/v_A, while the subsystem relaxes on the time scale L/v_A, we may consider the subsystem in steady state (if such a state exists). In this sense the subsystem constitutes a stationary driven reconnection configuration. By assuming up-down and right-left symmetry consideration can be restricted to a quadrant, for instance the upper left quadrant shown in Fig. 4, where the main parameters are indicated. The plasma along with its frozen-in magnetic field is injected from above and after reconnection ejected to the left. While u_∞, B_∞ are given, the internal parameters u_0, B_0 of the current layer and the dimensions δ, Δ are determined by the reconnection process.

A series of simulation runs (Biskamp, 1986) performed for different values of η and the (imposed) reconnection rate E, but identical boundary profile functions yields the following scaling laws of the cur-

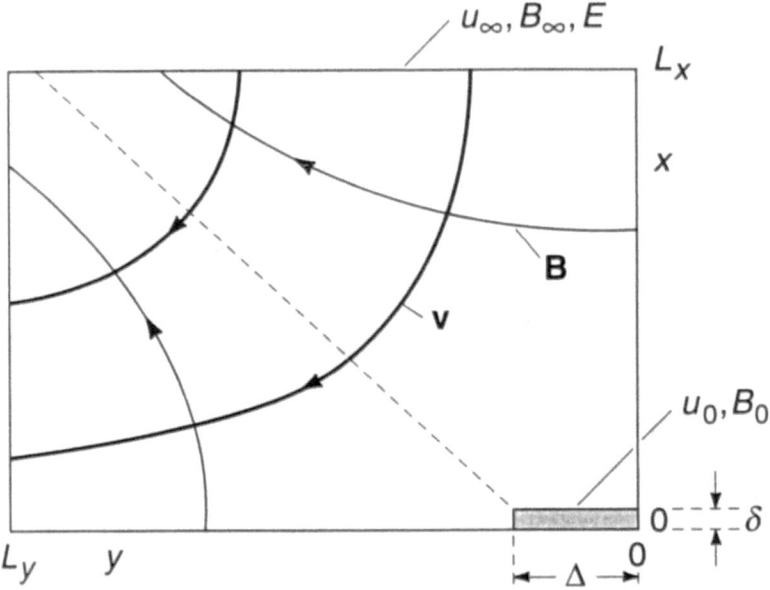

Figure 4. Computational box for driven reconnection studies.

rent sheet parameters

$$B_0 \sim E^2/\eta , \tag{25}$$
$$u_0 = E/B_0 \sim \eta/E , \tag{26}$$
$$\Delta \sim E^4/\eta^2 , \tag{27}$$
$$\delta \sim E\eta^0 . \tag{28}$$

Hence increasing E or decreasing η leads to a rapid stretching of the sheet length Δ as illustrated in Fig. 5. The width δ does not decrease but increases with E contrary to expectation, the reason being the increase (pile-up) of the magnetic field B_0 in front of the layer and the corresponding decrease of u_0, since $u_0 B_0 = E$.

The scaling laws (25) - (28) reflect an important property of the Sweet-Parker sheet. Consider the average inertia force along the sheet. Since the velocity increases about linearly $v_y \simeq B_0 y/\Delta$, we have

$$\overline{v_y \partial_y v_y} \simeq B_0^2/2\Delta \simeq E/2\delta , \tag{29}$$

using mass conservation $u_0 \Delta = B_0 \delta$ and Ohm's law $u_0 B_0 = E$. Hence the scaling (28) implies that the inertia force is invariant to changes of E and η, in particular remains finite for $\eta \to 0$.

When the sheet length Δ reaches the system size L as in Fig. 5c, Δ cannot increase further, and the scaling laws (25)-(28) are no longer

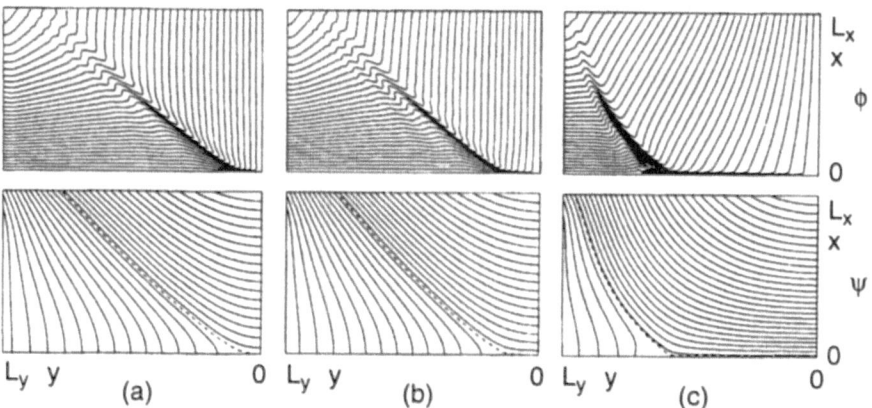

Figure 5. Flux function ψ and stream-function ϕ for three simulation runs with different values of η: (a) $\eta = \eta_0$, (b) $\eta = \eta_0/2$, (c) $\eta = \eta_0/4$, with identical boundary conditions.

valid. Instead we have $\Delta = L$. From $u_0 B_0 = E = \eta B_0/\delta$ one obtains $u_0 = \eta/\delta$, $B_0 = E\delta/\eta$. Using mass conservation we find

$$B_0 \sim (E^2 L/\eta)^{1/3} \,, \tag{30}$$

$$\delta \sim (\eta^2 L/E)^{1/3} \,. \tag{31}$$

In a closed system only a finite amount of free energy is available, hence B_0 is limited. This leads us back to the original sheet relations (22), (23) indicating $u_0 \sim \delta \sim \eta^{1/2}$.

3.3. PETSCHEK'S QUASI-IDEAL RECONNECTION MODEL

For two decades Petschek's slow-shock model (Petschek, 1964) was the generally accepted theory of magnetic reconnection. In recent years it has, however, been realized, that the model is not valid in the usually adopted framework of resistive MHD in the limit of small resistivity. Petschek's configuration, illustrated in Fig. 6, is characterized by two pairs of slow shocks standing back to back against the upstream flow, which they deviate by roughly 90° into the downstream cone between the shocks. Current density and vorticity are concentrated in the shocks and the central current sheet. The shocks derive their properties from the slow magnetosonic wave, a compressible mode, which survives with finite phase velocity in the incompressible limit, $\omega^2 = k_\|^2 v_A^2 = k^2 B_n^2$, where B_n is the component normal to the wave front. Hence the flow can be supersonic with respect to this mode for any given speed, if the angle between wave front and magnetic field is sufficiently small.

Petschek's configuration is a solution of the ideal MHD equations valid outside the singularities. The jump conditions at the shocks determine the position of the latter, i.e. the downstream angle α in terms of the upstream velocity. The reconnection rate is essentially independent of the resistivity, which has been the most attractive feature of the model. The ideal solution must, however, be matched to the resistive solution in the current sheet, which is very difficult and has previously been essentially ignored. Petschek *assumes* a current sheet of dimensions $\Delta \sim \delta \sim \eta$ adjusting automatically to the external solution. This is, however, not true. Numerical solutions of the full resistive problem show that Δ does not shrink with η, but becomes macroscopically large following the scaling laws (25)-(28), which changes the external configuration fundamentally and leads to a slow Sweet-Parker type reconnection rate $E = O(\eta^{1/2})$. Petschek's model may, however, be applicable, if the conditions in the reconnection region are alleviated. This is achieved by a locally enhanced effective resistivity $\eta_{eff} = O(1)$ due to turbulence generated for instance by the high current density in the sheet (Sato and Hayashi, 1979). Macroscopic current sheet formation can also be avoided, if collisionless reconnection processes are dominant as discussed in section 6.

4. Paradigms of resistive reconnection

The natural configuration of a magnetized plasma in the absence of nearby boundary surfaces or other types of forcing is a circular plasma tube or pinch (Fig. 7a), for instance a coronal loop. External forces may change this configuration primarily in two ways, either by stretching into an elongated sheet-like state (Fig. 7b) as in the case of the solar wind stretching the earth dipole field into the magnetotail configuration, or by twisting (Fig. 7c) such as performed by photospheric convection on coronal loops. Both processes lead to energization of the configuration, which gives rise to relaxation processes, if certain thresholds are exceeded. The stretched configuration becomes unstable to the tearing mode breaking up into several plasma tubes (Fig. 7d), which then coalesce into a single tube. In the twisted case the plasma tube kinks into a helical configuration (Fig. 7e), which is then carried back into a less twisted straight tube by the resistive kink mode.

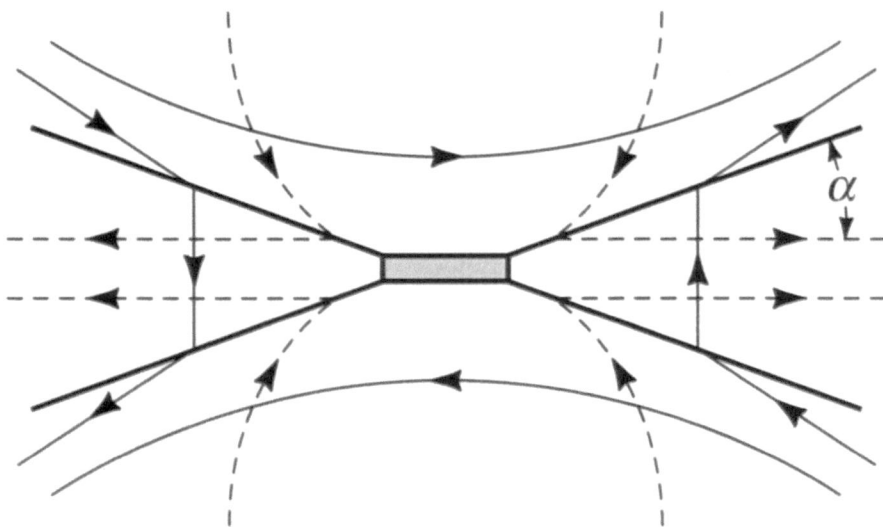

Figure 6. Schematic drawing of Petschek's slow-shock configuration.

The basic mechanism driving these relaxation processes is the attractive force between parallel currents. Tearing, coalescence and (resistive) kinking involve large-scale reconnection, which I will consider more in detail in the following. We should of course keep in mind that the sequence of events sketched in Fig. 7 is highly idealized ignoring 3D effects, for instance two flux tubes in general encounter obliquely under a finite angle. Nevertheless, these processes play a fundamental role in analyzing and understanding nonlinear plasma dynamics, serving as paradigms of reconnecting systems.

4.1. THE TEARING INSTABILITY

The tearing mode is probably the most frequently invoked instability in the context of magnetic reconnection. Nevertheless, its peculiar properties, in particular the slow nonlinear evolution, do not seem to be well known. I first recapitulate briefly the linear theory (Furth et al., 1963). Consider a plane configuration $\mathbf{B}_0 = (0, B_{0y}(x), B_{0z}(x))$, where the poloidal field reverses sign at $x = 0$, so that in the vicinity of this surface $B_{0y}(x) \simeq B'_{0y}x = \psi''_0 x$. In the presence of a sinusoidal perturbation one has

$$\psi(x,y) = \psi''_0 x^2/2 + \psi_1(x, t)\cos ky, \qquad (32)$$

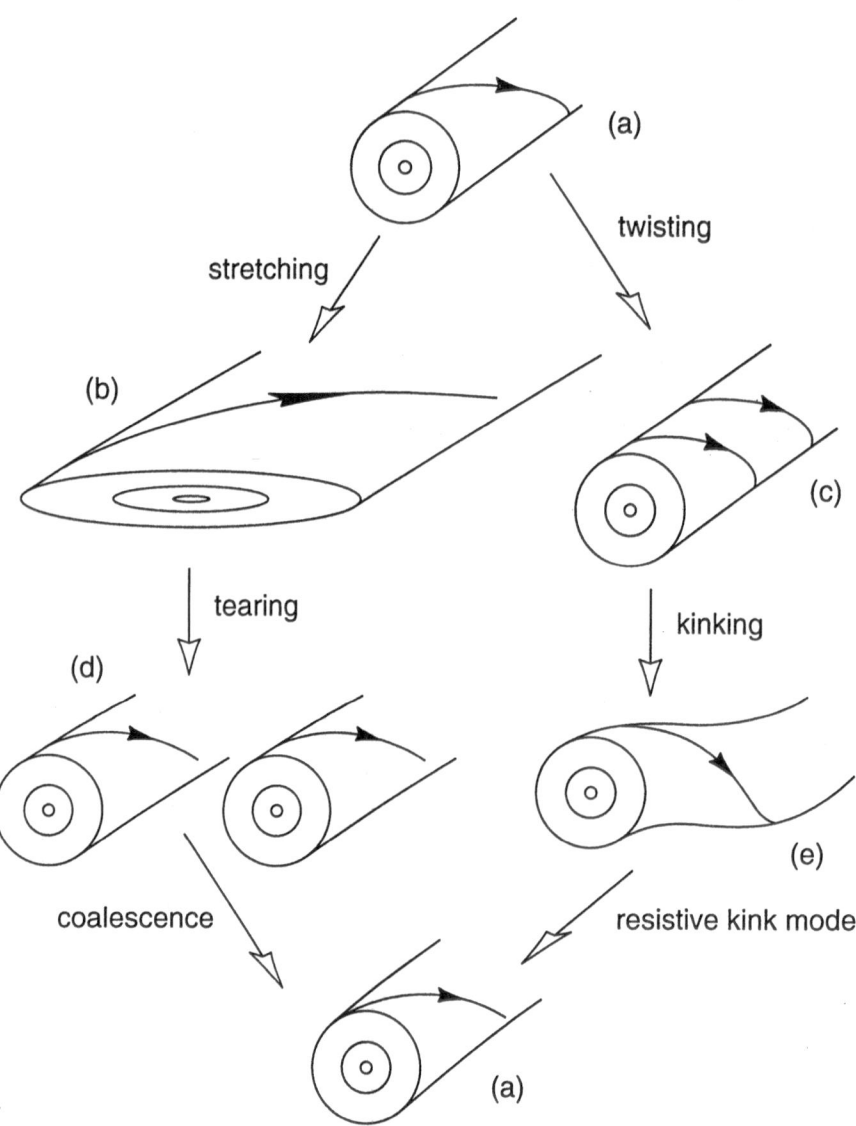

Figure 7. A circular pinch, forced externally, tends to relax back into a circular shape by different reconnection processes.

which consists of an alternating sequence of X- and O-points. The perturbation is characterized geometrically by the island size w, the distance between the separatrix branches at the O-point,

$$w = 4\sqrt{\psi_1/\psi_0''}, \tag{33}$$

ignoring the x-dependence of ψ_1. Linearizing eqs. (16), (17) and neglecting the viscous term one obtains

$$\partial_t \nabla^2 \phi_1 - \mathbf{B}_0 \cdot \nabla j_1 - \mathbf{B}_1 \cdot \nabla j_0 = 0 , \tag{34}$$

$$\partial_t \psi_1 - \mathbf{B}_0 \cdot \nabla \phi_1 = \eta \nabla^2 \psi_1 , \tag{35}$$

where for eigenmodes $\psi_1, \phi_1 \propto e^{\gamma t}$. Since the highest order derivative in eq. (35) is multiplied by the small parameter η, we have to solve a boundary layer problem. We distinguish between the narrow resistive layer, where $B_{0y} \simeq 0$, such that in eq. (35) the resistive term is important, and the external region, where B_{0y} is finite. In the resistive layer, $\mathbf{B}_0 \cdot \nabla = ik B_{0y} = ixF$ with $F = k B_{0y}'$, and $\nabla^2 \simeq \partial_x^2$, such that eqs. (34), (35) become (index i for "inner" solution)

$$\gamma \phi_i'' = ixF\psi_i'' , \tag{36}$$

$$\gamma \psi_i = ixF\phi_i + \eta \psi_i'' . \tag{37}$$

In addition we assume that $\psi_i(0)$ is finite, hence $\psi_i = $ const. inside the narrow layer, which is called the constant-ψ approximation. With

$$\psi_i'' \simeq \Delta_i' \frac{\psi_i}{\delta} , \quad \phi_i'' \simeq \frac{\phi_i}{\delta^2}, \tag{38}$$

where δ is the layer width and Δ_i' is the change of ψ_i' across the layer,

$$\Delta_i' = [\psi_i'(\delta) - \psi_i'(-\delta)]/\psi_i , \tag{39}$$

we can evaluate eqs. (36), (37) approximately

$$\gamma \simeq \eta^{3/5} (\Delta_i')^{4/5} F^{2/5} , \tag{40}$$

$$\delta \simeq \eta^{2/5} (\Delta_i')^{1/5} F^{-2/5} . \tag{41}$$

One can show that instability arises, i.e. $\gamma > 0$, if $\Delta_i' > 0$. We can also see that the constant-ψ approximation is in fact valid, by comparing the time $\tau_\delta = \delta^2/\eta \sim \eta^{-1/5}$ for a magnetic perturbation to diffuse across the layer (and hence assume a finite value $\psi_i(0)$) with the growth time $\gamma^{-1} \sim \eta^{-3/5}$, $\tau_\delta \ll \gamma^{-1}$.

In the external region, where B_{0y} is finite, not only the resistive term in eq. (35) but also the inertia term in eq. (34) can be neglected, while leaves us with the linearized equilibrium equation,

$$\psi_e'' - \left[k^2 + \frac{j_0'(x)}{B_{0y}(x)} \right] \psi_e = 0 \qquad (42)$$

In general the external solution $\psi_e(x)$ is not smooth at $x = 0$. By asymptotically matching the layer solution to the external one, we identify Δ_i' in eq. (40) with the jump Δ_e' of the derivative of the external solution at $x = 0$

$$\Delta_i' = \Delta_e' = [\psi_e'(0_+) - \psi_e'(0_-)]/\psi_e(0) = \Delta' , \qquad (43)$$

which determines the growth rate γ. In general $\Delta' > 0$ for $ka < 1$, where a is the gradient scale of the current profile $j_0(x)$.

With $\gamma \sim \eta^{3/5}$ the tearing mode has the appearance of a relatively fast reconnection process, not much different from the Sweet-Parker scaling $\eta^{1/2}$, which is in fact reached for the maximum growth rate at $ka \sim \eta^{-1/4}$. The exponential growth is, however, nonlinearly slowed down already at microscopic island width $w \sim \delta$. The main nonlinear effect is the flattening of the current profile $j_0(x)$ over the island width, which implies that in the linear equation (37) δ is to be replaced by w

$$\partial_t \psi_i \simeq \eta \frac{\Delta'}{\delta} \psi_i \;\rightarrow\; \eta \frac{\Delta'}{w} \psi_i \;\text{ for } w > \delta .$$

Hence the nonlinear growth is slow (Rutherford, 1973)

$$w \simeq \eta \Delta' t \qquad (44)$$

corresponding to a reconnection rate $E \sim \eta$, which is consistent with the absence of a macroscopic current sheet. This result is, however, strictly valid only in a periodic, i.e. closed configuration. The tearing mode in a Sweet-Parker sheet of finite length develops on a faster time scale, see section 4.4.

4.2. RESISTIVE KINK INSTABILITY

The kink instability is the most fundamental MHD instability in a plasma column, which occurs if the twist of the magnetic field exceeds a threshold given by $B_p L/B_z a > 2\pi$, where L is the length and a the radius of the column. The kink mode corresponds to a rigid helical shift of the column pushing against the surrounding plasma and magnetic field, which generates a current sheet at the so-called resonant surface $r = r_s$, where $B_p L / B_z r_s = 2\pi$. In its simplest form the resistive kink instability can be discussed in the same framework as the tearing mode in section 4.1. In the narrow layer around r_s eqs. (36), (37) apply with $x = r - r_s$, but since the instability is much faster, the constant-ψ approximation does not hold, such that $\psi_i'' \sim \psi_i / \delta^2$. One thus obtains

$$\gamma \simeq \eta^{1/3} F^{2/3}, \; \delta \simeq \eta^{1/3} F^{-2/3}. \tag{45}$$

Hence γ is larger than for the tearing mode, eq.(40). Nonlinearly the evolution of the resistive kink instability is determined by the reconnection in a current sheet of macroscopic length (Waelbroeck, 1989; Biskamp, 1991).

Conventionally, the kink mode is studied in a periodic system, practically speaking a toroidal plasma column, since primary interest in MHD instabilities has been for application in laboratory plasmas in fusion research. In twisted coronal loops with line-tying in the photosphere the kink mode does not develop as a uniform helical displacement, but is rather localized in the central part along the column, and so is the reconnection region (Amo et al., 1995). Because of this localization the reconnection process associated with the kink mode can be more rapid than in the periodic case.

4.3. COALESCENCE OF MAGNETIC FLUX BUNDLES

The most instructive example of fast reconnection is the coalescence of two flux bundles, for instance of two magnetic islands generated by the tearing mode. Consider for instance the corrugated sheet pinch equilibrium (Fadeev et al., 1965)

$$\psi(x, y) = B_\infty \, a \ln(\cosh\frac{y}{a} + \varepsilon \cos\frac{x}{a}), \tag{46}$$

which consists of a periodic chain of magnetic islands around the midplane $y = 0$ imbedded in an antiparallel field $B_x \to \pm B_\infty$ for $y \to \pm\infty$. The island width w is given by the parameter ε, $w/a \simeq 4\sqrt{\varepsilon}$ for $w \ll a$. The configuration (46) is unstable to pairwise attraction and coalescence of islands for any island width (Pritchett and Wu, 1979). The nonlinear process, which has been studied numerically (see, e.g. Biskamp and Welter, 1980) consists of an ideal MHD phase, where the islands are freely accelerated toward each other, which leads to flux compression and current sheet formation, and a slower reconnection phase. For intermediate values of η, a self-similar behavior is observed, with the scaling laws for the upstream quantities u_0, B_0, i.e. v_x and B_y taken in front of

D. BISKAMP

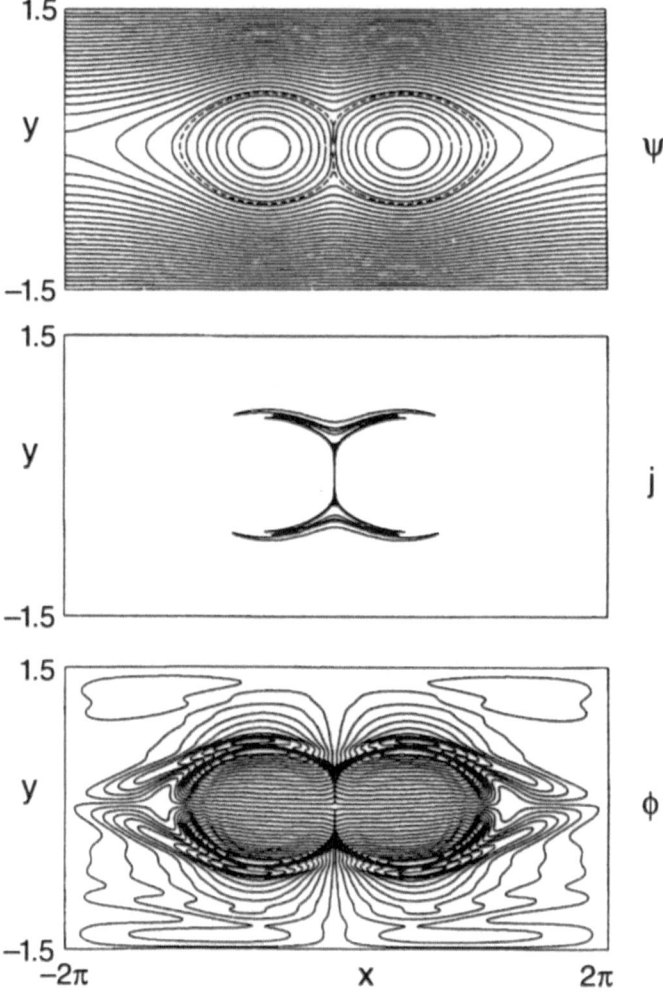

Figure 8. Coalescence of two magnetic islands starting from the equilibrium eq. (46).

the current sheet between the two islands, $u_0 \sim \eta^{1/3}$, $B_0 \sim \eta^{-1/3}$. As a consequence the reconnection rate $E = u_0 B_0$ is independent of η, which corresponds to the scaling laws (30), (31) for driven reconnection.

As mentioned in section 3.2, in a closed system such a behavior can only be valid for a certain η-range, since B_0 cannot exceed the maximum value B_m which would be obtained in the ideal case $\eta = 0$, when the attractive motion is reversed because of the repelling force due to field compression. Hence the scaling $B_0 \sim \eta^{-1/3}$ breaks down for $B_0 \simeq B_m$. For smaller values of η one finds a Sweet-Parker scaling $u_0 \sim \eta^{1/2}$, $B \simeq B_m$. The behavior during the coalescence is illustrated in Fig.

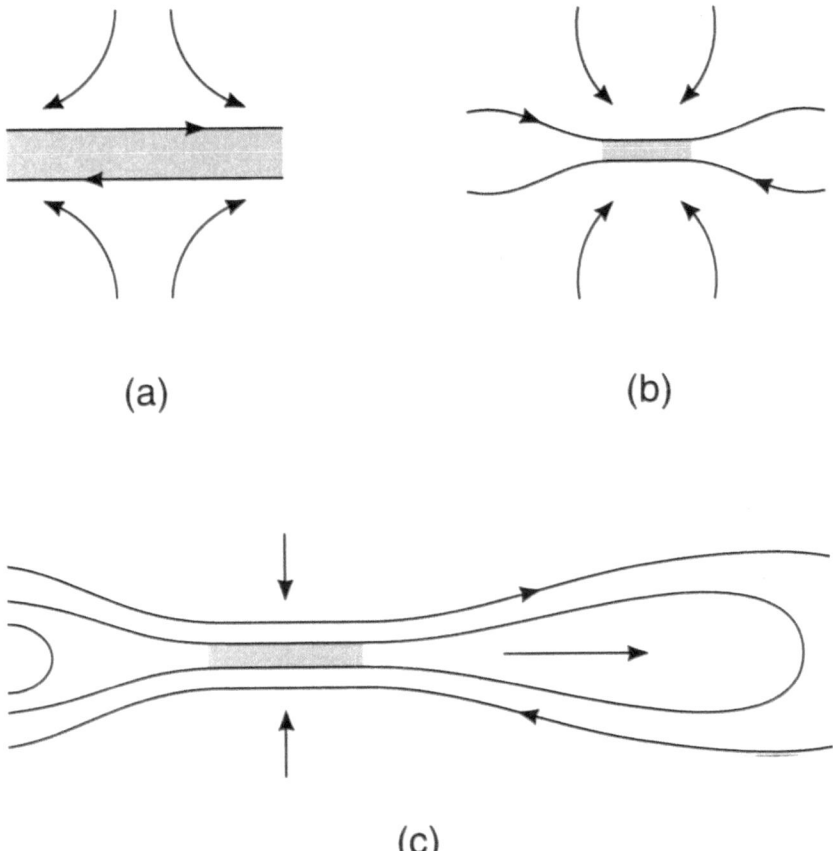

(a)   (b)

(c)

Figure 9. Local thinning of a current sheet leads to faster reconnection (b) and plasmoid generation (c).

8. The current sheet exhibits all the characteristic features discussed in section 3.1.

   These coalescence simulation as well as simulations of the nonlinear resistive kink instability prove (or at least give strong evidence) that the results of driven reconnection given in section 3.2 are representative of the behavior of resistive reconnection and that contrary to the basic assumption implied in Petschek's model, resistive reconnection takes place in macroscopic current sheets, as long as these are stable and a stationary configuration exists.

4.4. Tearing instability of a Sweet-Parker sheet and the generation of plasmoids

In a static sheet pinch, for which the tearing mode is usually considered, the instability condition $ka < 1$ (see section 4.1) implies that a sheet pinch of length Δ and width δ becomes unstable, if it accommodates more than, say, two marginally unstable wavelengths, $\Delta/\delta \simeq 2\lambda_c/\delta \sim 10$, $\lambda_c = 2\pi/k_c = 2\pi\delta$. The existence of apparently stable sheets of much larger Δ/δ as observed in numerical simulations indicates that in a Sweet-Parker sheet the strong inhomogeneous flow has a stabilizing effect (Bulanov et al., 1979; Syrovatskii, 1981). The velocity increases linearly along the sheet, $v_y(y) = \Gamma y$, where $\Gamma = v_A/\Delta = u/\delta = \eta/\delta^2$ using the properties of a Sweet-Parker sheet. The stability condition is roughly $\Gamma \simeq \gamma$, where γ is the tearing mode growth rate for a static configurationdiscussed in section 4.1. This result can easily be understood. While the tearing instability corresponds to a local current density condensation, the inhomogeneous flow tends to pull such current blobs apart. Hence the tearing mode can only be unstable, if the inhomogeneity is sufficiently weak. More quantitatively, the mode is stabilized, if the relative increase of the wavelength $\Delta\lambda/\lambda$ during one growth time exceeds about $1/4$

$$\left(v_y(y + \lambda) - v_y(y)\right)/\gamma_{max}\lambda = \Gamma/\gamma_{max} > \frac{1}{4}. \qquad (47)$$

Inserting the maximum growth rate $\gamma_{max} \propto \eta^{1/2}$ gives a condition for the maximum stable current sheet length. Tearing instability can hence be expected for

$$\Delta/\delta \gtrsim 10^2. \qquad (48)$$

The nonlinear dynamics of the tearing mode in a current sheet is in general quite different from that in a closed periodic configuration. Instead of a chain of magnetic islands typically only one island, a so-called plasmoid, is generated in the central part of the sheet, which is then convected along the sheet while still growing self-similarly in size. The basic instability mechanism is a local thinning of the current sheet, as illustrated in Fig. 9. While in the unperturbed sheet flux is piled up slowing down of the upstream flow (diverging flow pattern, Fig. 9a), a local thinning leads to field expansion and flow acceleration (converging flow, Fig. 9b), which increases the reconnection efficiency by reducing the local Lundquist number S_0, eq. (22). Note that both the field B_0 and the sheet length Δ in S_0 are reduced.

Contrary to the tearing mode in a periodic system the plasmoid growth and ejection is a nonequilibrium process. As illustrated in Fig. 9c the plasmoid is pushed away from the reconnection point by an imbalance of forces, which gives rise to a fast quasi-Alfvénic motion along the sheet. Reconnection occurs in a macroscopic current sheet and is much faster, $E \sim \eta^{1/2}$, than in the periodic nonlinear tearing mode with $E \sim \eta$, since the reconnected flux is carried away with the plasmoid pulling new flux from above and below into the reconnected region. Since plasmoid ejection takes place on the Alfvén time scale τ_A, while the supply of reconnected flux is much slower $\eta^{1/2}\tau_A$, the plasmoids tend to be rather narrow. For sufficiently small η the current sheet itself becomes unstable, giving rise to secondary plasmoids.

Plasmoids are believed to be important in various kinds of eruptive processes, notably in magnetospheric substorms and two-ribbon flares (Mikic et al., 1988; Biskamp and Welter, 1989b). Figure 10 gives a sequence of plasmoid states from a 2D compressible MHD simulation of the magnetotail (Hautz and Scholer, 1987). A large anomalous resistivity is assumed at the primary X-point at $x = 2.5$, which leads to large plasmoids. The first plasmoid is accelerated along the sheet in the direction of decreasing field intensity and pressure away from the earth (Fig. 10a). A long thin current sheet is generated between the dipole field on the left and the plasmoid (Fig. 10b), which becomes unstable generating a secondary plasmoid (Fig. 10c), which catches up and finally coalesces with the first large one (Fig. 10d). Interaction with the downtail plasma, which is still at rest, leads to shock formation and a blunt leading edge, giving the plasmoid a drop-like shape. Three-dimensional effects do not change the general picture (Birn and Hones, 1981; Otto et al., 1990). Kageyama et al. (1990) present global simulations of the interaction of the solar wind with the magnetosphere. A long tail is formed giving rise to continuous generation of plasmoids. It should, however, be mentioned that the basic problem in the theory of magnetospheric substorms is the identification of the reconnection mechanism, i.e. the relevant effect in Ohm's law, which has to be a primarily nondissipative process, since the magnetotail plasma is collisionless. I will discuss such processes in section 6. In addition, as in all eruptive processes, the problem of onset conditions arises, i.e. why is the configuration suddenly erupting after a long quiescent period.

5. MHD turbulence

A fundamental issue in magnetic reconnection theory is to explain the fast observed time scales, which for typical eruptive events, in particular in astrophysics, seem to be essentially independent of the value of the

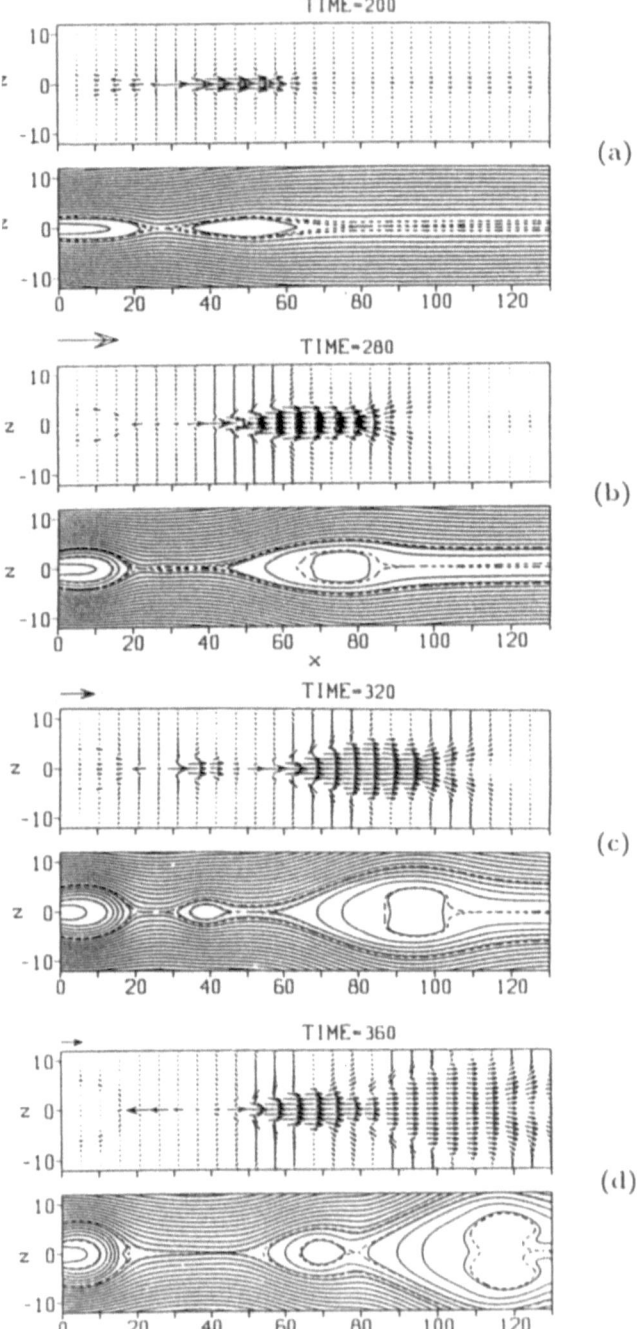

Figure 10. Compressible MHD simulation of plasmoid generation in the earth's magnetotail (from Hautz and Scholer, 1987).

collisional resistivity. As a consequence stationary current sheet reconnection, which appears to be the dominant mechanism at intermediate
R_m, are far too slow at almost collisionless plasma conditions, where R_m
is formally very large. It will be discussed in section 6 that under such
conditions nondissipative effects in Ohm's law are important. However,
also in the framework of resistive MHD fast η-independent reconnection is possible by a strongly nonstationary behavior in the reconnection
region. A first indication of such behavior has been given in the previous section, the tearing instability of the Sweet-Parker sheet leading
the plasmoid generation and ejection. It is not difficult to visualize the
transition to fully developed MHD turbulence for still larger R_m and
less symmetric systems. In this section I summarize the most important
properties of MHD turbulence (for more details see Biskamp, 1993).

5.1. GLOBAL AND SPECTRAL PROPERTIES

While nonmagnetic hydrodynamic turbulence is characterized by a single dimensionless parameter, the Reynolds number $Re = vL/\mu$, there
are two in the MHD case, Re and R_m, or R_m and the magnetic Prandtl
number $Pr_m = \mu/\eta$. Since for $Pr_m \lesssim 1$ the dependence on Pr_m is weak,
most studies of MHD turbulence assume $Pr_m = 1$, excluding the viscosity dominated regime $Pr_m \gg 1$.

The evolution of turbulence is strongly influenced by the ideal invariants of the system. For incompressible MHD, eqs. (13), (14), there are
three invariants, the total energy W, the cross-helicity K, and the magnetic helicity H^A, defined by

$$W = W^V \mid W^M = \tfrac{1}{2} \int (v^2 + B^2) d^3x , \tag{49}$$

$$K = \tfrac{1}{2} \int \mathbf{v} \cdot \mathbf{B} \, d^3x , \tag{50}$$

$$H^A = \tfrac{1}{2} \int \mathbf{A} \cdot \mathbf{B} \, d^3x , \tag{51}$$

where $\mathbf{A}$ is the vector potential, $\nabla \times \mathbf{A} = \mathbf{B}$. In 2D there are also three
(quadratic) invariants, the first two, W and K, are formally identical
with the 3D expressions, while H^A is replaced in 2D by the mean square
magnetic potential

$$H^\psi = \tfrac{1}{2} \int \psi^2 d^2x . \tag{52}$$

Including finite dissipation coefficients η, μ these invariants are in general found to decay in magnitude, but the decay rates $\Gamma^W = (dW/dt)/W$

etc., with

$$dW/dt = -\eta \int j^2 d^3x - \mu \int \omega^2 d^3x \,, \qquad (53)$$

$$dK/dt = -(\eta + \mu) \int \boldsymbol{\omega} \cdot \mathbf{j}\, d^3x \,, \qquad (54)$$

$$dH^A/dt = -\eta \int \mathbf{j} \cdot \mathbf{B}\, d^3x, \ \text{ or } dH^\psi/dt = -\eta \int B^2 d^2x \,, \qquad (55)$$

may be general largely different, in particular

$$\Gamma^W \gg \Gamma^A, \ \text{ or } \Gamma^W \gg \Gamma^\psi \text{ in 2D} \,, \qquad (56)$$

and

$$\Gamma^W \gg \Gamma^K. \qquad (57)$$

The different magnitudes of in the Γ's, called selective decay (Matthaeus and Montgomery, 1980) arise, because the strongest dissipation occurs at small scales and dW/dt contains higher derivatives than $dH^{A,\psi}/dt$, and because the integrand in dW/dt is positiv, while its sign varies in dK/dt leading to cancellations. The existence of several ideal invariants and their selective decays result in self-organization processes in decaying turbulence. On intermediate time scales $(\Gamma^W)^{-1} \ll t \ll (\Gamma^A)^{-1}$, or $(\Gamma^W)^{-1} \ll t \ll (\Gamma^\psi)^{-1}$ in 2D, the system relaxes to a state of minimum energy for a given value of H^A or H^ψ, which can be formalized by the variational principle

$$\delta \left[\int (v^2 + B^2) d^3x - \lambda \int \mathbf{A} \cdot \mathbf{B} d^3x \right] = 0 \,, \qquad (58)$$

the result of which is a static linear force-free field (Woltjer, 1958; Taylor, 1986)

$$\nabla \times \mathbf{B} - \lambda \mathbf{B} = 0 \,, \ \mathbf{v} = 0 \,, \qquad (59)$$

and in 2D

$$\delta \left[\int (v^2 + B^2) d^2x - \lambda^2 \int \psi^2 d^2x \right] = 0 \qquad (60)$$

with a similar linear field solution

$$\nabla^2 \psi + \lambda^2 \psi = 0 \,, \ \nabla^2 \phi = 0. \qquad (61)$$

Hence both in 3D and 2D the system relaxes to a static state $W^V/W^M \to 0$. The origin of this asymmetry between $\mathbf{B}$ and $\mathbf{v}$ is the lack of a conserved kinetic quantity equivalent to the magnetic ones H^A or H^ψ. The selective decay (57) leads to a different self-organization process, the alignment of $\mathbf{v}$ and $\mathbf{B}$ called dynamic alignment (Matthaeus and Montgomery, 1984). The appropriate variational principle

$$\delta \left[\tfrac{1}{2} \int (v^2 + B^2) d^3 x - \lambda \int \mathbf{v} \cdot \mathbf{B} d^3 x \right] = 0. \tag{62}$$

has the solutions (both in 3D and 2D)

$$\mathbf{v} = \pm \mathbf{B} , \tag{63}$$

which are called pure Alfvénic states. A dynamic mechanism of this relaxation has been developed by Dobrowolny et al. (1980). Strongly aligned states are often observed in the solar wind (e.g. Burlage and Turner, 1976). Which of the two relaxation processes dominates, the formation of force-free states (which in 3D is in alignment of $\mathbf{j}$ and $\mathbf{B}$) or the dynamic alignment of $\mathbf{v}$ and $\mathbf{B}$, depends on the initial values of H^A (or H^ψ) and K.

Quadratic ideal invariants are connected with detailed balance relations in wavenumber space meaning that the corresponding spectral quantities, e.g. $W_{\mathbf{k}}$, are conserved in elementary interactions between any triad of modes $\mathbf{k}, \mathbf{p}, \mathbf{q}$ with $\mathbf{k} + \mathbf{p} + \mathbf{q} = 0$, $W_{\mathbf{k}} + W_{\mathbf{p}} + W_{\mathbf{q}} = \text{const.}$ If such a quantity is injected into a certain spectral range $k \sim k_{in}$, it will diffuse in k-space. If this flow is primarily to large k, the quantity is said to exhibit a normal or direct cascade, and an inverse cascade, if the flow is to small k. Turbulence can be characterized by the cascade directions of its ideal invariants. Table 1 gives the cascade directions in MHD turbulence in 3D and 2D compared with the corresponding ones in Navier-Stokes turbulence, where in the latter W_k refers to the kinetic energy, H_k to the kinetic helicity, and Ω_k to the mean square vorticity. While the cascade properties in Navier-Stokes turbulence differ fundamentally in 3D and 2D, they are essentially identical for MHD turbulence indicating a stronger similarity between 3D and 2D than in the Navier-Stokes case. The presence of several ideal invariants with different cascade directions is generally believed to give rise to large-scale self-organization, i.e. the formation of large-scale coherent structures. In 2D Navier-Stokes turbulence these are large-scale velocity eddies (generated by isolated vortices), while in MHD turbulence these are large-scale magnetic structures.

In addition to the local mode interactions giving rise to the cascade flows MHD turbulence also exhibits nonlocal interactions in k-space corresponding to large scales directly affecting small scales, which is absent in Navier-Stokes turbulence. The magnetic field in the large-scale eddies containing most of the magnetic energy acts as a guide field $\mathbf{B}_0$ to the small-scale fluctuations $\widetilde{\mathbf{B}}, \widetilde{\mathbf{v}}$, which thus become strongly

Table I. Cascade directions in 3D and 2D for MHD and Navier-Stokes
turbulence.

	3D		2D	
MHD	W_k	direct	W_k	direct
	K_k	direct	K_k	direct
	H_k^A	inverse	H_k^ψ	inverse
Navier-Stokes	W_k	direct	W_k	inverse
	H_k	direct	Ω_k	direct

correlated,

$$\tilde{\mathbf{v}} \simeq \pm\tilde{\mathbf{B}} \,, \tag{64}$$

propagating as Alfvén waves along $\mathbf{B}_0$. (Note the difference between a global alignment eq. (63), which occurs only in a relaxed state, and the small-scale property (64), which is satisfied under normal turbulence conditions). The basic effect of (64) on the turbulent dynamics becomes evident, if we rewrite the MHD equations in terms of the Elsässer fields $\mathbf{z}^\pm = \mathbf{v} \pm \mathbf{B}$ (Elsässer, 1950)

$$\partial_t \mathbf{z}^\pm \mp \mathbf{v}_A \cdot \nabla \mathbf{z}^\pm + \mathbf{z}^\mp \cdot \nabla \mathbf{z}^\pm = \text{pressure and dissipation terms}\,, \tag{65}$$

where we have split off the large-scale field $\mathbf{B}_0 = \mathbf{v}_A$ to emphasize the Alfvén wave propagation properties. The crucial effect is that only waves propagating in opposite directions couple. Hence the interaction time between two eddies is $\tau \sim \tau_A = (kv_A)^{-1}$, which is short compared with the nonlinear scrambling time of hydrodynamic eddies $\tau \sim (k\tilde{v})^{-1}$, since $\tilde{v} \sim \tilde{B} \ll B_0$. This is called the Alfvén effect in MHD turbulence (Iroshnikov, 1964; Kraichnan, 1965), which strongly affects the energy transfer in the turbulent cascade process.

A characteristic property of high Reynolds-number turbulence is the existence of an inertial range in the energy spectrum defined by $k_{in} \ll k \ll k_d$, where k_{in} is the injection range, corresponding to the large energy-containing scales, and k_d is the dissipation range, the smallest scales excited. The angle-integrated energy spectrum W_k, $W = \int W_k dk$, can be "derived" by dimensional arguments. Assuming that the energy transfer rate ε, which is constant in the inertial range and equal to the energy dissipation rate, is only a function of k, W_k, v_A, and that due to the Alfvén effect $\varepsilon \propto \tau_A$, one obtains $\varepsilon \sim v_A^{-1} W_k^2 k^3$, or

$$W_k \sim (\varepsilon v_A)^{1/2} k^{-3/2} \,, \tag{66}$$

which is less steep than the well-known Kolmogorov law for hydrodynamic turbulence $W_k \sim \varepsilon^{2/3} k^{-5/3}$. The MHD spectrum (66) should apply to both 3D and 2D (in contrast to the Kolmogorov law, which is only valid in the 3D hydrodynamic case). In fact the spectrum (66) has been verified in 2D simulations (Biskamp and Welter, 1989a). The only high-R_m system of MHD turbulence, where the energy spectrum has been measured, is the solar wind. Here the spectral index appears to be close to 5/3 (e.g. Matthaeus et al., 1982), from which it has been concluded that in the solar wind the energy spectrum follows the Kolmogorov law instead of the flatter spectrum (66). However, only the magnetic part of the energy spectrum is measured, which in a refined theory is predicted to be somewhat steeper than $k^{-3/2}$, since $W_k^M - W_k^V = ak^{-2} > 0$ (Grappin et al., 1983).

Spectral laws can also be obtained for inversely cascading quantities in the range $k < k_{in}$. Assuming a constant magnetic helicity flux η_H one obtains the helicity spectrum $H_k^A \sim \eta_H^{2/3} k^{-2}$, and correspondingly in 2D $H_k^\psi \sim \eta_\psi^{2/3} k^{-7/3}$ for a constant mean square potential flux η_ψ. The latter law has been verified in numerical simulations (Biskamp and Bremer, 1994).

5.2. TURBULENT ENERGY DISSIPATION

In a magnetic configuration with a macro-current sheet the energy dissipation rate (53) is small, $\varepsilon = O(\eta^{1/2})$, as follows from the properties of a Sweet-Parker sheet. Since for sufficiently high R_m the system becomes turbulent, the question arises, whether the turbulent energy dissipation rate remains finite in the limit $R_m \to \infty$. This behavior is generally believed to be valid for Navier-Stokes turbulence, where a decrease of the viscosity simply leads to the excitation of smaller scales such that ε is essentially independent of μ. (This is, however, not true in the 2D Navier-Stokes case, where the energy dissipation is weak, $\varepsilon = O(\mu)$). In MHD turbulence the cascade and spectral properties are very similar in 2D and 3D, as discussed in the previous section. Numerical simulations of decaying 2D MHD turbulence have been performed for different Reynolds numbers (e.g. Biskamp and Welter, 1989a). Dissipation is found to be concentrated in micro-current sheets, distributed intermittently as seen in Fig. 11. Increasing R_m these micro-sheets become both smaller and more numerous , while the macro-state remains essentially unchanged. Quantitatively Fig. 12 gives the evolution of the energy dissipation rate $\varepsilon(t)$ for three cases starting from the same smooth initial conditions but with different values of ($\eta = \mu$). While in the initial nonturbulent phase $\varepsilon \propto \eta$, in the subsequent phase of fully developed turbulence $(t \gtrsim 1)\,\varepsilon$ is statistically identical in all three cases, i.e. ε is essentially independent of R_m. Such behavior is a

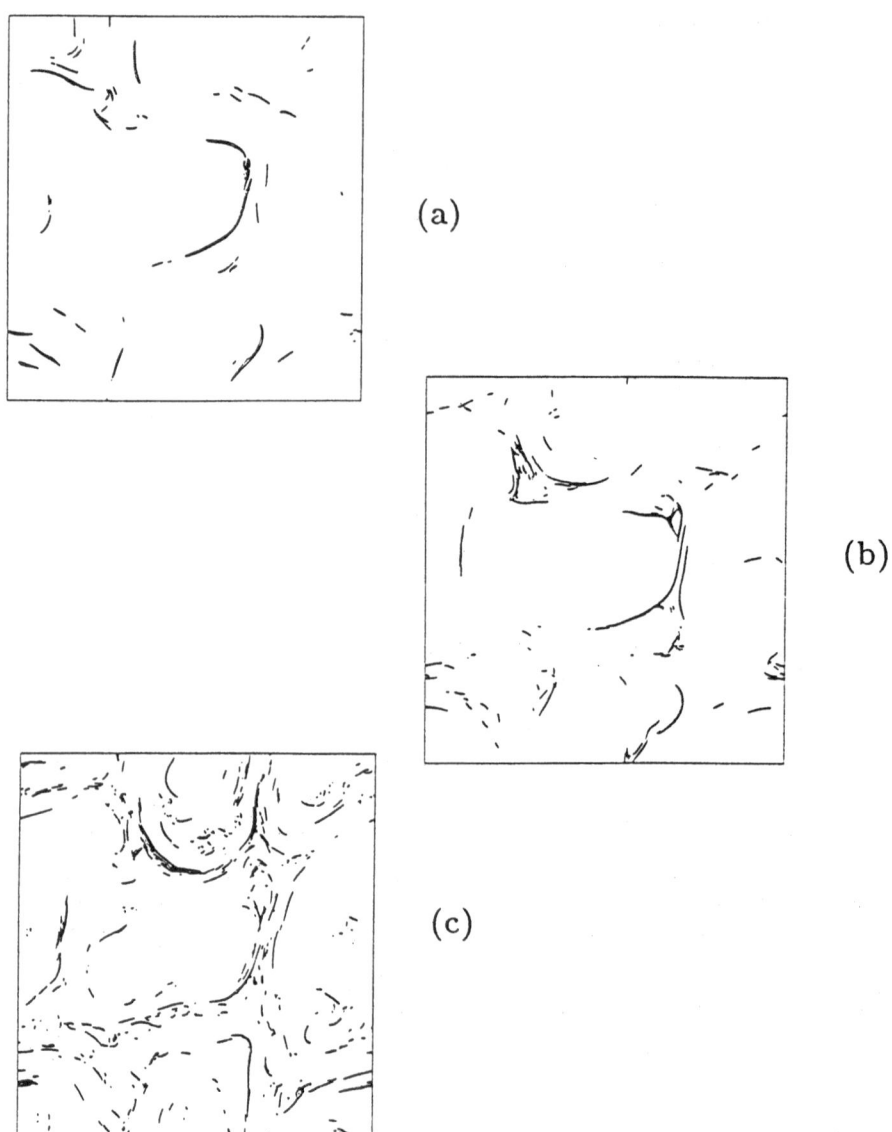

Figure 11. Distribution of micro-current sheets in 2D MHD turbulence for three different Reynolds numbers: (a) $R_m \simeq 12000$, (b) $R_m \simeq 25000$; (c) $R_m \simeq 50000$, for identical macro-states.

forteriori also expected to hold for 3D MHD turbulence. In fact recent numerical simulations of 3D MHD turbulence indicate, that dissipation is concentrated in sheet-like current and vorticity structures (Politano et al., 1995). Hence also with respect to the dissipation properties 2D and 3D systems seem to be rather similar.

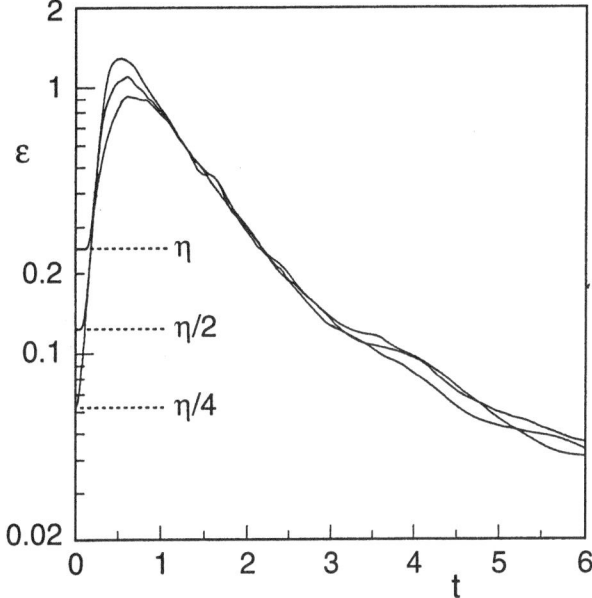

Figure 12. Energy dissipation rates $\varepsilon(t)$ in 2D decaying MHD turbulence of three simulation runs with different values of η.

The following picture of fast turbulent reconnection appears, illustrated in Fig. 13. MHD turbulence is preferentially excited in regions of strong current density (or current density gradients), i.e. around X-points, the natural loci of reconnection. Since in fully developed turbulence the energy dissipation rate is finite, independent of η, the magnetic field energy carried into the turbulent region is annihilated at a fast rate. This implies that the outside field is reconnected at the same rate.

5.3. THE TURBULENT DYNAMO EFFECT

While in the case of magnetized plasmas a 2D turbulence model may describe many features of the real 3D system, in primarily unmagnetized systems such as in stellar convection zones or accretion discs a fully 3D turbulence theory is required. In the case $W^M \ll W^V$ the analogy between the equation of motion (13) for $\boldsymbol{\omega}$, neglecting the Lorentz force, and the induction equation (14) for $\mathbf{B}$ indicates that $\mathbf{B}$ structures should be similar to $\boldsymbol{\omega}$ structures in Navier-Stokes turbulence, i.e. $\mathbf{B}$ should be concentrated in small-scale rope-like eddies. Concerning spectra one can argue that the magnetic energy spectrum

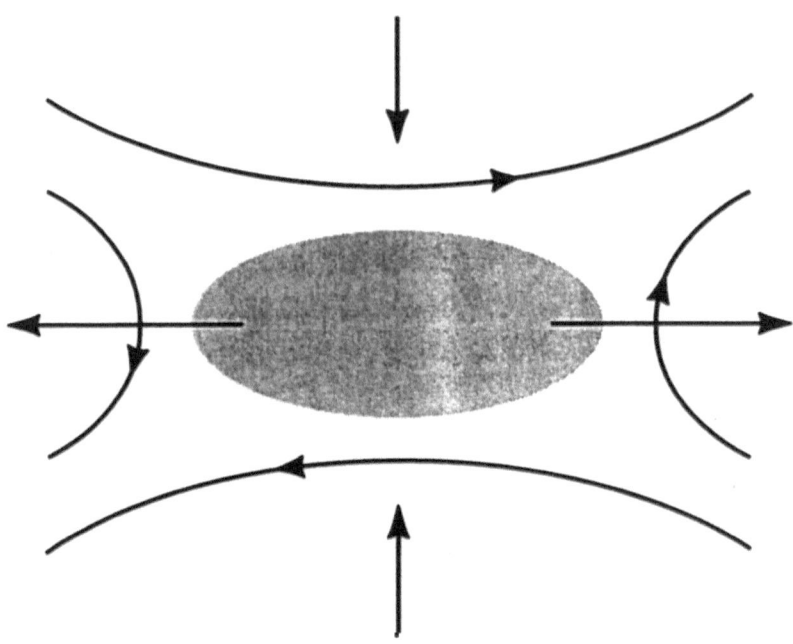

Figure 13. Schematic view of fast turbulent reconnection.

W_k^M should be similar to the vorticity spectrum $\Omega_k = k^2 W_k^V$

$$W_k^M \propto k^2 W_k^V \simeq \varepsilon^{2/3} k^{1/3}, \tag{67}$$

assuming a Kolmogorov law for the velocity spectrum W_k^V. Hence the magnetic spectrum for $W_k^M \ll W_k^V$ is fundamentally different from the $k^{-3/2}$ spectrum for MHD turbulence with $W^M \sim W^V$

The analogy between ω and $\mathbf{B}$ can, however, only give the shape of the magnetic spectrum not its magnitude, in particular, cannot determine, whether a weak seed field will grow or decay. Starting from a smooth seed field distribution, W^M will always grow initially owing to field line stretching and twisting much as vortex line stretching in Navier-Stokes turbulence. Because of mass and magnetic flux conservation a flux tube stretched by a factor A shrinks in diameter as $d \propto A^{-1/2}$, while the field intensity increases as $B \propto A$, such that the magnetic energy enclosed in the tube increases as $W^M \propto A^2$. If the tube diameter becomes small enough, resistive dissipation may dominate over the convective stretching effect leading to subsequent decay of W^M. This always happens in 2D, where only a temporary dynamo effect is possible (Biskamp and Welter, 1990). In 3D such behavior is possible for $Pr_m \ll 1$. In the opposite case $Pr_m > 1$, where the viscous dissipation scale is larger than the resistive one, magnetic energy increase may continue, until the Lorentz force becomes important leading to dynamic saturation. This saturation occurs for $W_k^M \sim W_k^V$ at the smallest scales. In this state the total magnetic energy is still small compared with the fluid energy because of the spectrum (67). The final saturation level depends on Pr_m, see 3D simulations by Yanase et al. (1991).

The picture of magnetoconvection given so far describes the process of small-scale field excitation. Flux ropes are swirled around in a spaghetti-like manner, until the magnetic tension is large enough to slow down the swirling motion. Magnetic tension may also be reduced by flux rope reconnection. Reconnection plays an important role in the build-up of *large-scale* fields - the proper topic of dynamo theory - which can already be seen from the kinematic theory in mean-field electrodynamics (Steenbeck et al., 1966). Here $\mathbf{v}$ and $\mathbf{B}$ are split into large-scale mean parts and small-scale fluctuating parts, and the equation for $\widetilde{\mathbf{B}}$ is linearized

$$\partial_t \widetilde{\mathbf{B}} - \nabla \times (\mathbf{v}_0 \times \widetilde{\mathbf{B}}) = \nabla \times (\mathbf{v} \times \mathbf{B}_0) \, , \qquad (68)$$

which obviously violates exact flux conservation. Averaging over small scales yields the well-known equation of turbulent dynamo theory

$$\partial \mathbf{B}_0 - \nabla \times (\mathbf{v}_0 \times \mathbf{B}_0) = -\nabla \times \mathbf{E} \, , \qquad (69)$$

where

$$\mathbf{E} = -\langle \widetilde{\mathbf{v}} \times \widetilde{\mathbf{B}} \rangle \simeq \alpha \mathbf{B}_0 + \beta \nabla \times \mathbf{B}_0 , \qquad (70)$$

$$\alpha = \tfrac{1}{3}\tau \langle \widetilde{\mathbf{v}} \cdot \nabla \times \widetilde{\mathbf{v}} \rangle, \ \beta = \tfrac{1}{3}\tau' \langle \widetilde{\mathbf{v}}^2 \rangle \, , \qquad (71)$$

and $\tau \simeq \tau'$ are velocity correlation times. While β is positive representing a turbulent resistivity, which leads to a field energy decrease, α, which can have both signs depending on the kinetic helicity, may lead to field amplification. We thus find that dynamo action requires sufficiently complicated flows, which in particular should not be reflectionally symmetric, excluding dynamo action in 2D. Since the dynamics and statistics of the velocity field entering the coefficients α, β are assumed to be independent of $\mathbf{B}_0$, eq. (70) is only valid for sufficiently weak fields. For finite $\mathbf{B}_0$ small-scale fluctuations become Alfvén waves with $\widetilde{\mathbf{v}} \times \widetilde{\mathbf{B}} \simeq 0$. As a consequence the α-term is reduced. There is, however, a different process, leading to a nonlinear turbulent dynamo effect (Pouquet et al., 1976). Since small-scale fields $\widetilde{\mathbf{B}}$ are in general connected with small-scale magnetic helicity H_k^A, the inverse cascade of H_k^A leads to the build-up of large-scale magnetic fields. 3D simulations showing this effect have been performed by Meneguzzi et al. (1981).

In inhomogeneous systems strong large-scale fields may be generated by spatially separating the region of primarily kinetic turbulence from

the adjacent region, into which large-scale magnetic flux is transport-
ed and thus accumulated. Such a process seems to occur in the solar
convection zone, where strong fields, which could be responsible for the
solar dynamo effect, are concentrated in the stable overshoot region at
the bottom of the convection zone (Nordlund et al., 1992).

6. Quasi-collisionless reconnection

In many hot, dilute plasmas, both in the laboratory and in space,
Coulomb collisions are very weak, such that the nondissipative terms in
R, eq. (9), are larger than the dissipative ones. This happens if typical
resistive (or viscous) scales $\delta \sim \eta^{1/2}$ become smaller than the length
scales characterizing the collisionless terms d_i, the ion inertia scale, or
$d_i\beta = \rho_s$, the ion Larmor radius. From eq. (9) we see that for magnetized plasmas ($\beta < 1$) the $\mathbf{j} \times \mathbf{B}$ term, called the Hall term, is formally
the most important collisionless effect. We should, however, note that
the Hall term by itself cannot give rise to reconnection, since it simply restates the property of flux conservation in more precise form,
the field being frozen to the electron flow instead of the mass flow.
In fact it appears that also the remaining two nondissipative terms in
R associated with electron pressure and electron inertia do not per-
mit quasi-stationary reconnection. Hence some amount of dissipation
is required, which, however, need not be due to Coulomb collisions but
can also be caused by collisionless dissipation such as Landau damping.
The important point is, that the reconnection dynamics, in particular
the time scales, no longer depend on these weak dissipation processes.

   In the low-β regime, to which I restrict myself, the pressure term
in **R** is negligible. The inertia term, though formally very small, has a
somewhat different quality and will hence be kept. The nondissipative
terms change the dispersion properties of the plasma at small scales.
Inserting **E** into Faraday's law and linearizing the latter together with
the equation of motion (6) gives the dispersion relation

$$\omega^2 = k_{\parallel}^2 v_A^2 \frac{1 + k^2 d_i^2}{(1 + d_e^2 k^2)^2} \,. \tag{72}$$

While at large scales $kd_i < 1$ we recover the dispersion-free Alfvén
wave $\omega^2 = k_{\parallel}^2 v_A^2$, waves become strongly dispersive at smaller scales,
for $d_i^{-1} < k < d_e^{-1}$, called the whistler mode,

$$\omega^2 = (\Omega_e d_e^2)^2 k_{\parallel}^2 k^2, \tag{73}$$

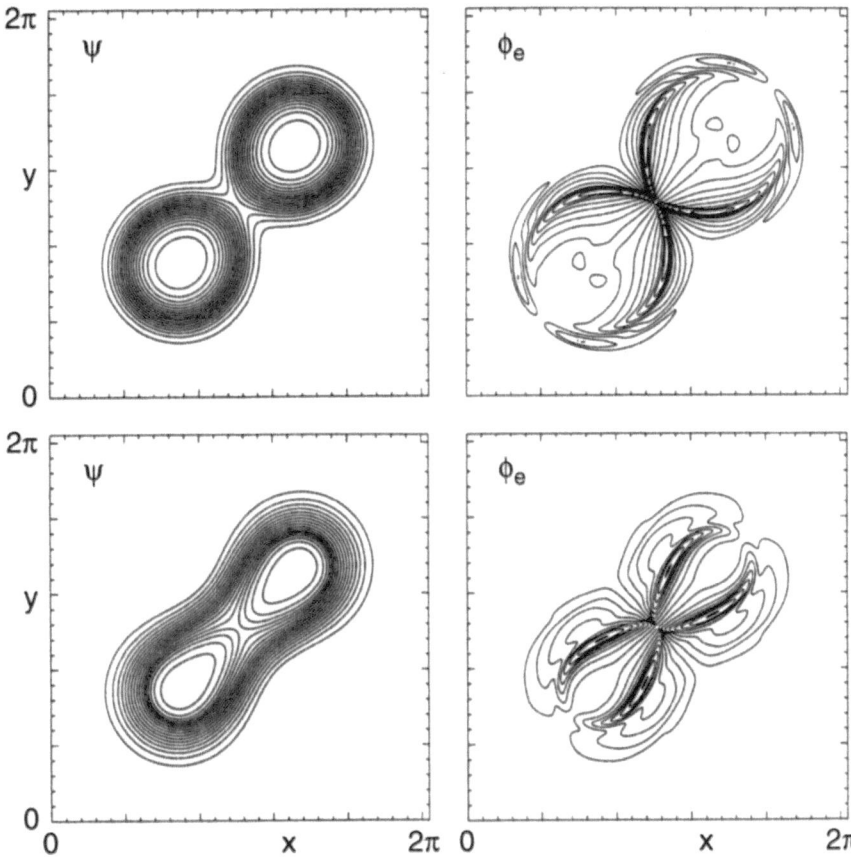

Figure 14. Coalescence of flux bundle in EMHD. Contours of ψ and ϕ_e, at two different times.

and for $kd_e > 1$, called the electron cyclotron mode,

$$\omega^2 = \Omega_e^2 k_{||}^2 / k^2. \tag{74}$$

While for the whistler mode the group velocity increases with k, $\partial\omega/\partial k \propto k$, reaching a maximum value $\sim \Omega_e d_e$, it drops to zero for the electron cyclotron mode, due to electron inertia. Note that the whistler is independent of both ion and electron inertia.

6.1. ELECTRON MAGNETOHYDRODYNAMICS

Let us first restrict consideration to scales smaller than d_i, where the ion dynamics can be neglected, the ions forming an immobile charge-neutralizing background. The dynamics is determined by the electron velocity $\mathbf{v}_e = -\mathbf{j}/ne$ and the selfconsistent magnetic field, following the equations

$$\partial_t(\mathbf{B} - d_e^2\nabla^2\mathbf{B}) - \nabla \times [\mathbf{v}_e \times (\mathbf{B} - d_e^2\nabla^2\mathbf{B})] = -\mu_e\nabla^{(4)}\mathbf{B}, \qquad (75)$$

$$\mathbf{v}_e = -\alpha\nabla \times \mathbf{B}, \qquad (76)$$

where $\alpha = c/4\pi ne$ is the Hall parameter. Equations (75), (76) are called electron magnetohydrodynamics (EMHD) (e.g. Kinsep et al., 1990). Linearization gives the whistler dispersion relation (72) (in the limit $kd_i > 1$). Since neglecting the ion dynamics corresponds to $m_i \to \infty$, i.e. $t_A \to \infty$, we have to use a different time unit, a convenient one being the whistler time $t_w = L^2/\Omega_e d_e^2(= t_A/d_i)$, which formally corresponds to setting $\alpha = 1$ in eq. (76). On the right side of eq. (75) electron viscosity is chosen as the relevant dissipation process, since resistivity (= electron friction) can in general not prevent the formation of singular gradients.

Whistler modes are destabilized by a strong current density gradient (Mikhai-lovskii, 1974), where the most unstable modes are perpendicular to $\mathbf{j}$, i.e. to $\mathbf{B}$ in a low-β plasma. This suggests to study whistler-related processes in the poloidal plane perpendicular to $\mathbf{B}_0 = B_0\hat{\mathbf{z}}$. The 2D EMHD equations can be written (as in 2D incompressible MHD, note that $\nabla \cdot \mathbf{v}_e \propto \nabla \cdot \mathbf{j} = 0$) in terms of a flux function ψ, $\mathbf{B}_p = \hat{\mathbf{z}} \times \nabla\psi$ and a stream-function ϕ_e, which is the axial field fluctuation, $\mathbf{v}_e = \hat{\mathbf{z}} \times \nabla\phi_e = \hat{\mathbf{z}} \times \nabla B_z$:

$$\partial_t(\psi - d_e^2 j) + \mathbf{v}_e \cdot \nabla(\psi - d_e^2 j) = -\mu_e\nabla^2 j, \qquad (77)$$

$$\partial_t(\phi_e - d_e^2\omega_e) + \mathbf{v}_e \cdot \nabla(\phi_e - d_e^2\omega_e) + \mathbf{B}_p \cdot \nabla j = -\mu_e\nabla^2\omega_e, \qquad (78)$$

$$j = \nabla^2\psi, \quad \omega_e = \nabla^2\phi_e. \qquad (79)$$

Let us consider stationary reconnection in the framework of eqs. (77), (78). For $d_e = 0$ and $\mu_e = 0$ and stationary conditions the equations reduce to

$$\mathbf{v}_e \cdot \nabla\psi = -E, \qquad (80)$$

$$\mathbf{B}_p \cdot \nabla j = 0 \qquad (81)$$

with $\partial_t\psi = E = $ const., which have the similarity solution

$$\psi = \tfrac{1}{2}(x^2 - a^2 y^2), \qquad (82)$$

$$\phi_e = \tfrac{1}{2}\frac{E}{a}\ln\left|\frac{x+ay}{x-ay}\right|. \tag{83}$$

The solution corresponds to an upstream flow converging toward the X-point and downstream flow diverging away from it. Finite dissipation is needed to smooth the flow singularity along the separatrix $x = \pm ay$. In the vicinity of the neutral point electron inertia must be retained and ψ and ϕ_e deviate from the solution (82), (83). A micro-current layer of length $\Delta \sim (Ed_e^2)^{1/3}$ and width $\delta \sim d_e$ is driven by the applied field E. One can show, that the integrated current in this layer goes to zero for $d_e \to 0$. Thus the magnetic field at the X-point retains its structure (82) for any E, the latter depending only on the external configuration, i.e. on the free magnetic energy of the system, in contrast to the resistive MHD case, where reduction of the smallness parameter η leads to a macroscopic current sheet and small $E = O(\eta^{1/2})$. To illustrate these results let me present some numerical simulations of the 2D EMHD equations. I consider the coalescence of two magnetic flux bundles as shown in Fig. 14. The conspicuous feature is, that flux surfaces appear to be pulled into the X-point instead of pushed against it as in the MHD case (see the MHD coalescence system, Fig. 8). This behavior is caused by the structure of the flow with streamlines converging (i.e. velocity increasing) toward the X-point, which is consistent with the solution (83). The reconnection rate is found to be independent of d_e and also of the dissipation coefficient μ_e, as long as the latter is not too large. For very small μ_e quasi-singular substructures inside the current layer are generated corresponding to cusp-like electron flow (or current) profiles. The collimated outflow finally becomes Kelvin-Helmholtz unstable generating turbulence, which is convected out from the reconnection region into the downstream cone. The turbulence gives a natural mechanismfor a finite energy dissipation rate in the limit $\mu_e \to 0$. Turbulence generation is found to be much more efficient than the generation of Alfvén turbulence in the case of resistive current sheets.

While the 2D EMHD model (77), (78) gives a good approximation for laminar quasi-stationary processes, they cannot adequately describe the dynamics, when turbulence gets in. In 3D turbulence is more violent because of modes with finite k_z excited by the z-component of the current density at the X-point. 3D simulations of unstable current layers in the framework of EMHD have recently been performed (Drake et al., 1994) indicating excitation of rather isotropic turbulence, which broadens the current layer to a width exceeding d_e. The effect of the turbulence can be described by a turbulent electron viscosity $\mu_{eff} = a\Omega_e d_e^2$, where the numerical coefficient is $a \sim 0.1$.

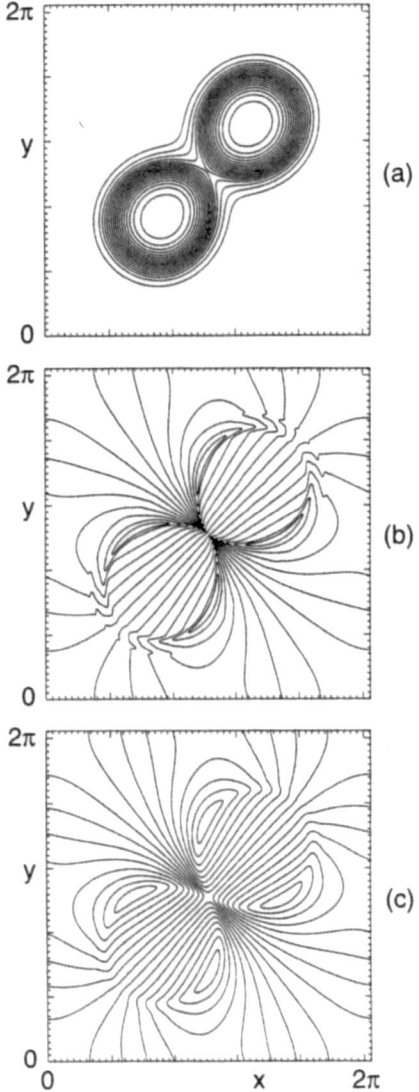

Figure 15. Coalescence of flux bundles as in Fig. 14, but including the ion motion.

6.2. ION CONTROLLED QUASI-COLLISIONLESS RECONNECTION

At spatial scales exceeding d_i the ion motion can no longer be neglected. Here ions and electrons move essentially together, which justifies
neglecting the Hall term in the MHD approximation. I include the ions
in a simple model assuming their motion to be incompressible. In the
2D case the ion velocity is hence given in terms of a stream-function

ϕ_i, $\mathbf{v}_i = \hat{\mathbf{z}} \times \nabla \phi_i$, $\omega_i = \nabla^2 \phi_i$. Adding the equations of motion for ions and electrons we obtain

$$d_i^2(\partial_t \omega_i + \mathbf{v}_i \cdot \nabla \omega_i) + d_e^2(\partial_t \omega + \mathbf{v}_e \cdot \nabla \omega_e) - \mathbf{B} \cdot \nabla j$$

$$= \mu_i \nabla^2 \omega_i + \mu_e \nabla^2 \omega_e. \qquad (84)$$

Neglecting the axial velocity of the ions compared with that of the electrons, the axial current density $j_z = j$ remains unchanged and hence eq. (77) for ψ. The poloidal current density is, however, modified by the ion motion, $\mathbf{j}_p = -\hat{\mathbf{z}} \times \nabla B_z = \hat{\mathbf{z}} \times \nabla(\phi_i - \phi_e)$. Hence eq. (78) becomes

$$\partial_t(\phi_e - \phi_i - d_e^2 \omega_e) + \mathbf{v}_e \cdot \nabla(\phi_e - \phi_i - d_e^2 \omega_e) + \mathbf{B} \cdot \nabla j = -\mu_e \nabla^2 \omega_e. \qquad (85)$$

In this equation the term $B_z \nabla \cdot \mathbf{v}_e$ has been neglected by assuming exact incompressibility $\nabla \cdot \mathbf{v}_i = 0$. While this is rigorous in EMHD since $\mathbf{v}_e = -\mathbf{j}$ and $\nabla \cdot \mathbf{j} = 0$, including the ion motion this approximation requires that $B_z \nabla \cdot \mathbf{v}_i$ is small, i.e. that $\nabla \cdot \mathbf{v}_i$ is small independent of the presence of a strong axial field, which can be proved only for a large background pressure, i.e. for high-β, since finite $dp_i/dt = -\gamma p_i \nabla \cdot \mathbf{v}_i$ enforces $\nabla \cdot \mathbf{v}_i \simeq 0$ if p_i is large. The low-β behavior is presently under investigation.

Numerical simulations of flux tube coalescence for $d_i < 1$ using eqs. (77), (84), (85) (Biskamp et al., 1995) show that the reconnection rate E is independent of the electron parameters d_e, μ_e, as could be expected from the EMHD results presented in section 6.1. E depends only on d_i, i.e. on the ion mass. Figure 15 gives a typical state during the coalescence process with $d_i = 0.1, d_e = 0.015$. It can be seen by comparing ϕ_e and ϕ_i contours, that at finite distances from the X-point (and the separatrix) ion and electron flow patterns are very similar. They differ, however, in the vicinity of the X-point, where ion spatial scales ($\sim d_i$) are much larger than those of the electrons ($\sim d_e$). For $|\mathbf{x}| < d_i$ the electron flow again shows the converging pattern as in the EMHD case.

For not too small values of d_i, $d_i \gtrsim 0.1$, one finds that E has an Alfvénic scaling. For smaller values of $d_i, d_i < 0.1$ the reconnection configuration assumes the shape of an elongated sheet of width d_i corresponding to $E \sim d_i v_A$, which for $d_i \ll 1$ is slower than Alfvénic, but still much larger than $E \sim d_e v_A$, as would be obtained neglecting the Hall term, which would imply forcing also the ions into a narrow channel of width d_e. In addition the elongated current layer is prone to plasmoid generation as in the resistive case, which further increases the reconnection rate.

7. Conclusions

I have presented an overview of the present status of the theory of magnetic reconnection in plasmas. In the conventional framework of resistive MHD reconnection at intermediate magnetic Reynolds numbers R_m takes place in quasi-stationary current sheets, so-called Sweet-Parker sheets, of macroscopic length giving rise to reconnection rates $E = O(\eta^{1/2})$, which are far too slow if extrapolated to the large R_m values in many astrophysical systems. This extrapolation is, however, not legitimate, since at high R_m reconnection occurs in a turbulent way. Fully developed MHD turbulence seems to allow finite reconnection rates in the limit $\eta \to 0$. The effect of the turbulence can be modelled by assuming a localized anomalous resistivity, as done in many simulations of substorms or flares.

Primarily nondissipative reconnection processes dominate in hot, dilute plasmas, where collisional effects are negligibly small. In fact parameters such as the Reynolds number lose their relevance to characterize the dynamics in quasi-collisionless systems because of the presence of much more effective collisionless processes. Hence one has to broaden the theoretical framework by including nondissipative terms in Ohm's law, above all the Hall term and electron inertia. It is found that reconnection rates are independent of electron inertia and also of the value of the residual dissipation coefficients, depending only on the ion mass. Small-scale whistler turbulence is easily excited in the reconnection region giving rise to anomalous electron viscosity.

I have not touched on kinetic effects, in particular collisionless dissipation due to resonant wave-particle interactions such as Landau or cyclotron damping. Inverse Landau damping can give rise to microinstabilities. A well-known example is the ion-sound instability driven by a sufficiently high current density in a plasma with temperature ratio $T_e/T_i > 1$, which generates small-scale electrostatic turbulence. Little is known about the efficiency of particle acceleration to high energies by reconnection processes. Ion acceleration is mainly bulk plasma acceleration up to the Alfvén speed. Electrons may be accelerated by the inductive field E in the reconnection region, but simple estimates predict that the period electrons stay in the region where they can be freely accelerated, are short (Biskamp and Schindler, 1971), in particular if the reconnection region is turbulent. The acceleration has probably a rather low efficiency depending strongly on the local coherence properties of the magnetic configuration, much in contrast to the robustness of the diffusive shock acceleration.

I have also not talked about sudden onset conditions of reconnection processes. This is a general problem which always arises, when trying to explain an eruptive process by a linear instability. It appears that in many cases the fast excitation of small-scale turbulence is a probable mechanism acting as a quasi-instantaneous switch-on of a high resistivity. With the advent of extremely powerful parallel computers high resolution numerical simulations in fully 3D geometry will shed more light on these still unexplored areas.

References

Amo, H., T. Sato, and A. Kageyama, 1994, Intermittent energy bursts and recurrent topological change of a twisted magnetic flux tube, National Institute for Fusion Science, Res. Report NIFS-309, Nagoya, Japan

Braginskii, S. I., 1965, Transport processes in a plasma, Reviews of Plasma Physics, ed. M. A. Leontovich (Consultants Bureau, New York), Vol. 1, 205-311

Birn, J., and E. W. Hones, Jr., 1981, Three-dimensional computer modeling of dynamic reconnection in the geomagnetic tail, J. Geophys. Res. **86**, 6802 - 6808

Biskamp, D., 1986, Magnetic reconnection via current sheets, Phys. Fluids **29**, 1520-1531

Biskamp, D., 1991, Algebraic nonlinear growth of the resistive kink instability, Phys. Fluids **B3**, 3353-3356

Biskamp, D., 1993, Nonlinear Magnetohydrodynamics (Cambridge University Press, Cambridge)

Biskamp, D., and U. Bremer, 1993, Dynamics and statistics of inverse cascade processes in 2D magnetohydrodynamic turbulence, Phys. Rev. Lett. **72**, 3819 - 3822

Biskamp, D., and K. Schindler, 1971, Instability of two-dimensional collisionless plasmas with neutral points, Plasma Phys. **13**, 1013-1026

Biskamp, D., and H. Welter, 1980, Coalescence of magnetic islands, Phys. Rev. Lett. **44**, 1069-1072

Biskamp, D., and H. Welter, 1989a, Dynamics of decaying two-dimensional magnetohydrodynamic turbulence, Phys. Fluids B **1**, 1964-1979

Biskamp, D., and H. Welter, 1989b, Magnetic arcade evolution and instability, Solar Phys. **120**, 49-77

Biskamp, D., and H. Welter, 1990, Magnetic field amplification and saturation in two-dimensional magnetohydrodynamic turbulence, Phys. Fluids **B 2**, 1781-1793

Biskamp, D., E. Schwarz, and J. F. Drake, 1995, Ion-controlled collisionless magnetic reconnection, submitted to Phys. Rev. Lett.

Bulanov, S. V., J. Sakai, and S. I. Syrovatskii, 1979, Tearing mode instability in approximately steady MHD configurations, Sov. J. Plasma Phys. 5, 157-163

Burlage, L. F., and J. M. Turner, 1976, Microscale Alfvén waves in the solar wind at 1 AU, J. Geophys. Res. **81**, 73-77

Cowley, S. W. H., 1975, Magnetic field line reconnection in a highly-conducting incompressible fluid: properties of the diffusion region, J. Plasma Phys. 14, 475-490

Dobrowolny, M., A. Mangeney, and P. Veltri, 1980, Fully developed anisotropic hydromagnetic turbulence in interplanetary space, Phys. Rev. Lett. **45**, 144-147

Drake, J. F., R. G. Kleva, and M. E. Mandt, 1994, Structure of thin current layers: Implications for magnetic reconnection, Phys. Rev. Lett. **73**, 1251-1254

Elsässer, W. M., 1950, The hydromagnetic equations, Phys. Rev. **79**, 183

Fadeev, V. M., I. F. Kvartskhava, and N. N. Komarov, 1965, Self-focusing of local plasma currents, Nucl. Fusion 5, 202-209

Furth, H. P., J. Killeen, and M. N. Rosenbluth, 1963, Finite resistivity instabilities of a sheet pinch, Phys. Fluids **6**, 459-484

Grappin, R., A. Pouquet, and J. Léorat, 1983, Dependence of MHD turbulence spectra on the velocity-magnetic field correlation, Astron. Astrophys. **126**, 51-58

Hautz, R., and M. Scholer, 1987, Numerical simulations on the structure of plasmoids in the deep tail, Geophys. Res. Lett. **14**, 969-972

Iroshnikov, P. S., 1964, Turbulence of a conducting fluid in a strong magnetic field, Sov. Astron. **7**, 566-571

Kageyama, A., K. Watanabe, and T. Sato, 1990, Global simulation of the magnetosphere with a long tail: The formation and ejection of plasmoids, National Institute of Fusion Studies, Res. Report NIFS-49, Nagoya, Japan

Kingsep, A. S., K. V. Chukbar, and V. V. Yan'kov, 1990, Electron magnetohydrodynamics, Reviews of Plasma Physics, ed. B. B. Kadomtsev (Consultants Bureau, New York), Vol. **16**, 243-288

Kraichnan, R. H., 1965, Inertial range spectrum in hydromagnetic turbulence, Phys. Fluids **8**, 1385-1387

Matthaeus, W. H., M. L. Goldstein, and C. Smith, 1982, Evaluation of magnetic helicity in homogeneous turbulence, Phys. Rev. Lett. **48**, 1256-1259

Matthaeus, W. H., and D. Montgomery, 1980, Selective decay hypothesis at high mechanical and magnetic Reynolds numbers, Ann. NY Acad. Sci. **357**, 203

Matthaeus, W. H., and D. Montgomery, 1984, Dynamic alignment and selective decay in MHD, in: Statistical Physics and Chaos in Fusion Plasmas, eds. W. Horton and L. Reichl (Wiley, New York), pp. 285-291

Meneguzzi, M., U. Frisch, and A. Pouquet, 1981, Helical and nonhelical turbulent dynamos, Phys. Rev. Lett. **47**, 1060-1064

Mikic, Z., D. C. Barnes, and D. D. Schnack, 1988, Dynamical evolution of a solar coronal magnetic arcade, Astrophys. J. **328**, 830-847

Mikhailovskii, A. B., 1974, Theory of Plasma Instabilities (Consultants Bureau, New York), Vol. 2, p. 63

Nordlund, A. A. Brandenburg, R. L. Jennings, M. Rieutord, J. Ruokolainen, R. F. Stein, and I. Tuominen, 1992, Dynamo action in stratified convection with overshoot, Astrophys. J. **392**, 647-652

Otto, A., K. Schindler, and J. Birn, 1990, Quantitative study of the nonlinear formation and acceleration of plasmoids in the earth's magnetotail, J. Geophys. Res. **95**, 15023-15037

Parker, E. N., 1963, The solar flare phenomenon and the theory of reconnection and annihilation of magnetic fields, Astrophys. J. Suppl. Ser. **8**, 177-211

Petschek, H. E., 1964, Magnetic field annihilation, AAS/NASA Symp. on the Physics of Solar Flares, ed. W. N. Hess (NASA, Washington, DC) pp. 425-437

Politano, H., A. Pouquet, and P. L. Sulem, 1995, Current and vorticity dynamics in three-dimensional MHD turbulence, Phys. Plasmas **2**, 2931-2939

Pouquet, A., U. Frisch, and J. Léorat, 1976, Strong MHD helical turbulence and the nonlinear dynamo effect, J. Fluid Mech. **77**, 321-354

Pritchett, P. L., and C. C. Wu, 1979, Coalescence of magnetic islands, Phys. Fluids **22**, 2140-2146

Rutherford, P. H., 1973, Nonlinear growth of the tearing mode, Phys. Fluids **16**, 1903-1908

Sato, T., and T. Hayashi, 1979, Externally driven magnetic reconnection and a powerful magnetic energy converter, Phys. Fluids **22**, 1189-1202

Steenbeck, M., F. Krause, and K. H. Rädler, 1966, Berechnung der mittleren Lorentz-Feldstärke $\langle v \times B \rangle$ für ein elektrisch leitendes Medium in turbulenter, durch Coriolis Kräfte beeinflusster Bewegung. Z. Naturforsch. **21a**, 369-376

Sweet, P. A., 1958, The production of high energy particles in solar flares, Nuovo Cimento Suppl. **8**, Ser. X, 188-196

Syrovatskii, S. I., 1971, Formation of current sheets in a plasma with a frozen-in strong magnetic field, Sov. Phys. - JETP **33**, 933-940

Syrovatskii, S. I., 1981, Pinch sheets and reconnection in astrophysics, Annu. Rev. Astron. Astrophys. **19**, 163-229

Taylor, J. B., 1986, Relaxation and magnetic reconnection in plasmas, Rev. Mod. Phys. **53**, 741-763

Waelbroeck, F. L., 1989, Current sheets and nonlinear growth of the $m = 1$ kink-tearing mode, Phys. Fluids **B 1**, 2372-2380

Woltjer, L., 1958, A theorem of force-free magnetic fields, Proc. Natl. Acad. Sci. (Washington) **44**, 489-492

Yanase, S., J. Mizushima, and S. Kida, 1991, Coherent structures in MHD turbulence and turbulent dynamo, in Turbulence and Coherent Structures, ed. O. Métais and M. Lesieur (Kluiver, Dordrecht), pp. 569-583

Don Melrose

Particle Acceleration and Nonthermal Radiation in Space Plasmas

D.B. Melrose
Research Centre for Theoretical Astrophysics
School of Physics, University of Sydney

Abstract. Acceleration processes for fast particles in astrophysical and space plasmas are reviewed with emphasis on stochastic acceleration by MHD turbulence and on acceleration by shock waves. Radiation processes in astrophysical and space plasmas are reviewed with emphasis on plasma emission from the solar corona and electron cyclotron maser emission from planets and stars.

Key words: Acceleration of particles, plasma emission, electron cyclotron maser emission

1. Introduction

Astrophysical and space plasmas are composed of thermal electrons, thermal ions, suprathermal distributions of particles of different kinds, and various forms of low-frequency and high-frequency plasma turbulence. The suprathermal particles include relativistic particles (e.g., cosmic rays), and can also include beams or magnetically trapped distributions of lower energy electrons and ions. The low-frequency turbulence includes large-scale flow motions, and low-frequency MHD (magnetohydrodynamic) waves. There may also be shock waves or other dynamical structures such as current sheets, electrostatic shocks or double layers. The high-frequency turbulence consists of waves in various wave modes (ion sound waves, whistler waves, Langmuir waves and so on), and is often referred to as *plasma turbulence*. The plasma turbulence is trapped in the plasma, in the sense that it is composed of waves that cannot propagate out of the plasma. The suprathermal particles and plasma turbulence interact with each other, exchanging energy and momentum. Such particle-wave interactions are central to our understanding of the physics of diffuse plasmas in space and in the laboratory.

Acceleration mechanisms and nonthermal radiation mechanisms in plasmas are discussed in these lectures. An *acceleration mechanism* is a process that increases the energy of nonthermal particles. Several different acceleration processes are thought to be important in space plasmas, and eight of these are described below. The discussion is biased

Astrophysics and Space Science **242**: 209–246, 1997.

toward acceleration of electrons, rather than ions, because nonthermal electrons are involved in nonthermal radiation mechanisms. A *radiation mechanism* is defined as a process that leads to waves that can escape from the plasma. Waves in only two wave modes can escape to infinity: the o-mode at frequencies $\omega > \omega_p$, ω_p is the plasma frequency, and the x-mode at frequencies $\omega > \omega_x$, where

$$\omega_x = \frac{1}{2}\Omega_e + \frac{1}{2}[4\omega_p^2 + \Omega_e^2]^{1/2} \tag{1.1}$$

is the cutoff frequency for the x-mode, with Ω_e the electron cyclotron frequency. A radiation mechanism must produce waves in either or both these wave modes. Nonthermal radiation mechanisms may be classified as incoherent or coherent: an *incoherent* mechanism involves nonthermal particles radiating independently of each other, and a *collective* or *coherent* radiation mechanism involves either wave-wave processes transferring energy from plasma turbulence into the escaping radiation, or nonthermal particles causing the escaping radiation to be generated through an instability. The two most important coherent radiation mechanisms are plasma emission and electron cyclotron maser emission.

2. Possible acceleration mechanisms

Acceleration mechanisms that are thought to be important for fast particles in astrophysical and space plasmas include the following:
  1) Stochastic acceleration by MHD waves
  2) Diffusive shock acceleration
  3) Resonant acceleration by MHD waves
  4) Resonant acceleration by Langmuir waves
  5) Shock drift acceleration
  6) Acceleration during magnetic reconnection
  7) Runaway acceleration by a parallel electric field
  8) Acceleration by potential double layers
The first two mechanisms are modern-day versions of second-order Fermi acceleration (Fermi 1949) and first-order Fermi acceleration (Fermi 1954), respectively. Both involve energy flowing from macroscopic motions (MHD turbulence and shocks, respectively) into the fast particles. Both also require effective resonant scattering of the fast particles. Mechanisms 3) and 4) involve direct energy transfer to the particles during resonant interactions. Mechanisms 5) and 6) involve acceleration by quasi-static electric fields that are perpendicular to a strongly inhomogeneous magnetic field. The final two mechanisms involve acceleration by quasi-static electric fields parallel to the magnetic field.

2.1. RESONANT WAVE-PARTICLE INTERACTIONS

The important wave-particle interactions are *resonant*, in the sense that the particle and the wave stay in phase for an anomalously long time. Such interactions allow effective exchange of momentum between particles and waves, leading to a variety of consequences, such as resonant scattering of the particles, acceleration of the particles, and generation of the plasma turbulence. Gyroresonant scattering is a particularly important concept in space plasma physics. Fast particles can be scattered very efficiently by waves, so that their distribution remains nearly isotropic. As a consequence, they diffuse along magnetic field lines rather than propagating according to simple orbit theory.

The resonance condition in the presence of a magnetic field is the Doppler condition

$$\omega - s\Omega - k_{\parallel}v_{\parallel} = 0, \tag{2.1}$$

where s is an integer, $\Omega = |q|B/m\gamma$ is the relativistic gyrofrequency, and $k_{\parallel}$, $v_{\parallel}$ are the components of the wavevector, $\mathbf{k}$, and the particle velocity, $\mathbf{v}$, respectively, parallel to $\mathbf{B}$.

Gyroresonant interactions between waves and a distribution of particles causes the distribution function to evolve according to (cf. Appendix A for details)

$$\frac{df(\mathbf{p})}{dt} = \frac{1}{\sin\alpha}\frac{\partial}{\partial\alpha}\left\{\sin\alpha\left[D_{\alpha\alpha}(\mathbf{p})\frac{\partial}{\partial\alpha} + D_{\alpha p}(\mathbf{p})\frac{\partial}{\partial p}\right]f(\mathbf{p})\right\}$$

$$+ \frac{1}{p^2}\frac{\partial}{\partial p}\left\{p^2\left[D_{p\alpha}(\mathbf{p})\frac{\partial}{\partial\alpha} + D_{pp}(\mathbf{p})\frac{\partial}{\partial p}\right]f(\mathbf{p})\right\}, \tag{2.2}$$

The expressions for the diffusion coefficients in (2.2), cf. Appendix A, involve integrals over the distribution of waves. The integrands of the relevant integrals are in the ratios

$$D_{\alpha\alpha} : (D_{p\alpha} = D_{\alpha p}) : D_{pp} = (\Delta\alpha)^2 : \Delta\alpha\Delta p : (\Delta p)^2$$

$$\frac{\Delta\alpha}{\Delta p} = \frac{\omega\cos\alpha - k_{\parallel}v}{\omega p\sin\alpha} \tag{2.3}$$

The coefficient $D_{\alpha\alpha}$ describes diffusion in pitch angle, and is the basis for a description of resonant scattering. The coefficient D_{pp} describes diffusion in momentum or energy, and is the basis for a description of resonant acceleration.

Pitch-angle scattering corresponds to a situation where the dominant term on the right hand side of (2.2) is the first term, involving a double derivative with respect to pitch angle, α. Thus, resonant scattering leads to predominantly pitch-angle scattering whenever the phase speed of wave is much less than the speed of the particle, so that one has $|\omega\cos\alpha| \ll |k_\parallel v|$, $|p\Delta\alpha| \gg |\Delta p|$, and hence $p^2 D_{\alpha\alpha} \gg D_{pp}$ in (2.3). For MHD waves with phase speed close to the Alfvén speed, v_A, pitch-angle scattering applies for $v \gg v_A$. In treating pitch-angle scattering one can often justify retaining only the terms $s = \pm 1$ in the sum over harmonics. In most cases the waves involved in resonant scattering are MHD waves of much higher frequency than those typically associated with macroscopic motions, and these higher frequency waves need to be generated through the resonant interaction with the particles. The waves can be generated by the particles themselves provided that the distribution function is sufficiently anisotropic, that is, provided that $\partial f/\partial\alpha$ is large enough. In all cases where waves are generated by an anisotropic distribution of particles, the scattering of the particles by the waves tends to reduce the anisotropy, and so tends to reduces the wave growth. The energy density required for the resonant waves to provide effective pitch-angle scattering is only a small fraction of the energy density in the resonant particles, so that effective momentum transfer can occur with little energy transfer between the particles and the waves.

The requirement that the speed of the particle exceed the phase speed of the wave leads to a threshold condition for particles to be scattered effectively. For ions resonating with MHD waves, this threshold requires $v > v_A$. The condition for electrons (rest mass m_e) to resonate with MHD waves is more restrictive than for ions, and requires that the momentum satisfy $p > (m_p/m_e)m_e v_A$, where m_p is the mass of the proton (cf. Melrose 1986, p. 237, for more details). However, slower electrons can resonate with whistlers; in this case the threshold condition is $v > (m_p/m_e)^{1/2} v_A$. Provided that the relevant threshold is exceeded, scattering can be very effective for fast particles, and can be much more effective than for slower particles. The recognition that resonant scattering can be very effective was an early major success in the field of plasma astrophysics (e.g., Wentzel 1974)

Gyroresonant acceleration, which is acceleration mechanism 3) above, is due to gyroresonant interactions at $|s| \gg 1$. This acceleration is described by the final term in (2.2), after averaging over pitch angle. In principle, damping of MHD turbulence due to this process can allow the energy in MHD turbulence to be transferred to the energetic particles. This mechanism has been suggested for relativistic electrons in a radiogalaxy (Lacombe 1977) and for energetic particles on the Sun (Barbosa 1979; Eichler 1979). Stochastic acceleration competes with gyroresonant acceleration and the former is usually more important.

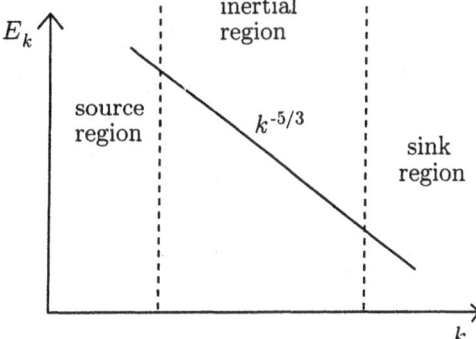

Figure 1. The energy per unit wave number in magnetic turbulence is plotted schematically as a function of wavenumber, showing the Kolmogorov spectrum in the inertial range, between the source at small k and the sink at large k.

2.2. GENERATION OF MHD TURBULENCE

The MHD turbulence needed for stochastic acceleration of fast particles requires an energy source to maintain it. The turbulent energy must arise from large-scale mass motion, which can lead to an instability, e.g., the Kelvin–Helmholtz instability. (Large scale mass motions can also lead to shock waves, and such energy is also then available for the acceleration of particles.) The instability that converts energy in directed fluid motions into wave energy leads to waves with relatively long wavelengths (small k), typical of the size of the system. The formation of the spectrum of MHD turbulence is thought to involve nonlinear wave-wave interactions that leads to a *cascade of turbulent energy* that transfers turbulent energy from long wavelengths to short wavelengths (cf. Zhou & Matthaeus 1990 for a qualitative explanation), as illustrated in Figure 1. The short-wavelength (large-k) limit of the turbulent spectrum is determined by the dominant dissipation mechanism. One may define a characteristic k-value, for each specific damping mechanism for the MHD waves, as the k at which the damping time is equal to the transfer time (from smaller to larger k) due to the wave-wave interactions, that is, to the turnover time in the cascade. The large-k limit of the spectrum is determined by the smallest such characteristic k. At this k effectively all the power injected into the turbulent spectrum at long scalelengths is transferred to the particles that cause the damping at this smallest characteristic k.

Under quite modest conditions, the smallest characteristic k can be due to damping by fast particles. Then the dominant dissipation mechanism is stochastic acceleration of the fast particles and virtually all the energy in the turbulent cascade at the longest scalelengths ultimately ends up in the fast particles. All other dissipation mechanisms are then unimportant, because they would become effective only at larger k, where there is no turbulent energy. (As noted below, only the magnetosonic component is damped by Fermi-type acceleration; however, the wave-wave processes in the cascade mix up the Alfvén and magnetosonic components, so that all the turbulent energy, and not just the magnetosonic component in it, is ultimately available for stochastic acceleration of fast particles.)

In principle, the waves in the turbulent spectrum could also lead to resonant scattering of fast particles. However, in practice the typical wavelength of the turbulence is much greater than is required for this purpose. The waves that are involved in resonant scattering have wavelengths of order the gyroradii of the scattered particles, which is typically much shorter than the wavelengths in the turbulent spectrum, and these short-wavelength waves must be generated by the anisotropic distribution of the particles themselves. However, the growth time for the instability that generates the resonant Alfvén waves increases rapidly with increasing $k \propto p$. The growth time becomes too long to be effective for cosmic rays at relatively modest energies, and appears to be far too long to be effective for energies $\gtrsim 10^{15}\,\mathrm{eV}$. Such higher energy cosmic rays can be scattered by MHD waves generated by turbulent motions, as in the turbulent spectrum illustrated in Figure 1. For example, resonant scattering in the interstellar medium of cosmic rays with energy $\gtrsim 10^{15}\,\mathrm{eV}$ requires waves with $k^{-1} \sim 0.3\,\mathrm{pc}$. The presence of such a turbulent spectrum of MHD waves in the interstellar medium is inferred from scintillations of radio sources (e.g., Rickett 1990).

2.3. STOCHASTIC ACCELERATION

Historically, the ideas underlying *stochastic acceleration* were first discussed by Fermi (1949), and it is often referred to as Fermi acceleration. A stochastic acceleration mechanism involves three ingredients: an energy changing mechanism for the particles, a scattering mechanism for the particles, and a statistical theory that links these two. In the original Fermi (1949) mechanism for the acceleration of galactic cosmic rays, the energy changing mechanism and the scattering process were combined, and attributed to reflection of charged particles from moving magnetized interstellar clouds. However, it was later realized that small very frequent changes in energy due to interaction with MHD waves can be much more effective in accelerating energetic particles (for historical references, cf. Melrose 1990; Axford 1994).

The energy changing mechanisms in MHD turbulence include magnetic pumping, reflection from moving magnetic compressions and transit-time acceleration (Kulsrud & Ferrari 1971). For example, the idea

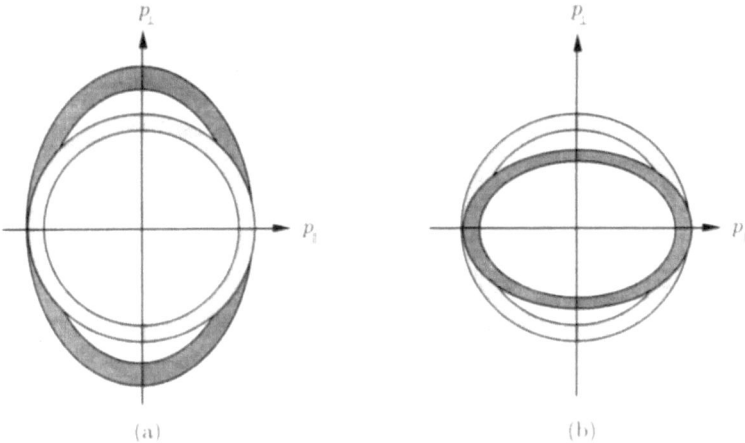

Figure 2. The effect of magnetic pumping on a distribution of particles is illustrated schematically. The lightly shaded region is a section of an isotropic distribution that is deformed into the darkly shaded region by (a) a compression, and (b) a rarefaction. As $B(t)$ increases the distribution function becomes anisotropic; the circles are deformed by being elongated along the $p_\perp$-axis, and as $B(t)$ decreases, the contours of constant f become shortened along the $p_\perp$-axis. In both cases scattering reduces the anisotropy, making the contours of constant f closer to circles.

in magnetic pumping is that in slow changes in the magnetic field strength, B, there is an adiabatic invariant

$$\frac{p_\perp^2}{B} = \text{constant}, \qquad (2.4)$$

where $p_\perp$ is the magnitude of the momentum component perpendicular to the ambient magnetic field **B**. The invariant (2.4) implies that the energy of a particle increases and decreases as B increases and decreases. Note that magnetic fields do no work, and the energy changes are to be attributed to the electric field due to the time-varying B. The stochastic process invoked is pitch-angle scattering. The need for this can be understood by considering the change in the distribution of particles during a cycle of compression and rarefaction of the B-field, as illustrated in Figure 2. In the absence of scattering the distribution function returns to its initial value at the end of one cycle, implying no net transfer of energy. Pitch-angle scattering tends to keep the distribution isotropic, and its inclusion implies a viscous-type drag which is such that the energy losses by the particles in the rarefaction phase are less than the energy gains during the compressive phase. For a more detailed discussion of magnetic pumping, cf. Kirk (1994).

Another interpretation of the energy-changing process in stochastic acceleration is in terms of the parallel (to **B**) component of the electric field, $E_\parallel$, in the wave (Achterberg 1981). From this viewpoint, magnetic pumping (or transit damping) is regarded as a resonant interaction, corresponding to $s = 0$ in the Doppler condition (2.1). In fact $E_\parallel$ causes changes only in the parallel component, $p_\parallel$, of the particle momentum. In describing the effects of interactions at $s = 0$, it is more convenient to write the diffusion in cylindrical coordinates $(p_\parallel, p_\perp)$, cf. equation (A.6) of Appendix A, rather than in spherical polar coordinates (p, α), as in equation (2.2). In cylindrical coordinates only the term involving $D_{\parallel\parallel}$ in nonzero on the right hand side of equation (A.6). The resulting evolution causes the distribution function to become anisotropic in such a way that the energy transfer to the particles suppresses itself (Achterberg 1981). Pitch-angle scattering is required to keep the distribution approximately isotropic, to avoid this self-suppression effect on the acceleration. If pitch-angle scattering is effective, one may average the diffusion equation, involving only the $D_{\parallel\parallel}$ term, over pitch angle. Then the diffusion effectively reduces to one in p alone, with the counterpart of the D_{pp} term in (2.2) determined by the average of $D_{\parallel\parallel}$ over pitch angle.

In general, MHD turbulence consists of waves in each of the three MHD wave modes: the fast, Alfvén and slow modes. Alfvén waves are shear or torsional waves; in the simplest approximation these waves have an electric field that is perpendicular to the ambient magnetic field, and they involve no compression of the gas or the ambient magnetic field. Hence (shear) Alfvén waves do not contribute to Fermi acceleration. For most of the plasmas of interest in plasma astrophysics, the Alfvén speed, v_A, is much greater than the sound speed, and then the fast mode is magnetosonic in character, with phase speed $\omega/k \approx v_A$; the slow mode is sonic in character and is of little interest. Magnetosonic waves do cause compression of B, and it is the magnetosonic component that leads to stochastic acceleration.

¿From either of the viewpoints discussed above, stochastic acceleration may be described in terms of an average diffusion in momentum space, where the average is over pitch angle. This corresponds to (e.g., Tverskoĭ 1967, 1968)

$$\frac{df(p)}{dt} = \frac{1}{p^2}\frac{\partial}{\partial p}\left[p^2 D(p)\frac{\partial f(p)}{\partial p}\right]. \tag{2.5}$$

For MHD (fast mode) turbulence with energy density W_M and mean frequency $\bar{\omega}$, the diffusion coefficient in (2.2) has the form (e.g., Kulsrud & Ferrari 1971; Achterberg 1981)

$$D(p) = \frac{\zeta\pi\bar{\omega}}{4}\frac{W_M}{B^2/2\mu_0}\frac{p^2 v_A}{v}, \tag{2.6}$$

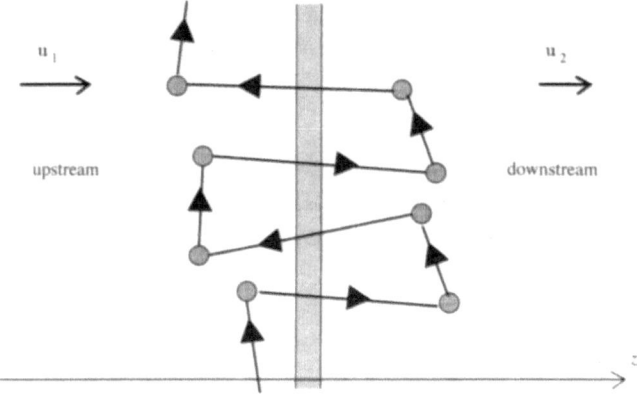

Figure 3. Diffusive shock acceleration is illustrated: the shaded vertical region is
the shock, the circular blobs denote idealized scattering centers, and the solid line
with arrows denotes the path of an idealized fast particle. The coordinate z and the
velocities u_1 and u_2 are shown for the case of a parallel shock.

where ζ is a factor of order unity that depends on the details of the
model. For comparison, the original Fermi mechanism, due to collisions
with clouds moving with speed $u \sim v_A$ separated by a mean distance
L, gives $D(p) = u^2 p^2 / 3Lv$.

In many specific applications, stochastic Fermi acceleration is generally out of favor compared with DSA (see below). (These two acceleration processes are related in the sense that acceleration by DSA due to
an ensemble of weak shocks reduces to stochastic Fermi acceleration.)
An exception is in the context of solar flares, where stochastic acceleration is considered a viable mechanism (for references, cf. Melrose 1990,
1992; Miller, Guessoum & Ramaty 1990).

2.4. DIFFUSIVE SHOCK ACCELERATION

Since it was first treated in detail in the late 1970s, diffusive shock
acceleration (DSA) (e.g., the reviews by Blandford & Eichler 1987;
Jones & Ellison 1991; Kirk 1994) has become the most widely accepted
acceleration mechanism for relativistic particles in astrophysical plasmas. It is a form of first-order Fermi acceleration (Fermi 1954) in the
sense that first order changes in energy do not cancel. The relevant
energy changes in DSA occur when a particle crosses the shock and is
scattered. The scattering centers (MHD waves causing pitch-angle scattering) are flowing with the plasma. The important point is that, when
viewed from a frame at rest in the plasma on one side of the shock, the
scattering centers on the other side of the shock are flowing toward the
shock, with a speed equal to the change in flow speed $|u_1 - u_2|$ across

the shock. Hence, on crossing the shock a particle is always scattered
head-on, and so gains energy as a result of the scattering, as illustrated
in Figure 3.

In simple analytic treatments of DSA, it is assumed that scattering
is effective on both sides of the shock, and causes spatial diffusion
along and across the magnetic field lines. The boundary conditions far
upstream (the pre-shock region), and far downstream (the post-shock
region), together with the boundary conditions at the shock itself, then
lead to a relation

$$f_2(p) = bp^{-b} \int_0^p dp'\, p'^{(b-1)} f_1(p'), \quad b = \frac{3u_2}{u_1 - u_2}, \tag{2.7}$$

for the downstream distribution $f_2(p)$ in terms of the upstream distribution $f_1(p)$. In (2.7), u_1 and u_2 ($u_1 > u_2$) are the flow speeds in the
upstream and downstream regions, respectively, in the frame in which
the shock is at rest cf. Figure 3.

If the injection spectrum is monoenergetic, $f_1(p) \propto \delta(p - p_0)$ say,
then (2.7) implies that the downstream spectrum is a power law at
$p > p_0$ of the form $f_2(p) \propto p^{-b}$. According to (2.7), the power law
index, b, is determined by the ratio u_1/u_2, which is equal to the inverse
of the density ratio, n_1/n_2, which determines the strength of the shock.
In terms of the mach number $M = u_1/v_{A1}$, where v_{A1} is the Alfvén
speed upstream of the shock, one finds

$$b = \frac{3r}{r-1} = \frac{3(\Gamma+1)M^2}{2(M^2-1)}, \quad r = \frac{(\Gamma+1)M^2}{2+(\Gamma-1)M^2}, \tag{2.8}$$

where r is the compression factor across the shock, $\Gamma = 5/3$ is the
adiabatic index of the gas and M is the Mach number of the shock.
The maximum strength of the shock is $r = 4$, for $M^2 \to \infty$, and
the minimum value of the spectral index implied by (2.8) is $b = 4$.
The observed cosmic ray spectrum has $b = 4.7$. This is interpreted
as being due to a balance between DSA by strong shocks, giving an
inferred power law index $b = 4.1$, and losses with a power law energy
dependence with index 0.6 inferred from the properties of secondary
cosmic rays (e.g., Kirk 1994).

Suggested applications of DSA include virtually all astrophysical
and space plasmas where shocks occur. The most widely favored applications include (a) energetic ions in interplanetary shocks and planetary bow shocks (e.g., Lee 1992), (b) energetic ions in solar flares
(e.g., Reames 1990), (c) galactic cosmic rays accelerated at supernova-
generated shocks in the interstellar medium (e.g., Axford 1992), and
(d) relativistic electrons in supernova remnants and other galactic and
extragalactic synchrotron sources (e.g., Achterberg 1990).

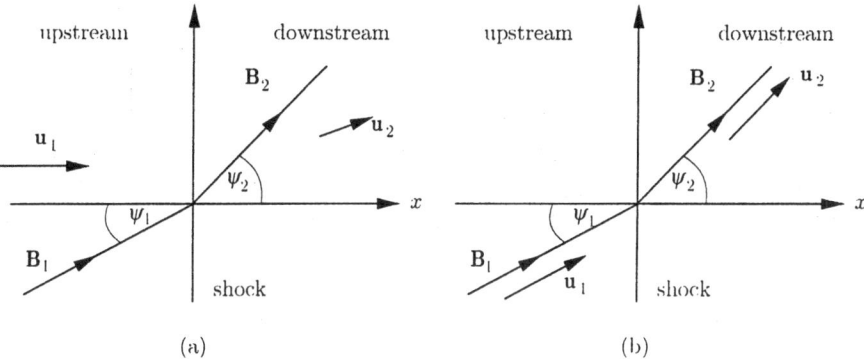

Figure 4. The changes across a shock front are illustrated in (a) the normal incidence frame (NIF) and (b) the de Hoffmann-Teller (HTF).

The injection of particles into DSA is usually assumed to consist of two components: a pre-existing component of energetic particles that is further accelerated by the shock, and thermal particles that are accelerated to form a nonthermal power-law distribution by the passage of the shock. The details of how the injection of thermal particles occurs is one of the weaker points of the theory. However, it is accepted that the pick-up of thermal ions occurs without significant preference for one ionic species over another. That is, the ionic composition of the fast ions produced by DSA reflects the ionic composition of the ambient plasma.

In most treatments of DSA, particles are injected artificially: through a δ-function injection in analytic treatments, and by a postulated effective scattering in numerical treatments. For ions the threshold $v > v_A$ for resonant scattering to be possible can be plausibly satisfied due to the presence of relative flow speed $> v_A$ from one side of the shock relative to the other. Thus thermal ions that simply convect through the shock would find themselves with $v \approx M v_A > v_A$ on the other side of the shock. Nevertheless, the need or otherwise for a specific injection mechanism to preaccelerate the ions remains somewhat controversial. For electrons the threshold condition $v > 43 v_A$ is not plausibly satisfied (for shocks with $M < 43$), and some form of preacceleration seems necessary (e.g., Achterberg 1990; Blandford 1994).

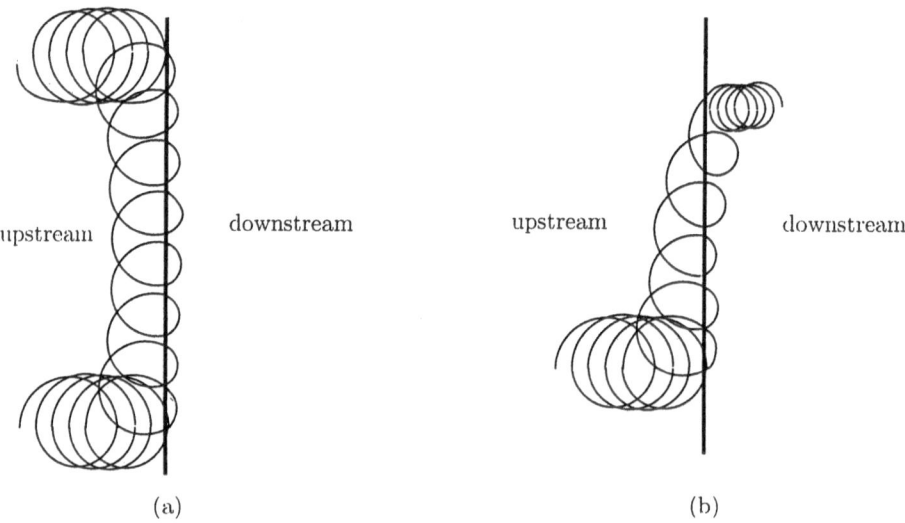

upstream   downstream   upstream   downstream

(a)   (b)

Figure 5. The path of an ion is indicated schematically as it encounters a shock, represented by the vertical line, and is (a) reflected or (b) transmitted. The gyroradius decreases abruptly from the upstream region to the downstream region due to the abrupt increase in B. This causes the orbit to drift along the shock front in the direction of the electric field.

2.5. SHOCK DRIFT ACCELERATION

Shock drift acceleration (SDA) (e.g., Webb, Axford & Terasawa 1983; Drury 1983; Kirk 1994) may be attributed to an electric field, $\mathbf{E} = -\mathbf{u} \times \mathbf{B}$, perpendicular to the magnetic field, due to the flow velocity, $\mathbf{u}$, not being along $\mathbf{B}$ on either side of the shock. This is illustrated in Figure 4a, which corresponds to the normal incidence frame (NIF) in which the inflow velocity is normal to the shock front. An ion that encounters a shock sees an abrupt increase in the magnetic field, which causes an abrupt decrease in its gyroradius. As a consequence, the ion drifts along the shock front and is either reflected or transmitted, cf. Figure 5. In either case, the ion gains energy due to its drift motion along the shock, which is in the direction of the electric field. Electrons have much smaller gyroradii, and for them the change in B occurs over a distance much greater than a gyroradius. They also drift along the shock (due to the grad B-drift) in the opposite sense to the ions, so that they also gain energy in being either reflected or transmitted at the shock.

The acceleration may be treated by transforming to the de Hoffmann-Teller frame (HTF), which is moving relative to the NIF along the shock front with velocity $u_0 = u_1 \tan \psi_1$, where ψ_1 is defined in Figure 4a. In

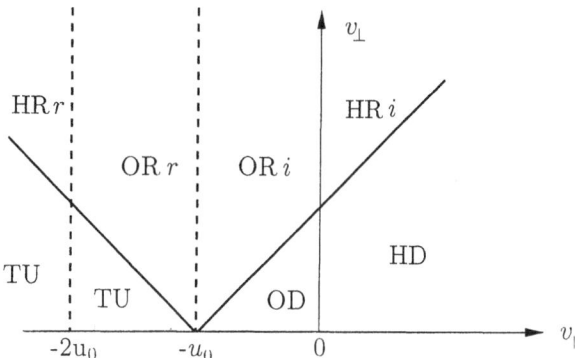

Figure 6. A schematic plot of the upstream velocity space in the NIF.

the HTF the flow velocity and the magnetic field are parallel on either side of the shock, cf. Figure 4b. There is no electric field in the HTF, and hence the energy of particles is conserved on either being reflected or transmitted at the shock. For electrons the adiabatic invariant (2.4) applies, and hence $(1 - \mu^2)/B$ is a constant, with μ the cosine of the pitch angle. Suppose that $B = B_1$ in the upstream plasma increases to $B = B_2$ in the downstream plasma. Electrons with $1 - \mu_1^2 > B_1/B_2$ are reflected and those with $1 - \mu_1^2 < B_1/B_2$ are transmitted.

In general, the treatment of the reflection or transmission of a particle at a shock is complicated by the fact that the details depend on the value of the gyrophase at the point where the particle crosses the shock (e.g., Drury 1983). For most purposes a simplified model that ignores this complication is adequate. One simply assumes that the speed, v', and cosine of pitch angle, μ', in the HTF are related to v and μ in the NIF by

$$v'_{\parallel} = v'\mu' = v_{\parallel} + u_0 = v\mu + u_0, \qquad v'^2_{\perp} = v'^2(1 - \mu'^2) = v^2_{\perp} = v^2(1 - \mu^2).$$
(2.9)

With this model, the reflection and transmission depends only on the initial position in (2-dimensional) velocity space in the upstream plasma, as indicated in Figure 6. The labels in Figure 6 correspond to overtaken particles that are transmitted downstream (OD), particles that encounter the shock head on and are transmitted downstream (HD), overtaken particles that are reflected downstream (OR), particles that encounter the shock head on and are reflected upstream (HR), and particles that are transmitted from downstream to upstream (TU).

The energy ratio $(\tfrac{1}{2}mv_r^2/\tfrac{1}{2}mv_i^2 = E_r/E_i)$ of the reflected (r) to the incident (i) particles can be inferred relatively simply from Figure 6. Consider two points, A and B, cf. Figure 7, that correspond to a particle

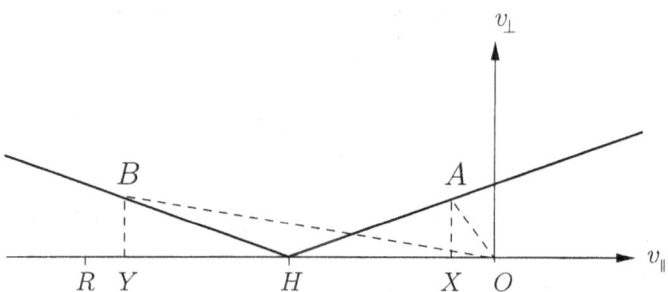

Figure 7. In velocity space in the NIF, the point O $(v_\parallel = 0, v_\perp = 0)$ is the origin, H $(v_\parallel = -u_0, v_\perp = 0)$ is the origin in the HTF frame, R $(v_\parallel = -2u_0,$ and $v_\perp = 0)$ is the reflection point corresponding to O. The slanting lines have the same interpretation as in Figure 6. The point A at $v_\parallel = -x$ and its reflection B at $v_\parallel = -(u_0 - x)$ maximize the ratio of the lengths OB to OA. The angles OHA and RHB are equal to $\alpha_c = \arcsin b$, the angle XOA is $\pi - \alpha_i$ and the angle YOB is $\pi - \alpha_r$.

before and after reflection. Then, if A is the point $v_\perp$, $v_\parallel = -x$ in velocity space, B is the point $v_\perp$, $v_\parallel = -(2u_0 - x)$. One has

$$\frac{E_r}{E_i} = \frac{v_r^2}{v_i^2} = \frac{(2u_0 - x)^2 + v_\perp^2}{x^2 + v_\perp^2}. \tag{2.10}$$

The maximum value of E_r/E_i occurs for the minimum value of $v_\perp^2$, which occurs on the slanted lines $(\alpha = \alpha_c, \sin^2 \alpha_c = B_1/B_2)$ that separate the regions where particles are reflected from the regions where they are transmitted, cf. Figure 7. For this case, inserting

$$v_\perp = v_i \sin \alpha_i = v_r \sin \alpha_r = (u_0 - x) \tan \alpha_c,$$
$$v_i \cos \alpha_i = -x, \qquad v_r \cos \alpha_r = -(2u_0 - x), \tag{2.11}$$

into (2.10) gives

$$\frac{E_r}{E_i} = \frac{(2u_0 - x)^2 + (u_0 - x)^2 \tan^2 \alpha_c}{x^2 + (u_0 - x)^2 \tan^2 \alpha_c}. \tag{2.12}$$

Toptygin (1980) suggested that the maximum energy ratio occurs for the minimum v_i, which corresponds to $x = u_0 \sin^2 \alpha_c$ in Figure 7. This gives $E_r/E_i = (4B_2 - 3B_1)/B_1$. However, modifying an argument due to Decker (1983), the actual maximum is found by maximizing (2.12) as a function of x. This maximum is for $x = u_0(1 - \cos \alpha_c)$, which gives

$$\left(\frac{E_r}{E_i}\right)_{\max} = \cot^2(\alpha_c/2) = \frac{1 + (1 - B_1/B_2)^{1/2}}{1 - (1 - B_1/B_2)^{1/2}}, \tag{2.13}$$

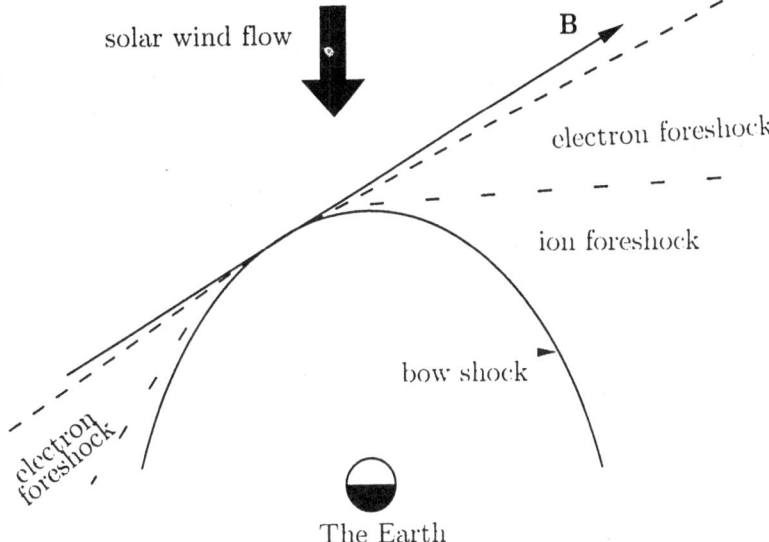

solar wind flow

B

electron foreshock

ion foreshock

bow shock

electron foreshock

The Earth

Figure 8. The electron foreshock is due to SDA on field lines nearly tangent to the bow shock; the dashed line indicates the leading edge of the electron foreshock region. There is also an ion foreshock, which is nearer to the bow shock due to the ions (being slower than the electrons) being swept back further by the solar wind.

which also applies to the relativistic case (Kirk 1994, p. 244). The magnetic compression ratio of the shock is determined by the Mach number M, the density compression ratio r and the angle ψ_1, cf. Figure 4:

$$\frac{B_2}{B_1} = \left(\left[\frac{(M^2 - 1)r}{M^2 - r} \right]^2 \sin^2 \psi_1 + \cos^2 \psi_1 \right)^{1/2}. \qquad (2.14)$$

The maximum density ratio is $r \to 4$ for $M^2 \to \infty$, and then (2.14) implies $B_2/B_1 \to 4$, provided that the angle ψ_1 is not too far from $\pi/2$. Thus for a nearly perpendicular strong shock, (2.13) implies $E_r/E_i \to$ 13.93. However, reflection is possible only for $v_i > M v_{A1} \tan \psi_1 \sin \alpha_c$, where v_{A1} is the Alfvén speed in the upstream plasma. In space plasmas one typically has $V_p \ll v_A \lesssim V_e$, where V_p and V_e are the thermal speeds of protons and electrons, respectively. Hence, the conditions for reflection with substantial energy gain are much more easily satisfied for electrons than for protons and other ions.

It is widely accepted that SDA of electrons is important at planetary bow shocks and at some interplanetary shocks. Reflection of electrons at the Earth's bow shock produces the *electron foreshock*, cf. Figure 8. The reflection is effective only near points where the magnetic field in the solar wind is tangent to the bow shock. The accelerated electrons stream out from this acceleration region along the field line, and are slowly swept toward the Earth by the solar wind flow. Slower ions are also reflected, and are swept back more strongly by the solar wind to form an ion foreshock. The electron beams formed in the foreshock excite Langmuir waves which produce plasma emission, as discussed further below. Reflection at some shocks nearer the Sun is thought to be the source of electrons that produce type II solar radio bursts through plasma emission. However, existing theories for such acceleration are not entirely satisfactory, in that they are either qualitative (e.g., Holman & Pesses 1983) or, at most, semiquantitative (Leroy & Mangeney 1984; Wu, Steinolfson & Zhou 1986).

2.6. ACCELERATION DURING MAGNETIC RECONNECTION

Magnetic reconnection occurs in a current sheet that separates two regions of opposite magnetic polarity. Acceleration of particles occurs for essentially the same reason as in SDA: there is an electric field that is perpendicular to the magnetic field (outside the current sheet) and particles drift such as to gain energy from this electric field. This has been demonstrated numerically by following the orbits of individual particles, which may become chaotic (e.g., Büchner & Zelenyi 1989; Kliem 1994).

Application of acceleration during reconnection to the production of energetic particles in solar flares, and to other flaring systems, is somewhat controversial. While there is wide acceptance that magnetic energy is released through reconnection, much of this energy is thought to go into heating the bulk of the plasma. There is no quantitative theory that allows one to discuss the acceleration of fast particles in a neutral sheet in a sufficiently general way to determine what fraction of the released magnetic energy that goes into fast particles.

2.7. ACCELERATION BY PARALLEL ELECTRIC FIELDS

Acceleration by an electric field, $E_{\parallel}$, parallel to the magnetic field is the simplest conceivable acceleration mechanism. However, plasmas are such good conductors that under most circumstances, currents flow quickly to screen out any $E_{\parallel}$.

There is clear evidence for parallel electric fields in the Earth's auroral zones (e.g., Mozer et al. 1980), and the favored interpretation is in terms of a large-scale potential drop that becomes localized in shock-like structures. Neither the nature of the large-scale potential drop (e.g., Lundin & Eliasson 1991), nor the detailed nature of the localized structures with $E_{\parallel} \neq 0$ is well understood. The shock-like structures have been interpreted in terms of electrostatic shocks (Mozer et al.

1980), double layers (e.g., the review by Raadu 1989) and ion phase-space holes (Boström *et al.* 1988). Electrons are accelerated to several keV in the auroral zones in so-called inverted-V events. Such electrons correlate strongly with auroral kilometric radiation (AKR), and it is accepted that the inverted-V electrons generate AKR through electron cyclotron maser emission, as discussed below. Analogous emission is also generated in Jupiter's magnetosphere, where the acceleration is thought to involve large-amplitude kinetic Alfvén waves generated by Io's motion through the magnetosphere (e.g., Goldstein & Goertz 1983).

Runaway acceleration by parallel electric fields has been suggested for $\lesssim 10\,$keV electrons in solar flares (e.g., Holman 1985). However, this is only one of several possible mechanisms that have been considered in the context of solar flares (e.g., Melrose 1992, 1994). One context in which acceleration by a parallel electric fields is not controversial is in the polar cap region of pulsars (e.g., Arons 1979; Michel 1991), where very strong electric fields are generated through the rapid rotation of a neutron star with a very strong magnetic field. Althought the details are uncertain, it is widely accepted that the acceleration of highly energetic ($\gtrsim 10^{14}\,$eV) primary particles that initiate an electron-positron pair cascade that populates the polar-cap zones with an outflowing plasma consisting of relativistic ($\sim 10^8\,$eV) pairs.

An important qualitative point concerns the source of the parallel electric fields. Despite the fact that the $E_{\parallel}$ is often attributed to a "potential drop" the fields are not potential fields. The electric field available to accelerate particle is inductive rather than capacitive. That is, the electric field is associated with a changing magnetic field, cf. the discussions by Schindler, Hesse & Birn (1991) in the context of magnetospheric physics, by Haerendel (1994) in the context of solar physics, and by Blandford (1994) in the context of pulsar physics.

3. Nonthermal emission processes

The nonthermal emission processes that are important in plasmas include:
  1) Synchrotron emission
  2) Gyrosynchrotron emission
  3) Bremsstrahlung
  4) Inverse Compton emission
  5) Plasma emission
  6) Electron cyclotron maser emission
  7) Coherent curvature emission
  8) Linear acceleration emission

These processes may be classified as incoherent or coherent. In the fore-going list the first four emission processes are incoherent, and the last four are coherent. The emphasis here is on two coherent emission processes: plasma emission and electron cyclotron maser emission, which are known to be important in space plasmas. The final two coherent emission processes have been invoked for pulsar radio emission (e.g., Melrose 1993).

3.1. INCOHERENT EMISSION PROCESSES

The important distinguishing features between thermal emission, non-thermal incoherent emission and coherent emission are the spectrum, the brightness temperature and the polarization. Thermal emission (in the Rayleigh-Jeans limit) has a spectrum of the form $I(\omega) \propto \omega^2 T$, where $I(\omega)$ is the specific intensity and T is the temperature. Nonthermal emission has a spectrum which is often a power law $I(\omega) \propto \omega^{-\alpha}$ with $\alpha \gtrsim 0$. One may write $I(\omega) \propto \omega^2 T_B(\omega)$, thereby defining the brightness temperature, $T_B(\omega)$. In general, $T_B(\omega)$ is depends on frequency. Besides being a function of frequency, $T_B(\omega)$ is often much greater than the ambient temperature of the plasma.

There is a limit on the maximum brightness temperature for incoherent emission. To every emission process there is a corresponding absorption process, and this limits T_B (in energy units) to less than about the energy of the radiating particles. The rest energy of the electron corresponds to $T_B = 0.6 \times 10^{10}$ K. Hence, sources with $T_B \gg 10^{10}$ K cannot be due to incoherent emission from nonrelativistic or mildly relativistic electrons. Emission by nonrelativistic or mildly relativistic ions could produce a higher T_B than for electrons, but emission by ions is not thought to be important in practice. (This is due to the power radiated in gyroemission and related processes being inversely proportional to the square of the mass of the radiating particle.) Incoherent emission by highly relativistic electrons is dominated by synchrotron emission, which has a limit $T_B \lesssim 10^{12}$ K. As this limit is approached inverse Compton losses due to the relativistic electrons scattering the synchrotron photons become catastrophic (e.g., Blandford 1990).

High brightness temperatures, $T_B \gg 10^{10}$ K, are observed in the following sources: (a) solar type III bursts, with $T_B \lesssim 10^{13}$ K for bursts from the corona, and up to $T_B \sim 10^{16}$ K for bursts in the interplanetary medium, (b) auroral kilometric radiation (AKR) from the Earth and the giant planets, with $T_B \gtrsim 10^{18}$ K for bursts from Jupiter, (c) probably solar spike bursts, with $T_B \gtrsim 10^{18}$ K, (d) bursts from some flare stars, with $T_B \gtrsim 10^{15}$ K in some flares, and (e) radio emission from pulsars with $T_B \gtrsim 10^{25}$ K. For these sources, the high brightness temperature imply form of coherent emission.

Incoherent emission is usually weakly to moderately polarized and coherent emission is expected to be highly polarized. However, there is evidence, notably from solar radio bursts, for depolarization occurring as a propagation effect. Thus, whereas very high polarization ($\sim 100\%$) is taken to be evidence for coherent emission, lower degrees of polarization do not necessarily provide evidence against coherent emission.

3.2. QUALITATIVE DISCUSSION OF PLASMA EMISSION

Plasma emission is any indirect emission process in which (a) the exciting agency generates plasma turbulence which cannot escape directly from the plasma, and (b) this turbulence leads to escaping radiation through some secondary process. Plasma emission occurs in solar radio bursts, of which there are several types, including (e.g., McLean & Labrum 1985) type I bursts, which occur in storms that appear to be associated with ongoing reconnection in nonflaring regions, type II bursts, which are excited by flare-initiated shock waves, and type III bursts, which are due to electron streams accelerated either in flares or in storms. In the following discussion emphasis is given to type III bursts.

The first detailed theory for plasma emission from the solar corona was for type III bursts (Ginzburg & Zheleznyakov 1958). At the time the theory was proposed it was accepted that type III bursts involve emission at the fundamental ($\omega = \omega_p$) and second harmonic ($\omega = 2\omega_p$) of the plasma frequency, and that the emission is excited by a stream of electrons. A variant of their theory is outlined schematically in Figure 9. It consists of three stages: 1) generation of Langmuir turbulence through a streaming instability, 2) production of fundamental plasma emission by scattering of Langmuir waves into transverse waves by plasma particles, and 3) production of second harmonic emission through coalescence of two Langmuir waves to form a transverse wave. Since the theory was originally proposed, the details of each of these stages have been updated several times as ideas on the underlying plasma theory developed over the past three decades. There is now a vast literature on the theory of solar radio bursts, and there is a wide variety of detailed ideas on the specific processes that are important (e.g., Melrose 1991).

In the variant of the theory of plasma emission illustrated schematically in Figure 9 the scattering is attributed to ion sound waves, also called ion acoustic waves. These scattering processes may be described using weak turbulence theory. Qualitatively, the relevant processes involve three-wave interactions in which two waves beat to generate a third

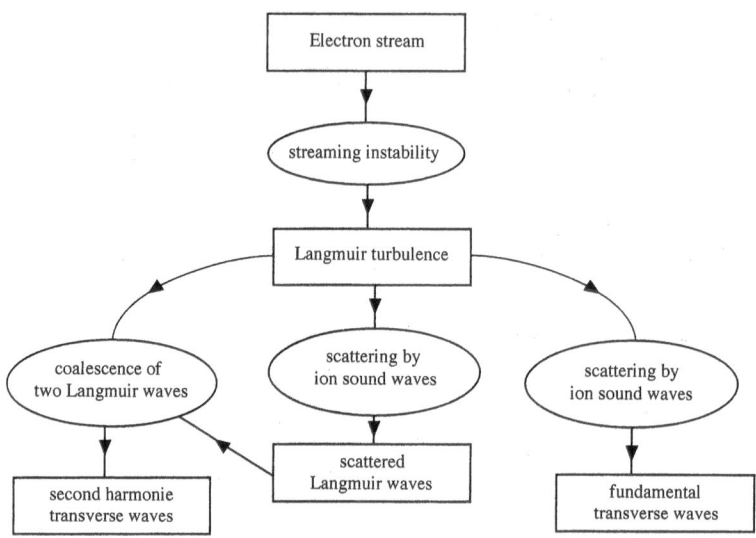

Figure 9. Flow diagram for a variant of the theory of Ginzburg & Zheleznyakov (1958) for the generation of plasma emission. In other variants the processes indicated that involve ion sound waves are replaced by other nonlinear plasma processes.

wave. Let the three initial waves be described by frequencies ω_1, ω_2, ω_3 and wave vectors $\mathbf{k}_1$, $\mathbf{k}_2$, $\mathbf{k}_3$. Then the specific beat process $1+2 \to 3$, in which waves 1 and 2 beat to form wave 3, satisfies the beat conditions

$$\omega_1 + \omega_2 = \omega_3, \quad \mathbf{k}_1 + \mathbf{k}_2 = \mathbf{k}_3. \tag{3.1}$$

One process that can lead to fundamental plasma emission is the beat $L+S \to T$, where L refers to a Langmuir wave, S to an ion sound wave and T to a transverse wave. The process corresponding to scattering of Langmuir waves is the beat $L + S \to L'$, where L' refers to the scattered Langmuir wave. These are wave coalescence processes. The inverse of a coalescence is a decay process. The decay processes $L \to T + S$ and $L \to L' + S$ are qualitatively similar to the corresponding coalescence processes, and lead to a slight frequency downshift, rather than a slight frequency upshift. Second harmonic emission results from the coalescence process $L + L' \to T$.

Each wave must satisfy the relevant dispersion relations. The dispersion relation for an arbitrary wave mode labeled M may be written in the form $\omega = \omega_M(\mathbf{k})$. The dispersion relations are, for Langmuir waves, ion sound waves and transverse waves, respectively,

$$\omega_L(\mathbf{k}) \approx \omega_p + 3k^2 V_e^2/2\omega_p, \quad \omega_S(\mathbf{k}) \approx k v_s, \quad \omega_T(\mathbf{k}) = (\omega_p^2 + k^2 c^2)^{1/2}, \tag{3.2}$$

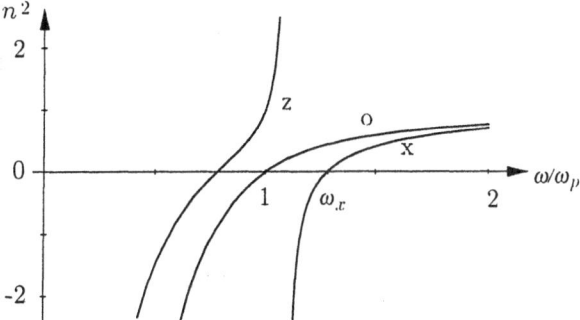

Figure 10. The squares of the refractive indices for magnetoionic modes are plotted for frequencies near the plasma frequency. The three modes are labeled as the o-mode, the x-mode and the z-mode. Langmuir waves may be regarded as thermally modified z-mode waves; their dispersion curve is an extension of the z-mode curve to $n^2 \gg 1$ at $\omega \approx \omega_p$.

where $V_e = (T_e/m_e)^{1/2}$ is the thermal speed of electrons and $v_s \approx V_e/43$ is the ion sound speed. In practice the dispersion relations can place severe restrictions on when specific three-wave interactions are allowed. How the ion sound waves are generated is not clear, but there is evidence that low-frequency turbulence (which may be ion sound waves or lower hybrid waves) is found to be associated with type III events the interplanetary medium (e.g., Lin *et al.* 1986).

There are other possible versions of the plasma emission processes. Of particular interest from a formal viewpoint is the possibility that strong-turbulence effects might play a role. This possibility is discussed separately below.

3.3. POLARIZATION OF PLASMA EMISSION

Simple theory predicts that the polarization of fundamental plasma emission should be 100% in the sense of the o-mode. The argument is as follows. The refractive indices of the ordinary and extraordinary wave modes in a cold plasma are illustrated schematically in Figure 10. The cutoff frequency for the x-mode is at $\omega = \omega_x$, cf. (1.1), with $\omega_x \approx \omega_p + \frac{1}{2}\Omega_e$ for $\omega_p \gg \frac{1}{2}\Omega_e$. Hence, in the range $\omega_p < \omega < \omega_x$ the only radiation that can escape is in the o-mode. Langmuir waves correspond to a resonance in the z-mode, cf. Figure 10. An important qualitative point is that the frequency of the Langmuir waves generated by type III electron streams is below ω_x. Specifically, for Langmuir waves the thermal correction in (3.2) to the plasma frequency is $3k^2V_e^2/2\omega_p \approx (3V_e^2/2v_b^2)\omega_p$, where v_b is the beam speed. This correction is relatively small, and in practice is less than the difference $\frac{1}{2}\Omega_e$ between ω_x and ω_p. The change in frequency due to the ion sound waves in a coalescence process is also small compared with $\frac{1}{2}\Omega_e$, and hence the beat frequency cannot exceed ω_x. Hence, it should be impossible to generated x-mode radiation in fundamental plasma emission, which then should be polarized 100% in the o-mode.

Qualitatively, this prediction is supported by the observations in vhat fundamental plasma emission is polarized in the sense of the o-mode. However, for type III bursts (and also for type II bursts) the polarization is never 100%. In contrast, for type I bursts the polarization is typically close to 100%. The degree of polarization is a maximum for bursts near the central meridian, and decreases toward the solar limb, with some bursts being unpolarized. This suggests that there is a depolarizing process that is more effective for type II and type III bursts than for type I bursts, and which becomes more effective as sources approach the solar limb. There is no entirely satisfactory explanation for this depolarization. Thus, as stated above, while a very high degree of polarization is usually indicative of a coherent emission process, the absence of very high polarization is not necessarily an argument against an emission process being coherent.

3.4. GROWTH OF THE LANGMUIR WAVES

The growth of Langmuir waves is attributed to an instability driven by a beam of electrons. In principle the instability may be either kinetic (random wave growth) or reactive (phase coherent growth). For type III electron streams the kinetic version appears to be appropriate. The kinetic instability is amenable to a relatively simple treatment in a one-dimensional model, where 'one-dimensional' refers to a three-dimensional theory in which the only Langmuir waves considered have $\mathbf{k}$ along the direction of the beam and the particle distribution function is integrated over the two momentum components orthogonal to the beam direction. The evolution is described by a pair of (quasilinear) equations. The wave growth is described by

$$\frac{dW(v_\phi)}{dt} = -\gamma_{\mathrm{L}}(v_\phi)W(v_\phi), \qquad \gamma_{\mathrm{L}}(v_\phi) = -\frac{\pi\omega_p}{n_e}v_\phi^2\frac{dF(v_\phi)}{dv_\phi}, \qquad (3.3)$$

where $v_\phi \approx \omega_p/k$ is the phase speed of the Langmuir waves and $W(v_\phi)dv_\phi$ is the energy density in the Langmuir waves in the range v_ϕ to $v_\phi+dv_\phi$. The evolution of the distribution of electrons is described by

$$\frac{dF(v)}{dt} = \frac{\partial}{\partial v}D(v)\frac{\partial F(v)}{\partial v}, \qquad D(v) = \frac{\pi\omega_p}{n_e m_e}vW(v). \qquad (3.4)$$

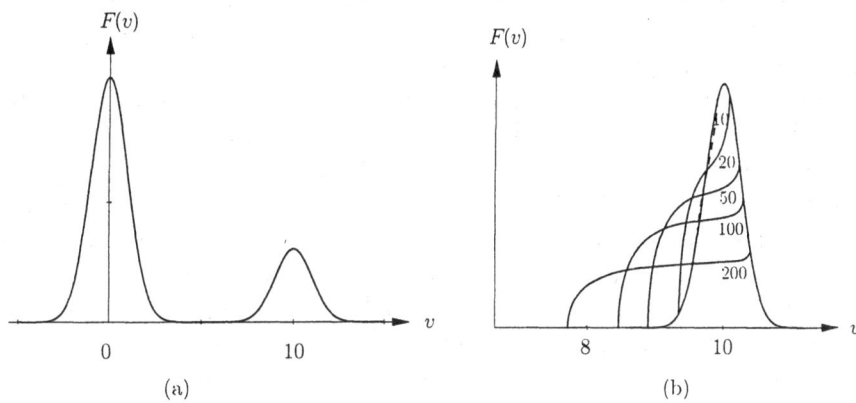

Figure 11. The evolution of the bump-in-tail instability is illustrated. (a) The initial distribution causes waves to grow for $v_\phi = v$ in the range where $dF(v)/dv$ is positive. (b) As the growth proceeds, the bump in the tail is eaten away to form a plateau $dF(v)/dv \approx 0$ that extends to lower v as time (denoted by numbers in units of the initial growth time) evolves. [After Grognard (1975, 1985).]

3.5. QUASILINEAR EVOLUTION

The quasilinear evolution of the distribution of electrons is illustrated in Figure 11. Waves with $v_\phi = v$ grow only on the rising portion of the distribution, with $dF(v)/dv > 0$. The back reaction to the wave growth eats away the bump in the distribution function, reducing $dF(v)/dv$ and tending to form a plateau, $dF(v)/dv \approx 0$, that extends to lower v as time evolves.

Quasilinear evolution tends to form a plateau distribution, and this involves a substantial energy loss by the stream to the Langmuir waves in ~ 100 growth times. However, type III streams are known to propagate to the orbit of the Earth and beyond without slowing down significantly. One is forced to conclude that the energy loss to Langmuir waves is much smaller than simple theory suggests, thus posing a problem as to how the streams avoid catastrophic energy losses. One idea on how this difficulty might be overcome is due to an interplay between two opposing effects. One effect is "fractionization": faster electrons outpace slower electrons, which tends to increase $dF(v)/dv$. Specifically, in the absence of quasilinear evolution, at a fixed point the first particles to arrive are necessarily the fastest particles, implying $dF(v)/dv > 0$ below a peak in the distribution, with the peak moving to lower v as time evolves. The second effect is quasilinear evolution, tending to reduce $dF(v)/dv$. Grognard (1984, 1985) developed a numerical code

to follow the evolution of the distribution function due to these two
effects. He found that a characteristic distribution function results from
this interplay between the two effects. Further, Grognard showed that
measured distribution functions for the type III electrons in the interplanetary medium are consistent with this characteristic distribution
function.

The apparent success of this simple theory of Grognard (1984) is
surprising. because there are several reasons why one would not expect
the one-dimensional quasilinear equations to be provide a valid model
for the evolution of type III streams. In particular, one expects the distribution of Langmuir waves to evolve due to wave-wave processes, as
discussed below, on a faster timescale than the "fractionization" occurs,
thereby invalidating the model. Another reason is that the observed
Langmuir waves are very strongly localized into "clumps" (e.g., Lin *et
al.* 1981), whereas the theory is based on the assumption that the system is locally homogeneous. However, it can be shown that the quasilinear equations continue to apply even for such an inhomogeneous
distribution of Langmuir waves (Melrose & Cramer 1989).

3.6. SATURATION OF THE BEAM INSTABILITY

The beam instability saturates as the distribution function becomes
sufficiently distorted that the growth rate is reduced substantially from
its initial value. Semi-quantitatively, saturation occurs when the energy density in the Langmuir waves becomes a significant fraction of
the initial energy density in the electron beam. A rough estimate of
the effective temperature of the Langmuir waves at saturation may be
made as follows. For simplicity assume that the saturation level of the
Langmuir waves is approximately constant, $W(v_\phi) = W_0$ say, for waves
in a range $v_b - \Delta v < v_\phi < v_b$ and zero otherwise, where v_b is the beam
speed. If the energy density of these waves is a fraction ζ of the initial
energy density in the beam, then we have

$$W_0 = \zeta n_1 m v_b^2 / 2\Delta v, \qquad (3.5)$$

where n_1 is the number density of electrons in the beam. Further assume
that the Langmuir waves are confined to a range $\Delta\Omega$ of solid angle. The
range of solid angles $\Delta\Omega$ can be estimated only in terms of a three-
dimensional theory for the instability. One simple estimate is to write
$\Delta\Omega = \pi(\Delta\theta)^2$, with $\Delta\theta = \Delta v/v_b$. Then the effective temperature T_L is
related to W_0 by, for $\Delta v \ll v_b$,

$$T_L = \frac{(2\pi)^3 v_b^4}{\omega_p^3 \Delta\Omega} W_0 = \frac{\zeta(2\pi)^2 n_1 m v_b^8}{\omega_p^3 (\Delta v)^3}. \qquad (3.6)$$

3.7. Weak Turbulence Theory

In weak turbulence theory an expansion of the basic equations is made in powers of the wave amplitude and the waves are treated in the random phase approximation. Consider waves with wavevector $\mathbf{k}$ in a wave mode M with dispersion relation $\omega = \omega_M(\mathbf{k})$. A distribution of such waves may be regarded as a collection of wave quanta with energy $\hbar\omega$ and momentum $\hbar\mathbf{k}$ and occupation number $N_M(\mathbf{k})$. The occupation number is a dimensionless quantity that corresponds to an energy $T_M(\mathbf{k})d^3\mathbf{k}/(2\pi)^3$ in the range $d^3\mathbf{k}/(2\pi)^3$, where

$$T_M(\mathbf{k}) = \hbar\omega_M(\mathbf{k})N_M(\mathbf{k}) \qquad (3.7)$$

is the effective temperature of the waves (in energy units). In a time-independent medium the transfer of radiation has the simple form, in the absence of emission or absorption, $N_M(\mathbf{k}) = $ constant or $T_M(\mathbf{k}) = $ constant along a ray, corresponding to conservation of wave quanta. In principle, measurement of the brightness temperature of radiation at the Earth provides direct information on the effective temperature of the radiation as it leaves the source, provided that the radiation is not scattered or absorbed significantly along the ray path from the source to the observer.

The kinetic equation for the waves in the mode M due to a three-wave process $L+S \leftrightarrow M$, with $M = T$ or L', may be written in terms of the probabilities, $u^{\pm}_{MLS}$, for the two three-wave processes, $L \pm S \to M$. The actual form of $u^{\pm}_{MLS}$ is given in Appendix B, but is not important in the following discussion. The two probabilities differ only in the sign in the beat conditions $\omega_L \pm \omega_S = \omega_M$, $\mathbf{k}_L \pm \mathbf{k}_S = \mathbf{k}_M$, which appear as the arguments of δ-functions in the probability. The kinetic equation is

$$\frac{dN_M(\mathbf{k})}{dt} = \sum_{\pm} \int \frac{d^3\mathbf{k}'}{(2\pi)^3} \frac{d^3\mathbf{k}''}{(2\pi)^3} u^{\pm}_{MLS}(\mathbf{k}, \mathbf{k}', \mathbf{k}'')$$
$$\times \{ N_L(\mathbf{k}')N_S(\mathbf{k}'') - N_M(\mathbf{k})[N_S(\mathbf{k}'') \pm N_L(\mathbf{k}')] \}, \qquad (3.8)$$

where the sum is over the two processes $L \pm S \to M$.

3.8. Saturation Model for Plasma Emission

A quantitative treatment of plasma emission based on the foregoing equations is cumbersome even when simplifying assumptions are made. However, there is one limiting case that may be treated simply and which may be understood in terms of simple physical arguments. This is the limit in which the three wave interaction saturates. Imagine the processes $L \pm S \to M$ with $N_S \gg N_L$ building up the level of waves in the mode M. If initially one has $N_M \ll N_L$ then the combination of terms inside the braces in (3.8) may be approximated by $N_L N_S$. As N_M increases the quantity in braces in (3.8) decreases, primarily due to the term $N_M N_S \gg N_M N_L$ increasing as N_M increases, and also because N_L decreases as N_M increases. The three-wave interaction is said to saturate when the quantity in braces approaches zero and the interaction ceases to cause N_M to increase. Such saturation occurs for

$$N_M = \frac{N_L N_S}{N_S \pm N_L} \approx N_L. \qquad (3.9)$$

In particular, for fundamental emission due to scattering by ion sound waves with $N_S \gg N_L$, the process saturates at $N_T = N_L$, and as the frequencies of the fundamental transverse and Langmuir waves are nearly equal, this saturation level corresponds to $T_T = T_L$. Saturation of the two stages in second harmonic emission leads to its saturation at the same level, at least to within a factor of two. Thus despite the highly nonthermal nature of these processes, this saturation level is what one might expect from a thermodynamic-type argument.

A saturation model for plasma emission seems to account for the observed brightness temperatures for type III bursts in the interplanetary medium (Melrose 1989). As implied by (3.9), if the weak turbulence processes saturate due to processes involving ion sound (or other appropriate low frequency) turbulence, then all secondary wave distributions tend to saturate at the same effective temperature T_L, at least to within a factor of two or so. Thus if the plasma emission processes saturate, then the observed brightness temperature should be of order T_L, and (3.6) then implies a predicted functional relation between the properties of the beam of electrons and the brightness of the observed plasma emission. Such a prediction is roughly consistent with the relatively small range of brightness temperatures, $\sim 10^{15}\,\mathrm{K}$, observed for type III bursts in the interplanetary medium.

It may be concluded that saturation models for plasma emission, despite the simple and almost naïve assumptions made, seem to provide a reasonable basis for a semiquantitative treatment of plasma emission.

3.9. STRONG TURBULENCE AND PLASMA EMISSION

Strong turbulence effects, by definition, cannot be treated using weak turbulence theory. Various strong turbulence effects that are relevant to plasma emission include parametric instabilities, modulational instabilities and Langmuir collapse (e.g., Goldman 1984). The growth of Langmuir turbulence is very fast, and an important question is what limits this growth. Possibilities include saturation of the instability as the free energy is exhausted, as described by the quasilinear equations, and various nonlinear saturation mechanisms that involve strong-turbulence effects. The data do not provide a clear indication of which processes are involved. What is clear is that the instability develops strongly in local regions, leading to clumps of Langmuir turbulence, with a relatively low level of Langmuir turbulence between the clumps.

Qualitatively, in the case of a three-wave interaction important differences between weak (random phase) and strong (phase-coherent) cases relate to the required initial level of the waves, and the nature of the subsequent evolution of the wave distributions. In weak turbulence theory a three-wave process causes the turbulence to evolve only when two excited wave distributions beat together to generate the third wave distribution. Both these two wave distributions must be present to drive the three-wave interaction, and the evolution is a monotonic one from the initial values toward the saturation values. In the phase-coherent case, only one wave distribution needs to be highly excited initially. This distribution acts as a pump that generates the other two wave distributions. The system then typically evolves in a periodic or quasiperiodic manner, with energy flows from the pump to the daughter wave, and then back to the initial pump. A variety of modulation and parametric instabilities have been explored in connection with type III bursts. For example, one such instability was suggested by Chian & Alves (1988) as a mechanism for the generation of fundamental plasma emission.

Langmuir collapse is related to self-focusing of laser light in nonlinear optics: a uniform distribution of turbulence breaks up into localized regions in which the energy density in the turbulence increases to very high values on a very short timescale. Radiation during Langmuir collapse is usually treated as a weak turbulence process that has no significant effect on the collapse. The role that strong turbulence effects may play remains unclear because of the paucity of definitive observational tests for specific theories.

Some progress has been made in developing a semiquantitative theory that includes the clumpiness of the growth of the Langmuir turbulence and some strong turbulence effects in the emission. In particular, Robinson & Cairns (1993) have shown that a *stochastic growth theory* can account satisfactorily for the volume emissivities and brightness temperatures of type III events in the solar wind.

3.10. ELECTRON CYCLOTRON MASER EMISSION

The absorption corresponding to electron cyclotron emiss on can be negative, leading to electron cyclotron maser emission (ECME). The most favorable case for negative absorption is for x-mode waves at $\theta \approx \pi/2$ at the fundamental $s = 1$ in a plasma with $\omega_p \ll \Omega_e$. The required source of free energy is a distribution of electrons with $\partial f/\partial p_\perp > 0$ at small $p_\perp$. The discussion of electron cyclotron maser emission here is centered around explaining why this case is the most favorable, and outlining relevant applications.

An important preliminary point is that ECME strongly favors the x-mode. The reason is that the electric vector in the x-mode rotates in the same sense as electrons gyrate, providing a strong coupling between the x-mode and gyrating electrons. The electric vector in the o-mode rotates in the opposite sense, and the coupling between the o-mode and gyrating electrons is much weaker than for the x-mode.

Historically, Twiss (1958) was the first to recognize that ECME is possible in principle. However, Twiss restricted his attention to a special case in which there is no Doppler shift ($k_\| v_\| = 0$) in the resonance condition (2.1), which is equivalent to assuming perpendicular propagation ($k_\| = 0$ or $\theta = \pi/2$). Twiss noted that ECME can be driven by a distribution with $\partial f(\mathbf{p})/\partial p_\perp > 0$. However, with $k_\| v_\| = 0$, (2.1) implies $\omega = \Omega_e/\gamma < \Omega_e < \omega_x$, so that any emission at $s = 1$ in the x-mode cannot escape from the plasma. In the case considered by Twiss, emission in the o-mode is identically zero, and hence Twiss' theory of ECME is not applicable to astrophysical phenomena. An alternative form of ECME was developed by Melrose (1976), who made the non-relativistic approximation, and required $k_\| v_\| > 0$ to Doppler shift the emission to above ω_x, so that generation of escaping x-mode emission is possible. This upshift is possible only for $\Omega_e \gg \omega_p$, requiring that the emission occur in a low density plasma, this requirement led to a successful prediction that $\omega_p \ll \Omega_e$ must be satisfied in the source region for AKR. However, this form of ECME can be driven only by a gradient in $p_\|$ (cf. Melrose 1986, p. 191 for an explanation), and it was subsequently found that the distribution of electrons associated with AKR does not satisfy the necessary conditions. A further alternative theory was proposed by Wu & Lee (1979), and this is now the favored version of ECME. This theory is similar to Twiss' theory in that the driving term is a $p_\perp$ gradient, but with the Doppler shift retained to boost the emission frequency (for the x-mode at $s = 1$) to above ω_x.

3.11. THE RESONANCE ELLIPSE

The Doppler condition (2.1) may be interpreted as determining the velocities ($v_\perp$ and $v_\|$) of all the particles that resonate with a given wave (given ω and $k_\|$) at a given harmonic s. If the resonance condition is plotted in velocity space ($v_\perp$-$v_\|$ space), for fixed s, ω, θ, it defines an

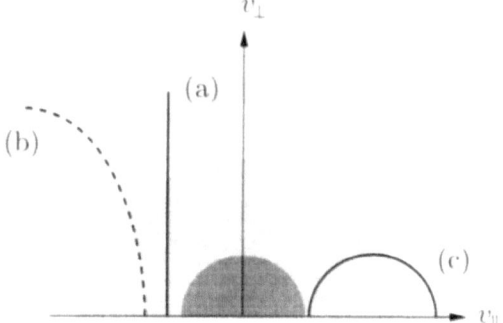

Figure 12. Examples of resonance ellipses in the non-relativistic region of velocity space. Thermal electrons occupy the darkly shaded region, and nonthermal electrons occupy the lightly shaded region. Curves (a) corresponds to the nonrelativistic approximation, (b) to an intermediate case and (c) to nearly perpendicular propagation.

ellipse, called a *resonance ellipse*. Some examples of resonance ellipses are illustrated in Figure 12. In the special case of perpendicular propagation, relevant to Twiss' (1958) version of ECME, the ellipse reduces to a circle centered on the origin and of radius $v/c = (1 - \omega^2/s^2\Omega_e^2)^{1/2}$. In the nonrelativistic case, relevant to parallel-driven ECME, the ellipse reduces to a vertical line in the nonrelativistic region of velocity space, cf. curve (a) in Figure 12. In the case considered by Wu & Lee (1979), a *semirelativistic* approximation to the resonance condition is relevant, and it corresponds to replacing $\Omega = \Omega_e/\gamma$ by $\Omega_e(1 - v^2/2c^2)$ in (2.1). Thus the resonance condition in the semirelativistic approximation is

$$\omega - s\Omega_e(1 - \tfrac{1}{2}v_\perp^2/c^2 - \tfrac{1}{2}v_\parallel^2/c^2) - \omega[1 - n_M(\omega, \theta)(v_\parallel/c)\cos\theta] = 0, \quad (3.10)$$

where n_M is the refractive index for mode M. This corresponds to a resonance ellipse that is a circle with its center displaced from the origin, cf. curve (c) in Figure 12.

3.12. PERPENDICULAR-DRIVEN ECME

The effect of a distribution of electrons in causing waves at given ω and θ to damp or grow is determine by an integral around the appropriate resonance ellipse. To see this, first note the general form of the gyromagnetic absorption coefficient, cf. (A.3),

$$\gamma_M(\mathbf{k}) = - \sum_{s=-\infty}^{\infty} 2\pi \int_{-\infty}^{\infty} dp_\parallel \int_0^{\infty} dp_\perp p_\perp$$

$$\times w_M(\mathbf{p}, \mathbf{k}, s)\hbar \left(\frac{s\Omega}{v_\perp} \frac{\partial}{\partial p_\perp} + k_\parallel \frac{\partial}{\partial p_\parallel} \right) f(p_\parallel, p_\perp). \quad (3.11)$$

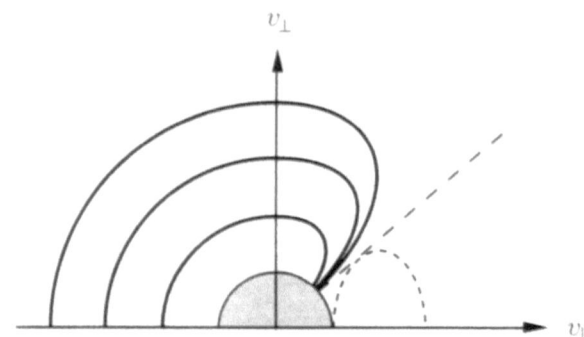

Figure 13. A loss-cone distribution with loss-cone angle indicated by the dashed line. The resonance ellipse illustrated favors maser action.

One of the integrals in (3.11) can be performed over the δ-function, and the other can then be expressed and an integral around the resonance ellipse. There are two possible driving terms for ECME in (3.11), the $p_\perp$ gradient and the $p_\parallel$ gradient. The semi-relativistic approximation (3.10) defines an ellipse that is close to the origin (where nonrelativistic particles are located) only for nearly perpendicular propagation, $\theta \approx \pi/2$, and then $k_\parallel$ is intrinsically small. In this case only the $p_\perp$-gradient term in (3.11) is significant.

Varying ω and θ causes the resonance ellipse to change in size and causes its center to move along the $v_\parallel$-axis. Even small changes in ω and θ alter the ellipse significantly from its optimum shape and location. As a consequence, ECME is confined to a narrow range of frequencies and to a narrow range of angles. The restrictions on the angular distribution of the radkation imply emission on the surface of a hollow cone with a half-angle $\theta \sim \arccos(v/c)$, where v is the typical speed of the electrons that cause the wave growth.

3.13. REQUIREMENTS FOR ECME

There are two important requirements that need to be satisfied for ECME to be a source of escaping radiation.

One requirement is that there be a distribution of electrons with available free energy in the appropriate form, specifically with $\partial f / \partial p_\perp > 0$ at small $p_\perp$. A favorable case is a loss-cone distribution, in which $f(\mathbf{p})$ is zero (or very small) inside the loss cone, $\alpha < \alpha_0$ say, and is nonzero outside the loss cone. Such a distribution has $f(\mathbf{p}) = 0$ at $p_\perp = 0$, with $f(\mathbf{p})$ increasing as a function of $p_\perp$ up to a maximum on the loss cone, as illustrated in Figure 13. Consider a resonance ellipse that is entirely within the loss cone. Then $\partial f / \partial p_\perp$ is positive everywhere around the resonance ellipse, and hence the absorption coefficient (3.11) is necessarily negative, implying wave growth. The circle (c) in Figure 12 illus-

trates this case; the circle is chosen so that it lies just outside the region
where the thermal electrons are located so that thermal gyromagnetic
absorption is absent (this resonance ellipse does not pass through the
region where the thermal electrons are located, and so no thermal electrons can resonate with waves with this ω, θ). Loss cones form naturally
in a magnetic trap where electrons are confined due to the magnetic
mirror effect. Thus this first requirement is expected to be satisfied for
magnetically trapped electrons under a wide variety of circumstances.

The other requirement is that there be waves with the dispersion
properties needed (i) to define the most favorable resonant ellipse, as
in the case illustrated in Figure 13, and (ii) to allow direct escape from
the plasma. The most favorable case for emission is at $s = 1$ in the
x-mode above the cutoff frequency at $\omega = \omega_x$, cf. Figure 10. Negative
absorption at $s = 1$ for radiation in the x-mode that can escape occurs
when $|\theta - \pi/2|$ is small but nonzero.

The evaluation of the absorption coefficient for ECME needs to be
performed numerically in practice. By making appropriate simplifying
assumptions one can estimate the maximum growth rate (the modulus
of the maximum negative value of the absorption coefficient). For a
loss-cone distribution of suprathermal electrons with mean speed $\langle v \rangle$
and loss-cone angle α_0, the maximum growth is estimated to be

$$|\gamma_x| \approx \frac{\pi \omega_p^2}{\Omega_e} \frac{n_1}{n_e} \frac{c^2}{\langle v \rangle^2 \sin \alpha_0}, \qquad (3.12)$$

where n_1 is the number density that would be required to fill the loss
cone.

3.14. APPLICATIONS OF ECME

ECME is the accepted emission mechanism for the Earth's auroral kilometric radiation (AKR), Jupiter's decametric radio emission (DAM)
and analogous radiation from the other giant planets, and is favored
for solar spike bursts and very bright emission from flare stars.

AKR consists of bursts of cyclotron emission from regions above
the auroral zones of the Earth. AKR correlates with inverted-V electron events that occur in a density cavity in the auroral zone. The
inverted-V electrons are accelerated downward by a potential drop of
several kilovolts along the magnetic field lines. The acceleration of the
electrons depopulates the auroral zone, thereby producing the cavity
and the necessary condition $\omega_p \ll \Omega_e$ for ECME to be possible. The
distribution of electrons has a loss-cone feature that could drive the
ECME. One simple explanation for the formation of the loss-cone distribution is as follows (Wu & Lee 1979). Below the acceleration region

conservation of the adiabatic invariant $v^2 \sin^2 \alpha / B$ implies that $\sin^2 \alpha$ increases as B increases leading to the magnetic mirror effect. Electrons with sufficiently large initial pitch angle, α, reflect and electrons with smaller α precipitate into the atmosphere. This leaves a deficiency of reflected electrons with small α, implying an upward directed loss cone anisotropy, as illustrated in Figure 13. However, although there is some qualitative support for this model in the measured distribution functions, there are other features in the distribution function that could be relevant to ECME. One is the presence of a distribution that is "trapped" by the magnetic mirror on the lower side, and by the electric potential that accelerates the electrons downward on the upper side (Louarn et al. 1990). At present it is not clear which feature in the distribution of inverted-V electrons causes the emission of AKR. This is because the time required to measure the distribution function is long compared with the timescale over which the development of the instability should modify the distribution. Thus, to identify the feature that drives the instability observationally, a faster rate for measuring the distribution function is required.

Jupiter's DAM is also attributed to ECME. The ECME theory provides a natural explanation for the bizarre radiation pattern for DAM, in which the emission is confined to a narrow range of angles, $\sim 1°$, on a cone with a half-angle $\sim 80°$ (Dulk 1967).

ECME is also the favored mechanism for solar spike bursts (e.g., Benz 1986), which are one special class of solar radio burst, and for very bright radio emission from some flare stars. However, there is a serious problem in the application of ECME to the Sun and stars: simple estimates suggest that the radiation cannot escape because it is strongly absorbed at the second harmonic layer, where ω is locally equal to $2\Omega_e$. How this difficulty might be overcome is unclear. One suggestion is that the ECME is generated at a higher harmonic, $s \geq 2$, or in the o-mode, so that the strong absorption of the x-mode at the second harmonic layer is avoided. However, the growth rate is largest for $s = 1$ in the x-mode, and a more slowly growing instability can dominate the emission only when the instability at $s = 1$ in the x-mode is suppressed. Another suggestion involves 'windows' through which the radiation can escape. The simplest is a window at small angles of propagation where the absorption coefficient at $s = 2$ varies as $\sin^2 \theta$. Another window near $\theta = \pi/2$ exists for the o-mode; partial conversion of x-mode radiation into o-mode radiation as this window is approached can allow escape of the resulting o-mode component (Robinson 1989). Yet another suggestion is that the radiation incident on the absorbing layer from below with $\omega < 2\Omega_e$ can modify the distribution function of the electrons there, allowing re-emission at $\omega > 2\Omega_e$ (McKean, Winglee & Dulk

1989). These latter two suggestions imply very inefficient escape, and if either of them is the correct explanation, an implication is that there must be a very large unobserved power in ECME generated in flares. Until this problem with the escape of ECME is understood, a serious doubt remains about the viability of ECME for spike bursts and for radio emission from stellar corona.

Appendix

A. Kinetic Theory for Particle-Wave Interaction

It is convenient to adopt a semiclassical description in which the waves are regarded as a collection of wave quanta with energy $\hbar\omega$, momentum $\hbar\mathbf{k}$ and occupation number $N(\mathbf{k})$. A wave in a particular mode, M, is described by its dispersion relation $\omega = \omega_M(\mathbf{k})$, its unimodular polarization vector, $\mathbf{e}_M(\mathbf{k})$, and the ratio, $R_M(\mathbf{k})$, of electric to total energy in the waves, all of which are found by solving the relevant wave equation (e.g., Melrose & McPhedran 1991).

The interaction between a particle, with charge q mass m and momentum components $p_\parallel$ and $p_\perp$ in a uniform magnetic field $\mathbf{B}$, and a wave in the mode M may be described by the probability per unit time that the particle emit a wave in the range $d^3\mathbf{k}/(2\pi)^3$. The probability of emission at the sth harmonic is

$$w_M(\mathbf{p}, \mathbf{k}, s) = \frac{2\pi q^2 R_M(\mathbf{k})}{\varepsilon_0 \hbar \omega_M(\mathbf{k})} \left| \mathbf{e}_M^*(\mathbf{k}) \cdot \mathbf{V}(\mathbf{k}, \mathbf{p}; s) \right|^2 \delta(\omega_M(\mathbf{k}) - s\Omega - k_\parallel v_\parallel),$$

$$\mathbf{V}(\mathbf{k}, \mathbf{p}; s) = \left(v_\perp \frac{s}{k_\perp R} J_s(k_\perp R), -i\eta v_\perp J_s'(k_\perp R), v_\parallel J_s(k_\perp R) \right),$$

$$\Omega = \frac{\Omega_0}{\gamma}, \quad \Omega_0 = \frac{|q|B}{m}, \quad \eta = \frac{q}{|q|}, \quad R = \frac{v_\perp}{\Omega} = \frac{p_\perp}{|q|B}. \quad (A.1)$$

This probability contains the Doppler condition (2.1) through the δ-function.

A.1. The Absorption Coefficient

The kinetic equation for the waves is

$$\frac{dN_M(\mathbf{k})}{dt} = \alpha_M(\mathbf{k}) - \gamma_M(\mathbf{k}) N_M(\mathbf{k}), \quad\quad\quad (A.2)$$

where the $\alpha_M(\mathbf{k})$ is an emission coefficient that describes the effect of spontaneous emission, and the other term defines the absorption coefficient, $\gamma_M(\mathbf{k})$. The absorption coefficient reduces to

$$\gamma_M(\mathbf{k}) = - \sum_{s=-\infty}^{\infty} \int d^3\mathbf{p} \, w_M(\mathbf{p}, \mathbf{k}, s) \hat{D}_s f(\mathbf{p}). \qquad (A.3)$$

The differential operator $\hat{D}_s$ may be rewritten in terms of either cylindrical polar coordinates, or spherical polar coordinates, p, α in momentum space, with $p_\parallel = p \cos\alpha$, $p_\perp = p \sin\alpha$:

$$\hat{D}_s = \hbar \left(\frac{s\Omega}{v_\perp} \frac{\partial}{\partial p_\perp} - k_\parallel \frac{\partial}{\partial p_\parallel} \right) = \frac{\hbar\omega}{v} \left(\frac{\partial}{\partial p} + \frac{\cos\alpha - n_M(v/c)\cos\theta}{p\sin\alpha} \frac{\partial}{\partial\alpha} \right).$$
$$(A.4)$$

A simple way of deriving the coefficients in (A.4) is to note that one has $\hat{D}_s = (\Delta p)\partial/\partial p + (\Delta\alpha)\partial/\partial\alpha$, with $\Delta p = \hat{D}_s p$, $\Delta\alpha = \hat{D}_s\alpha$.

Cyclotron emission and absorption may be treated in an approximate way by noting that the argument of the Bessel functions in (A.1) is small. The power series expansion

$$J_\nu(z) = \sum_n \frac{(-)^n}{n!\Gamma(\nu + n + 1)} \left(\frac{z}{2} \right)^{2n+\nu} \qquad (A.5)$$

then converges rapidly and only the leading term need be retained.

The quasilinear diffusion equation for magnetized particles is

$$\frac{df(\mathbf{p})}{dt} = \frac{1}{p_\perp} \frac{\partial}{\partial p_\perp} \left\{ p_\perp \left[D_{\perp\perp}(\mathbf{p}) \frac{\partial}{\partial p_\perp} + D_{\perp\parallel}(\mathbf{p}) \frac{\partial}{\partial p_\parallel} \right] f(\mathbf{p}) \right\}$$

$$+ \frac{\partial}{\partial p_\parallel} \left\{ \left[D_{\parallel\perp}(\mathbf{p}) \frac{\partial}{\partial p_\perp} + D_{\parallel\parallel}(\mathbf{p}) \frac{\partial}{\partial p_\parallel} \right] f(\mathbf{p}) \right\}$$

$$= \frac{1}{\sin\alpha} \frac{\partial}{\partial\alpha} \left\{ \sin\alpha \left[D_{\alpha\alpha}(\mathbf{p}) \frac{\partial}{\partial\alpha} + D_{\alpha p}(\mathbf{p}) \frac{\partial}{\partial p} \right] f(\mathbf{p}) \right\}$$

$$+ \frac{1}{p^2} \frac{\partial}{\partial p} \left\{ p^2 \left[D_{p\alpha}(\mathbf{p}) \frac{\partial}{\partial\alpha} + D_{pp}(\mathbf{p}) \frac{\partial}{\partial p} \right] f(\mathbf{p}) \right\}. \qquad (A.6)$$

Explicit expressions for the diffusion coefficients are given by

$$D_{QQ'}(\mathbf{p}) = \sum_{s=-\infty}^{\infty} \int \frac{d^3\mathbf{k}}{(2\pi)^3} w_M(\mathbf{p}, \mathbf{k}, s) \, \Delta Q \, \Delta Q' \, N_M(\mathbf{k}), \quad (A.7)$$

with Q, Q' identified as $p_\perp$, $p_\parallel$, p, or α, and with $\Delta Q = \hat{D}_s Q$. One finds

$$\Delta p_\perp = \frac{s\Omega}{v_\perp}, \quad \Delta p_\parallel = \hbar k_\parallel, \quad \Delta\alpha = \frac{\hbar(\omega\cos\alpha - k_\parallel v)}{pv\sin\alpha}, \quad \Delta p = \frac{\hbar\omega}{v}.$$
$$(A.8)$$

Appendix

A. Kinetic Theory for Three-Wave Interactions

The kinetic equation for the three-wave interactions $L + S \leftrightarrow M$ (upper sign), and $L \leftrightarrow M + S$ (lower sign), are

$$
\frac{dN_M(\mathbf{k})}{dt} = \int \frac{d^3k'}{(2\pi)^3} \frac{d^3k''}{(2\pi)^3} u^{\pm}_{MLS}(\mathbf{k}, \mathbf{k}', \mathbf{k}'')
$$
$$
\times \{N_L(\mathbf{k}')N_S(\mathbf{k}'') - N_M(\mathbf{k})[N_S(\mathbf{k}'') \pm N_L(\mathbf{k}')]\}. \qquad (B.1)
$$

The two probabilities $u^{\pm}_{MLS}$ differ only in the sign in the beat conditions $\omega_L \pm \omega_S = \omega_M$, $\mathbf{k}_L \pm \mathbf{k}_S = \mathbf{k}_M$, which appear as the arguments of δ-functions in the following expression:

$$
u^{\pm}_{MPS}(\mathbf{k}, \mathbf{k}', \mathbf{k}'') = \frac{e^2 \hbar \omega_p^2}{2\varepsilon_0 m_e^2 V_e^2} \frac{\omega_S(\mathbf{k}'') R_M(\mathbf{k}) R_P(\mathbf{k}')}{\omega_M(\mathbf{k})\omega_P(\mathbf{k}')}
$$
$$
\times |\mathbf{e}_M^*(\mathbf{k}) \cdot \mathbf{e}_P(\mathbf{k}')|^2 (2\pi)^4 \delta^3(\mathbf{k} - \mathbf{k}' \mp \mathbf{k}'') \, \delta(\omega_M(\mathbf{k}) - \omega_P(\mathbf{k}') \mp \omega_S(\mathbf{k}'')), \qquad (B.2)
$$

with $P = L$ for Langmuir waves. The beat conditions (3.1) play the role of a resonance condition and are contained in the δ-functions in (B.2).

For transverse waves the factor involving the polarization vectors is summed over the two final states of polarization or averaged over the initial states of polarization. The various possible cases then give

$$
|\mathbf{e}_M^*(\mathbf{k}) \cdot \mathbf{e}_P(\mathbf{k}')|^2 = \begin{cases} |\kappa_L \cdot \kappa_{L'}|^2 & M = L, \, P = L', \\ |\kappa_L \times \kappa_T|^2 & M = L, \, P = T, \\ \frac{1}{2}|\kappa_T \times \kappa_L|^2 & M = T, \, P = L, \\ \frac{1}{2}(1 + |\kappa_L \cdot \kappa_{L'}|^2) & M = T, \, P = T'. \end{cases} \qquad (B.3)
$$

The kinetic equation for the generation of second harmonic emission due to the process $L + L' \rightarrow T$ is

$$
\frac{dN_T(\mathbf{k})}{dt} = \int \frac{d^3k'}{(2\pi)^3} \frac{d^3k''}{(2\pi)^3} u^{\pm}_{TLL}(\mathbf{k}, \mathbf{k}', \mathbf{k}'')
$$
$$
\times \{N_L(\mathbf{k}')N_{L'}(\mathbf{k}'') - N_T(\mathbf{k})[N_L(\mathbf{k}') + N_{L'}(\mathbf{k}'')]\}. \qquad (B.4)
$$

The probability for the process $L + L' \rightarrow T$ of second harmonic plasma emission, after summing over the two states of polarization of the transverse waves, is

$$
u_{TLL'}(\mathbf{k}_T, \mathbf{k}_L, \mathbf{k}_{L'}) = \frac{e^2 \hbar}{32\varepsilon_0 m_e^2 \omega_p} \frac{(k_L^2 - k_{L'}^2)^2}{k_T^2} |\kappa_L \times \kappa_{L'}|^2
$$
$$
\times (2\pi)^4 \delta^3(\mathbf{k}_T - \mathbf{k}_L - \mathbf{k}_{L'}) \, \delta(\omega_T(\mathbf{k}_T) - \omega_L(\mathbf{k}_L) - \omega_L(\mathbf{k}_{L'})). \qquad (B.5)
$$

References

Achterberg, A.: 1981, 'On the nature of small amplitude Fermi acceleration', *Astron. Astrophys.* **97**, 259–264

Achterberg, A.: 1990, 'Particle acceleration near astrophysical shocks', in W. Brinkman, A.C. Fabian & F. Giovanelli *Physical Processes in Hot Cosmic Plasmas*, Kluwer (Dordrecht), pp. 67–80

Arons, J.: 1979, 'Some problems of pulsar physics', *Space Sci. Rev.* **24**, 437–510

Axford, W.I.: 1992, 'Particle acceleration on galactic scales', in G.P. Zank & T.K. Gaisser (eds) *Particle Acceleration in Cosmic Plasmas*, AIP (New York), pp. 45–56

Axford, W.I.: 1994, 'The origins of high-energy cosmic rays', *Astrophys. J. Suppl.* **90**, 937–944

Barbosa, D.D.: 1979, 'Stochastic acceleration of solar flare protons', *Astrophys. J.* **233**, 383–394

Benz, A.O.: 1986, 'Millisecond radio spikes', *Solar Phys.* **104**, 99–110

Blandford, R.D.: 1990, 'Physical processes in active galactic nuclei', in R.D. Blandford, H. Netzer & L. Woltjer (eds) *Active Galactic Nuclei*, Springer-Verlag (Berlin), pp. 161–275

Blandford, R.D.: 1994, 'Particle acceleration mechanisms', *Astrophys. J. Suppl.* **90**, 515–520

Blandford, R.D., & Eichler, D.: 1987, 'Particle acceleration at astrophysical shocks: a theory of cosmic ray origin', *Phys. Rep.* **154**, 1–75

Boström, R., Gustafsson, G., Holback, B., Holmgren, G., Koskinen, H., & Kintner, P.: 1988, 'Characteristics of solitary waves and weak double layers in the magnetospheric plasma', *Phys. Rev. Lett.* **61**, 82–85

Büchner, J., & Zelenyi, L.M.: 1989, 'Regular and chaotic charged particle motion in magnetotaillike field reversals 1. Basic theory of trapped motion', *J. Geophys. Res.* **94**, 11,821–11,842

Chian, A.C.-L., & Alves, M.V.: 1988, 'Nonlinear generation of the fundamental radiation of interplanetary type III radio bursts', *Astrophys. J.* **330**, L77–L80

Decker, R.B.: 1983, 'Formation of shock-spike events at quasi-perpendicular shocks', *J. Geophys. Res.* **88**, 9959–9973

Drury, L.O'C.: 1983, 'An introduction to the theory of diffusive shock acceleration of energetic particles in tenuous plasmas', *Rep. Prog. Phys.* **46**, 973–1027

Dulk, G.A.: 1967, 'Apparent changes in the rotation rate of Jupiter', *Icarus* **7**, 173–182

Eichler, D.: 1979, 'Particle acceleration in collisionless shocks: regulated injection and high efficiency', *Astrophys. J.* **229**, 419–423

Fermi, E.: 1949, 'On the origin of cosmic radiation', *Phys. Rev.* **75**, 1169–1174

Fermi, E.: 1954, 'Galactic magnetic field and the origin of cosmic radiation', *Astrophys. J.* **119**, 1–6

Ginzburg, V.L., & Zheleznyakov, V.V.: 1958, 'On the possible mechanisms of sporadic radio solar emission (Radiation in an isotropic plasma)', *Soviet Astron. AJ* **2**, 235–668

Goldman, M.V.: 1984, 'Strong turbulence of plasma waves', *Rev. Mod. Phys.* **56**, 709–735

Goldstein, M.L., & Goertz, C.K.: 1983, 'Theories of radio emissions and plasma waves', in A.J. Dessler (ed.) *Physics of the Jovian Magnetosphere*, Cambridge University Press, pp. 317–352

Grognard, R. J.-M.: 1975, 'Deficiencies of the asymptotic solutions commonly found in the quasilinear relaxation theory', *Aust. J. Phys.* **28**, 731–753

Grognard, R. J.-M.: 1984, 'Partial reconstruction of the initial conditions for streams of energetic electrons associated with a type III burst', *Solar Phys.* **94** 165–170

Grognard, R. J.-M.: 1985, 'Propagation of electron streams', in D.J. McLean & N.R. Labrum *Solar Radiophysics*, Cambridge University Press, pp. 253–286

Haerendel, G.: 1994, 'Acceleration from field-aligned potential drops', *Astrophys. J. Suppl.* **90**, 765–774

Holman, G.D.: 1985, 'Acceleration of runaway electrons and joule heating in solar flares', *Astrophys. J.* **293**, 584–594

Holman, G.D., & Pesses, M.E.: 1983, 'Solar type II radio emission and the shock drift acceleration of electrons', *Astrophys. J.* **267**, 837–843

Jones, F.C., & Ellison, D.C.: 1991, 'The plasma physics of shock acceleration', *Space Sci. Rev.* **58**, 259–346

Kirk, J.G.: 1994, 'Particle acceleration', in J.G. Kirk, D.B. Melrose, & E.R. Priest *Plasma Astrophysics*, Springer-Verlag (Berlin), pp. 225–314

Kliem, B.: 1994, 'Particle orbits, trapping, and acceleration in a filamentary current sheet model', *Astrophys. J. Suppl.* **90**, 719–728

Kulsrud, R.M., & Ferrari, A.: 1971, 'The relativistic quasilinear theory of particle acceleration by hydromagnetic turbulence', *Astrophys. Space Sci.* **12**, 302–318

Lacombe, C.: 1977, 'Acceleration of particles and plasma heating by turbulent Alfvén waves in a radiogalaxy', *Astron. Astrophys.* **54**, 1–16

Lee, M.A.: 1992, 'Particle acceleration in the heliosphere', in G.P. Zank & T.K. Gaisser (eds) *Particle Acceleration in Cosmic Plasmas*, AIP (New York), pp. 27–44

Leroy, M.M., & Mangeney, A.: 1984, 'A theory of energization of solar wind electrons by the Earth's bow shock', *Ann. Geophys.* **2**, 449–455

Lin, R.P., Potter, D.W., Gurnett, D.A., & Scarf, F.L.: 1981, 'Energetic electrons and plasma waves associated with a solar type III radio burst', *Astrophys. J.* **251** 364–373

Lin, R.P., Levedahl, W.K., Lotko, W., Gurnett, D.A., & Scarf, F.L.: 1986, 'Evidence for nonlinear wave-wave interactions in solar type III radio bursts', *Astrophys. J.* **308** 954–965

Louarn, P., Roux, A., de Féraudy, H., Le Quéau, D., André, M., & Matson, L.: 1990, 'Trapped electrons as a free energy source for the auroral kilometric radiation', *J. Geophys. Res.* **95**, 5983–5995

Lundin, R., & Eliasson, L.: 1991, 'Auroral energization processes', *Ann. Geophys.* **9**, 202–223

McKean, M.E., Winglee, R.M., & Dulk, G.A.: 1989, 'Propagation and absorption of electron-cyclotron maser radiation during solar flares', *Solar Phys.* **122**, 53–89

McLean, D.J., & Labrum, N.R.: 1985, *Solar Radiophysics*, Cambridge University Press

Melrose, D.B.: 1976, 'An interpretation of Jupiter's decametric radiation and the terrestrial kilometric radiation as direct amplified gyro-emission', *Astrophys. J.* **207**, 651–662

Melrose, D.B.: 1986, *Instabilities in Space and Laboratory Plasmas*, Cambridge University Press

Melrose, D. B.: 1989, 'The brightness temperatures of solar type III bursts', *Solar Phys.* **120**, 369–381

Melrose, D.B.: 1990, 'Particle acceleration processes in the solar corona', *Aust. J. Phys.* **43**, 703–752

Melrose, D.B.: 1991, 'Collective plasma radiation processes', *Ann. Rev. Astron. Astrophys.* **29**, 31–57

Melrose, D.B.: 1992, 'The Sun as a lab for particle acceleration mechanisms', in G.P. Zank & T.K. Gaisser (eds) *Particle Acceleration in Cosmic Plasmas*, AIP (New York), pp. 3–14

Melrose, D.B.: 1993, 'Coherent radio emission from pulsars', Chapter 7 in R.D. Blandford, A. Hewish, A.G. Lyne & L. Mestel *Pulsars as Physics Laboratories*, Oxford University Press, pp. 105–115

Melrose, D.B.: 1994, 'Turbulent acceleration in solar flares', *Astrophys. J. Suppl.* **90**, 623–630

Melrose, D. B., & Cramer, N.F.: 1989, 'Quasilinear relaxation of electrons interacting with an inhomogeneous distribution of Langmuir waves', *Solar Phys.* **123**, 343–365

Melrose, D.B., & McPhedran, R.C.: 1991, *Electromagnetic Processes in Dispersive Media*, Cambridge University Press

Michel, F.C.: 1991, 'Theory of Neutron Star Magnetospheres', (Chicago: Univ. Chicago Press)

Miller, J.A., Guessoum, N., & Ramaty, R.: 1990, 'Stochastic Fermi acceleration in solar flares', *Astrophys. J.* **361**, 701–708

Mozer, F.S., Cattell, C.A., Hudson, M.K., Lysak, R.L., Temerin, M., & Torbert, R.B.: 1980, 'Satellite measurements and theories of low altitude auroral particle acceleration', *Space Sci. Rev.* **27**, 155–213

Raadu, M.A.: 1989, 'The physics of double layers and their role in astrophysics', *Phys. Rep.* **178**, 27–97

Reames, D.V.: 1990, 'Energetic particles from impulsive solar flares', *Astrophys. J. Suppl.* **73**, 235–251

Rickett, B.J.: 1990, 'Radio propagation through the turbulent interstellar plasma', *Ann. Rev. Astron. Astrophys.* **28**, 561–605

Robinson, P.A.: 1989, 'Escape of fundamental electron cyclotron maser emission from the Sun and stars', *Astrophys. J.* **341**, L99–L102

Robinson, P.A., & Cairns, I.H.: 1993, 'Stochastic growth theory of type III solar radio emission', *Astrophys. J.* **418**, 506–509

Schindler, K., Hesse, M., & Birn, J.: 1991, 'Magnetic field-aligned electric potentials in nonideal plasma flows', *Astrophys. J.* **380**, 293–301

Toptygin, I.N.: 1980, 'Acceleration of particles by shocks in a cosmic plasma', *Space Sci. Rev.* **26**, 157–213

Tverskoĭ, B.A.: 1967, 'Contribution to the theory of Fermi statistical acceleration', *Sov. Phys. JETP* **25**, 317–325

Tverskoĭ, B.A.: 1968, 'Theory of turbulent acceleration of charged particles in a plasma', *Sov. Phys. JETP* **26**, 821–828

Twiss, R.Q.: 1958, 'Radiation transfer and the possibility of negative absorption in radio astronomy', *Aust. J. Phys.* **11**, 564–579

Webb, G.M., Axford, W.I., & Terasawa, T.: 1983, 'On the drift mechanism for energetic charged particles at shocks', *Astrophys. J.* **270**, 537–553

Wentzel, D.G.: 1974, 'Cosmic-ray propagation in the Galaxy: collective effects', *Ann. Rev. Astron. Astrophys.* **12**, 71–96

Wu, C.S., & Lee, L.C.: 1979, 'A theory of the terrestrial kilometric radiation', *Astrophys. J.* **230**, 621–626

Wu, C.S., Steinhofson, R.S., & Zhou, G.C.: 1986, 'The synchrotron-maser theory of type II solar radio emission: the physical model and generation mechanism', *Astrophys. J.* **309**, 392–401

Wu, C.S., & Lee, L.C.: 1979, 'A theory of the terrestrial kilometric radiation', *Astrophys. J.* **230**, 621–626

Zhou, Y., & Matthaeus, W.H.: 1990, 'Models of inertial range spectra of interplanetary magnetohydrodynamic turbulence', *J. Geophys. Res.* **95**, 14,881–14,892

Abraham Chian

Nonlinear Wave-Wave Interactions in Astrophysical and Space Plasmas

A. C.-L. Chian
National Institute for Space Research - INPE
P.O. Box 515, 12201-970 São José dos Campos - SP, Brazil

Abstract. We discuss nonlinear mode-mode coupling phenomena in cosmic plasmas. Four problems are considered: (1) nonlinear three-wave processes in the planetary magnetosphere involving the interaction of auroral Langmuir, Alfvén and whistler waves, (2) nonlinear three-wave processes in the solar wind involving the modulation of Langmuir and electromagnetic waves by ion-acoustic waves, (3) order and chaos in nonlinear four-wave processes in cosmic plasmas, and (4) regular and chaotic dynamics of the relativistic Langmuir turbulence and its application to pulsar and AGN emissions. The observational evidence in support of nonlinear wave-wave interactions in space and astrophysical plasmas is presented.

Key words: Wave-Wave Interactions, Planetary Magnetospheres, Solar Wind, Pulsars, AGNs, Chaos

1. Introduction

Astrophysical and space plasmas are very rich in nonlinear wave-wave interactions. There is increasing observational evidence rendering support for the occurrence of this class of nonlinear plasma phenomena in the cosmos. In this paper, four such processes are discussed. Firstly, we study the theory of the auroral LAW (Langmuir-Alfvén-whistler) events which indicate the occurrence of nonlinear three-wave coupling involving Langmuir, Alfvén and whistler waves in the planetary magnetosphere. This theory can also account for the observation of the leaked AKR (Auroral Kilometric Radiation). Secondly, we study the nonlinear modulation of Langmuir waves by ion-acoustic waves in the interplanetary medium and planetary foreshocks. Two types of nonlinear three-wave processes produced by the beam-excited Langmuir waves are considered: electrostatic Langmuir decay and electromagnetic Langmuir decay. This theory can explain the fine structures of intense Langmuir waves detected in the solar wind. Thirdly, we examine the nonlinear dynamical behavior of four-wave parametric coupling processes describing the resonant interaction of two wave triplets. As an illustration, we analyze the saturation state of the stimulated Langmuir modulational instability for which the ion-acoustic wave is a normal plasma mode. We show that in the absence of driving and damping such a Hamiltonian system exhibits transition from order to chaos. The phenomenon of chaos is of great importance for understanding the nonlinear plasma dynamics since it is an indication of the onset of turbulence. Finally, we investigate the regular and chaotic behaviors of Langmuir turbulence by solving numerically the weakly relativistic Zakharov equations. We apply the theoretical results to the interpretation of the variability in pulsars and AGN (Active Galactic Nuclei).

2. Nonlinear Three-Wave Process in the Planetary Magnetosphere

2.1. INTRODUCTION

Auroral Langmuir-Alfvén-whistler (LAW) events refer to the events observed in the auroral region of the planetary magnetosphere during which well correlated Langmuir (L), Alfvén (A) and whistler (W) waves are simultaneously detected (Chian et al. 1994a,b). These LAW events occur in association with the intense flux of low-energy (100 eV to 10 keV) field-aligned electron beams (Boehm et al. 1990). The close correlation of these three types of plasma waves suggests strongly that the nonlinear three-wave processes $W \rightleftharpoons L \pm A$ are operative. Presumably, these mode-mode coupling events are triggered by large-amplitude whistler or Langmuir waves driven by electron beams. The nonlinear interaction of whistler and Langmuir waves is effected through the ponderomotive coupling to low-frequency Alfvén waves.

Linear theories of parametric instabilities for $W \Rightarrow L \pm A$ and $L \Rightarrow W \pm A$ have been investigated separately by Chian et al. (1994a,b). In the linear theory the pump amplitude is assumed constant, thus neglecting the effect of pump depletion. In this section, we discuss a coherent nonlinear theory of the process $W \rightleftharpoons L \pm A$, taking into consideration the effect of pump depletion. In addition, the roles played by dissipation and frequency mismatch in the nonlinear wave dynamics are analyzed numerically. This nonlinear theory may contribute towards the understanding of the physical origins of fine structures of auroral whistler-mode emissions and amplitude modulations of auroral Langmuir waves.

2.2. DERIVATION OF COUPLED WAVE EQUATIONS

Consider all the interacting plasma waves propagating along the ambient magnetic field $\mathbf{B}_0 = B_0\hat{\mathbf{z}}$. The electric fields of whistler waves ($\mathbf{E}_W$), Alfvén waves ($\mathbf{E}_A$) and Langmuir waves ($\mathbf{E}_L$), written in the modula-

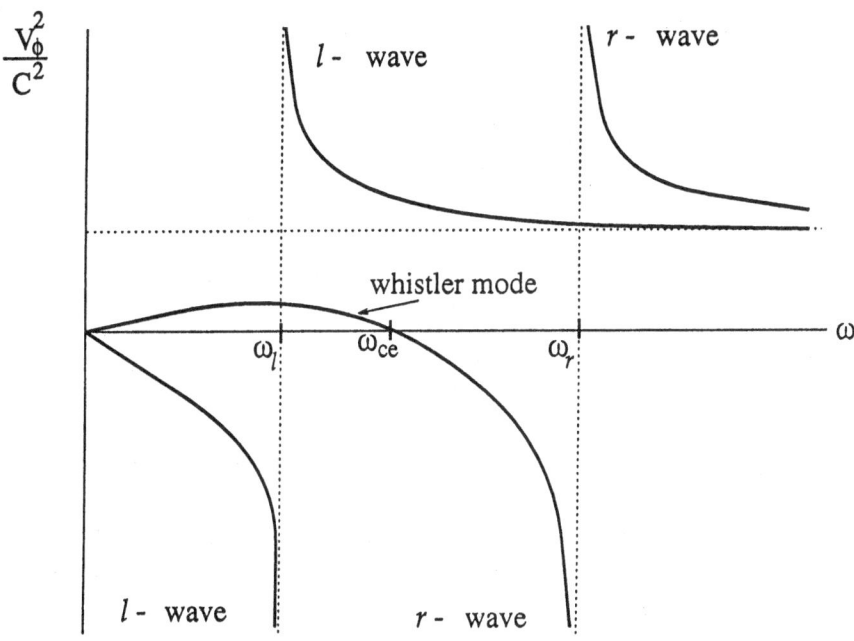

Figure 1. Dispersion curves of right-hand (r) and left-hand (l) electromagnetic cyclotron waves. V_ϕ is the wave phase velocity, $\omega_{r,l}$ are the cut-off frequency, $\omega_r = 1/2[\omega_{ce} + (\omega_{ce}^2 + 4\omega_{pe}^2)^{1/2}]$ and $\omega_l = 1/2[-\omega_{ce} + (\omega_{ce}^2 + 4\omega_{pe}^2)^{1/2}]$.

tional representation are

$$\mathbf{E}_W = \frac{1}{2}\mathcal{E}_W(\mathbf{z}, t)\exp[i(k_W z - \omega_W t)] + c.c., \qquad (1)$$

$$\mathbf{E}_A = \frac{1}{2}\mathcal{E}_A(\mathbf{z}, t)\exp[i(k_A z - \omega_A t)] + c.c., \qquad (2)$$

$$\mathbf{E}_L = \frac{1}{2}\mathcal{E}_L(\mathbf{z}, t)\exp[i(k_L z - \omega_L t)] + c.c., \qquad (3)$$

where $\mathcal{E}(\mathbf{z}, t)$ is a slowly varying complex amplitude with $|\partial_z^2\mathcal{E}| \ll |k\partial_z\mathcal{E}|$ and $|\partial_t^2\mathcal{E}| \ll |\omega\partial_t\mathcal{E}|$.

In the absence of coupling, the rapidly varying phases in (1)-(3) obey the following linear dispersion relations for whistler, Alfvén and Langmuir waves, respectively,

$$\omega_W^2 = c^2 k_W^2 + \frac{\omega_{pe}^2 \omega_W}{\omega_W - \omega_{ce}}, \tag{4}$$

$$\omega_A^2 = c_A^2 k_A^2, \tag{5}$$

$$\omega_L^2 = \omega_{pe}^2 + 3 v_{th}^2 k_L^2, \tag{6}$$

where the electron cyclotron frequency $\omega_{ce} = eB_0/m_e$, the electron plasma frequency $\omega_{pe} = (n_0 e^2/m_e \epsilon_0)^{1/2}$, the electron thermal velocity $v_{th} = (K_B T_e/m_e)^{\frac{1}{2}}$ and the Alfvén velocity $c_A = B_0/(n_0 m_i \mu_0)^{1/2}$.

The coherent nonlinear interactions $W \rightleftharpoons L + A$ occur if the wave triplet satisfies the following phase matching conditions

$$\omega_W \simeq \omega_L + \omega_A, \tag{7}$$

$$\mathbf{k}_W = \mathbf{k}_L + \mathbf{k}_A. \tag{8}$$

Fig. 1 displays linear dispersion curves of electromagnetic cyclotron waves propagating along the field lines. Since $\omega_W \leq \omega_{ce}$ and $\omega_A \ll \omega_{W,L}$, the process $W \rightleftharpoons L + A$ is possible if $\omega_L \sim \omega_{pe} \leq \omega_{ce}$. Two examples of wavevector kinematics for $W \rightleftharpoons L + A$ are illustrated in Fig. 2. In addition to the phase matching conditions, the wave triplet must also satisfy the conservation of wave helicity (Wilhelmsson 1969; Stenflo 1970; Chian et al. 1994a,b). Since whistler waves are right-hand circularly polarized and Langmuir waves are longitudinal waves (i.e., unpolarized), the Alfvén waves for $W \rightleftharpoons L+A$ are right-hand circularly polarized fast magnetosonic waves. Hence, in (1) and (2) we can write $\mathcal{E}_{W,A} \equiv \mathcal{E}_{W,A} \hat{\mathbf{r}}$ with $\hat{\mathbf{r}} \equiv (\hat{\mathbf{x}} + i\hat{\mathbf{y}})/\sqrt{2}$.

The governing equations are the two-fluid equations of motion

$$\partial_t \mathbf{v}_\alpha + (\mathbf{v}_\alpha \cdot \nabla)\mathbf{v}_\alpha = \frac{q_\alpha}{m_\alpha}(\mathbf{E} + \mathbf{v}_\alpha \times \mathbf{B}) - \nu_\alpha \mathbf{v}_\alpha - \frac{\gamma_\alpha K_B T_\alpha}{m_\alpha} \nabla \ln n_\alpha, \tag{9}$$

the equations of continuity

$$\partial_t n_\alpha + \nabla \cdot (n_\alpha \mathbf{v}_\alpha) = 0, \tag{10}$$

and Maxwell's equations

$$\nabla \times \mathbf{E} = -\partial_t \mathbf{B}, \tag{11}$$

$$\nabla \times \mathbf{B} = \mu_0 \left(\sum_\alpha q_\alpha n_\alpha \mathbf{v}_\alpha + \epsilon_0 \partial_t \mathbf{E} \right), \tag{12}$$

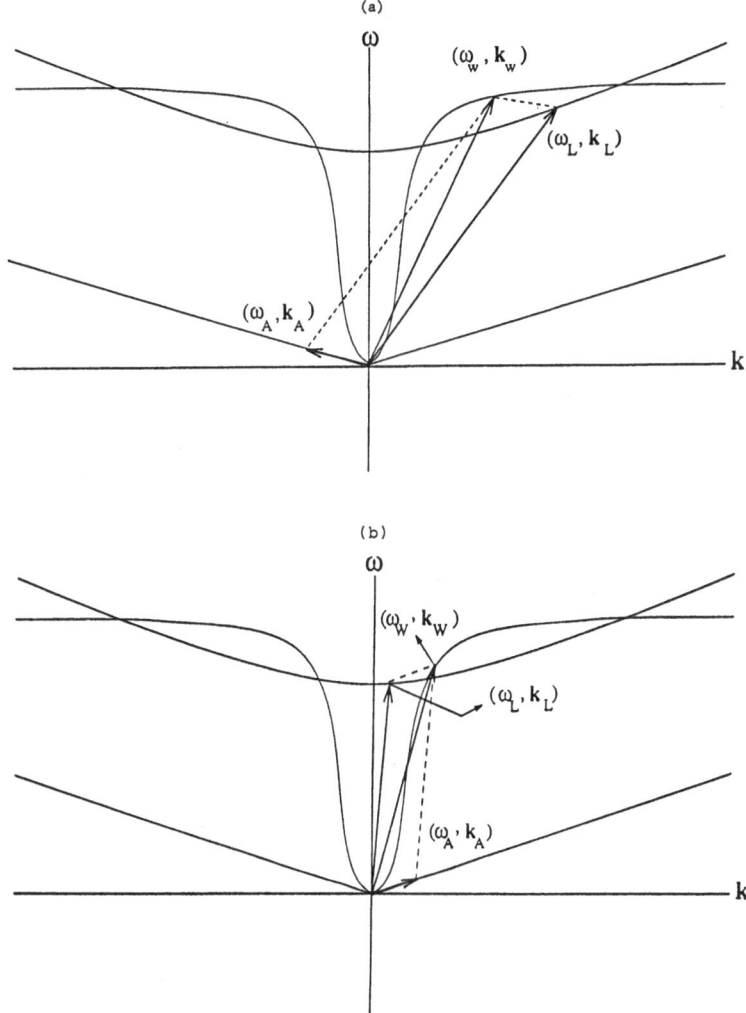

(a)

$(\omega_w, \mathbf{k}_w)$

$(\omega_L, \mathbf{k}_L)$

$(\omega_A, \mathbf{k}_A)$

(b)

$(\omega_W, \mathbf{k}_W)$

$(\omega_L, \mathbf{k}_L)$

$(\omega_A, \mathbf{k}_A)$

Figure 2. Examples of wavevector kinematics for $W \rightleftharpoons L + A$.

where $\alpha = (e, i)$, $q_e = -e$, $q_i = e$, ν is the damping frequency and γ is the ratio of specific heats. Equations (11) and (12) can be combined to yield a wave equation

$$\nabla^2 \mathbf{E} - \nabla(\nabla \cdot \mathbf{E}) - \frac{1}{c^2}\partial_t^2 \mathbf{E} = -\mu_0 \partial_t \sum_\alpha q_\alpha n_\alpha \mathbf{v}_\alpha. \qquad (13)$$

Dividing the interacting wave fields into two time-scales (Chian & Alves 1988; Chian 1991; Rizzato & Chian 1992; Chian et al. 1994a,b; Chian & Rizzato 1994): high-frequency fields, $\mathbf{E}_W$ and $\mathbf{E}_L$, oscillating near the electron plasma frequency; and low-frequency field, $\mathbf{E}_A$, oscillating below the ion cyclotron frequency $\omega_{ci} = eB_0/m_i$. Separating the physical variables in two time-scales, we have for the high-frequency scale

$$\mathbf{E}^h = \mathbf{E}_W + \mathbf{E}_L, \quad \mathbf{B}^h = \mathbf{B}_W, \tag{14}$$

$$n_e^h = n_L, \quad \mathbf{v}_e^h = \mathbf{v}_W + \mathbf{v}_L, \quad n_i^h = \mathbf{v}_i^h = 0; \tag{15}$$

and for the low-frequency scale

$$\mathbf{E}^l = \mathbf{E}_A, \quad \mathbf{B}^l = \mathbf{B}_A, \tag{16}$$

$$\mathbf{v}_e^l = \mathbf{v}_A, \quad \mathbf{v}_i^l = \mathbf{v}_i, \quad n_e^l = n_i^l = 0. \tag{17}$$

Introducing the above two time-scales into (9)-(10), for $W \rightleftharpoons L + A$, we obtain a system of three nonlinearly coupled wave equations

$$D_W \mathbf{E}_W = -ic_{LA} E_L \mathbf{E}_A, \tag{18}$$

$$D_A \mathbf{E}_A = ic_{WL} E_L^* \mathbf{E}_W, \tag{19}$$

$$D_L \mathbf{E}_L = ic_{WA} E_W E_A^* \hat{\mathbf{z}}, \tag{20}$$

where the dispersion operators are

$$D_W = -\omega_W^2 + c^2 k_W^2 + \frac{\omega_{pe}^2 \omega_W}{(\omega_W - \omega_{ce})} - i\nu_W \omega_W, \tag{21}$$

$$D_A = -\omega_A^2 + c_A^2 k_A^2 - i\nu_A \omega_A, \tag{22}$$

$$D_L = -\omega_L^2 + \omega_{pe}^2 + 3v_{th}^2 k_L^2 - i\nu_L \omega_L, \tag{23}$$

the coupling coefficients are

$$c_{WA} = \frac{e\omega_{pe}^2}{2m_e(\omega_W - \omega_{ce})} \left[\frac{k_A}{\omega_A} - \frac{k_W(\omega_W - \omega_{ce})}{\omega_W(\omega_A - \omega_{ce})} \right], \tag{24}$$

$$c_{WL} = \frac{\omega_A^2 c_A^2}{c^2 \omega_L^2} c_{WA}, \tag{25}$$

$$c_{LA} = \frac{\omega_W^2}{\omega_L^2} c_{WA}, \tag{26}$$

and the wave damping frequencies are

$$\nu_W = \frac{\omega_{pe}^2 \nu_e}{(\omega_W - \omega_{ce})^2}, \tag{27}$$

$$\nu_L = (\omega_{pe}^2/\omega_L^2)\nu_e, \tag{28}$$

$$\nu_A = \frac{\omega_{pe}^2 c_A^2 \nu_e}{c^2(\omega_A - \omega_{ce})^2} + \frac{\omega_{pi}^2 c_A^2 \nu_i}{c^2(\omega_A - \omega_{ci})^2}. \tag{29}$$

In deriving (18)-(20) we introduced harmonic spatial and temporal operators, $\nabla \to ik$ and $\partial_t \to -i\omega$, which are valid if $\mathcal{E}(z,t)$ in (1)-(3) is assumed to vary slowly in space-time in relation to the fast-varying phase.

In the case of resonant interactions, we can expand (18)-(20) about the linear solutions and identifying $k \to -i\partial_z$ and $\omega \to i\partial_t$, we then obtain the following set of three coupled wave equations describing the nonlinear space-temporal dynamics of $W \rightleftharpoons L + A$

$$(\partial_t + v_{gW}\partial_z + \nu_W')\mathcal{E}_W = \frac{c_{LA}\mathcal{E}_L\mathcal{E}_A \exp(-i\Delta t)}{2\omega_{pe}(1 + \omega_{ce}\omega_{pe}/2(\omega_{pe} - \omega_{ce})^2)}, \tag{30}$$

$$(\partial_t + v_{gA}\partial_z + \nu_A')\mathcal{E}_A = -\frac{c_{WL}\mathcal{E}_W\mathcal{E}_L^* \exp(i\Delta t)}{2c_A k_A}, \tag{31}$$

$$(\partial_t + v_{gL}\partial_z + \nu_L')\mathcal{E}_L = -\frac{c_{WA}\mathcal{E}_W\mathcal{E}_A^* \exp(i\Delta t)}{2(\omega_{pe}^2 + 3v_{th}^2 k_L^2)^{\frac{1}{2}}}, \tag{32}$$

where $v_g = \partial\omega/\partial k$ is the group velocity, $\nu' = \nu\omega/\partial_\omega D$ for the respective wave and $\Delta = \omega_W - \omega_A - \omega_L$ is the frequency mismatch. Equations(30)-(32) can also be derived from the coupled-mode formalism developed by Sjölund & Stenflo (1967) and Stenflo (1970). Note, however, that the system of equations (18)-(20) are valid for both resonant and non-resonant interactions; in contrast, the coupled-mode formalism is only valid for resonant interactions.

2.3. LINEAR THEORY

The linear stability theory for the parametric decay process $W \Rightarrow L + A$ with the whistler wave acting as the pump is governed by (19)-(20) or (31)-(32), assuming $\mathbf{E}_W = $ constant and $|\mathbf{E}_W| \gg |\mathbf{E}_A|, |\mathbf{E}_L|$ (Chian et al. 1994a). In the weak pump regime ($|\omega|^2 \simeq c_A^2 k_A^2$) wherein the Alfvén wave is a normal mode, the dispersion relation is

$$(\omega - \Delta + i\nu_A')(\omega + i\nu_L') + \frac{c_{WL}c_{WA}|E_W|^2}{4c_A k_A(\omega_{pe}^2 + 3v_{th}^2 k_L^2)^{\frac{1}{2}}} = 0, \tag{33}$$

the growth rate well above the threshold is

$$\Gamma = |E_W|\left[\frac{c_{WL}c_{WA}}{4c_A k_A(\omega_{pe}^2 + 3v_{th}^2 k_L^2)^{\frac{1}{2}}}\right]^{\frac{1}{2}} \tag{34}$$

and the threshold condition is

$$|E_W|^2 \geq \frac{4c_A k_A (\omega_{pe}^2 + 3v_{th}^2 k_L^2)^{\frac{1}{2}}}{c_{WL} c_{WA}} \nu_L' \nu_A'. \tag{35}$$

In the strong pump regime ($|\omega|^2 \gg c_A^2 k_A^2$) wherein the Alfvén wave is a quasi-reactive mode, the dispersion relation is

$$\omega^3 = -\frac{e^2 \omega_{pe} c_A^2 k^2 |E_W|^2}{8 m_e^2 c^2 (\omega_{pe} - \omega_{ce})^2}, \tag{36}$$

which can be solved to give

$$\omega = \frac{1}{2} \left[\frac{e^2 \omega_{pe} c_A^2 k_A |E_W|^2}{4 m_e^2 c^2 (\omega_{pe} - \omega_{ce})^2} \right]^{\frac{1}{3}} (1 + i\sqrt{3}). \tag{37}$$

The linear stability theory for the parametric fusion process $L + A \Rightarrow W$ with the Langmuir wave acting as the pump is governed by (18)-(19) or (30)-(31), assuming $\mathbf{E}_L =$ constant and $|\mathbf{E}_L| \gg |\mathbf{E}_W|, |\mathbf{E}_A|$ (Chian et al. 1994b). In the weak pump regime, the dispersion relation is

$$(\omega - \Delta + i\nu_A')(\omega + i\nu_L') + \frac{c_{WL} c_{LA} |E_L|^2}{4 c_A k_A \omega_{pe} [1 + \omega_{ce} \omega_{pe} / 2(\omega_{pe} - \omega_{ce})^2]} = 0, \tag{38}$$

the growth rate well above the threshold is

$$\Gamma = |E_L| \left[\frac{c_{WL} c_{WA}}{4 c_A k_A \omega_{pe} [1 + \omega_{ce} \omega_{pe} / 2(\omega_{pe} - \omega_{ce})^2]} \right]^{\frac{1}{2}}, \tag{39}$$

and the threshold condition is

$$|E_L|^2 \geq \frac{4 c_A k_A \omega_{pe} [1 + \omega_{ce} \omega_{pe} / 2(\omega_{pe} - \omega_{ce})^2]}{c_{WL} c_{WA}} \nu_W' \nu_A'. \tag{40}$$

In the strong pump regime, the dispersion relation is

$$\omega^3 = -\frac{e^2 \omega_{pe} c_A^2 k_A^2 |E_L|^2}{8 m_e^2 c^2 (\omega_{pe} - \omega_{ce})^2 [1 + \omega_{ce} \omega_{pe} / 2(\omega_{pe} - \omega_{ce})^2)]}, \tag{41}$$

and the solution is

$$\omega = \frac{1}{2} \left[\frac{e^2 \omega_{pe} c_A^2 k_A |E_L|^2}{16 m_e^2 c^2 (\omega_{pe} - \omega_{ce})^2 [1 + \omega_{ce} \omega_{pe} / (2(\omega_{pe} - \omega_{ce})^2)]} \right]^{\frac{1}{3}} (1 + i\sqrt{3}). \tag{42}$$

2.4. NONLINEAR THEORY

2.4.1. *General equations*

We now take into account the effect of pump depletion and study the nonlinear behavior of (30)-(32). After some standard algebraic manipulations, (30)-(32) can be put in the normalized form

$$\dot{A}_W = A_L A_A - \nu_W'' A_W, \tag{43}$$

$$\dot{A}_A = -A_W A_L^* - \nu_A'' A_A + i\delta A_A, \tag{44}$$

$$\dot{A}_L = -A_W A_A^* - \nu_L'' A_L, \tag{45}$$

with

$$A_W = \left[\frac{c_W L c_W A}{k^2 (v_{gL} - v)(v_{gA} - v)\partial_\omega D_L \partial_\omega D_A} \right]^{\frac{1}{2}} \mathcal{E}_W, \tag{46}$$

$$A_A = \left[\frac{c_{LA} c_W A}{k^2 (v_{gL} - v)(v_{gW} - v)\partial_\omega D_L \partial_\omega D_W} \right]^{\frac{1}{2}} \mathcal{E}_A \exp(i\Delta t), \tag{47}$$

$$A_L = \left[\frac{c_{LA} c_W L}{k^2 (v_{gW} - v)(v_{gA} - v)\partial_\omega D_W \partial_\omega D_A} \right]^{\frac{1}{2}} \mathcal{E}_L, \tag{48}$$

where the dot denotes differentiation with respect to the phase variable $\tau = k(z - vt)$, v and k are arbitrary wave velocity and wave vector respectively, $\nu'' = \nu'/[k(v_g - v)]$ and $\delta = \Delta/[k(v_g - v)]$.

It is convenient to rewrite (43)-(45) using $A_{W,A,L} = F_{W,A,L}^{1/2} \exp(i\varphi)$, where F and φ are real amplitude and phase, respectively, giving

$$\dot{F}_W = 2(F_W F_A F_L)^{\frac{1}{2}} \cos\varphi - \nu_W F_W, \tag{49}$$

$$\dot{F}_A = -2(F_W F_A F_L)^{\frac{1}{2}} \cos\varphi - \nu_A F_A, \tag{50}$$

$$\dot{F}_L = -2(F_W F_A F_L)^{\frac{1}{2}} \cos\varphi - \nu_L F_L, \tag{51}$$

$$\dot{\varphi} = \delta + \left[-\left(\frac{F_L F_A}{F_W} \right)^{\frac{1}{2}} + \left(\frac{F_W F_L}{F_A} \right)^{\frac{1}{2}} + \left(\frac{F_W F_A}{F_L} \right)^{\frac{1}{2}} \right] \sin\varphi, \tag{52}$$

where $\varphi = \varphi_W - \varphi_A - \varphi_L$ and we set $\nu = \nu''$ to simplify the notation.

2.4.2. *Nondissipative Case*

In the absence of wave damping ($\nu = 0$), a number of conserved quantities (constants of motion) can be obtained from (49)-(52). The first constant of motion is the Hamiltonian of the system

$$H = 2(F_W F_A F_L)^{\frac{1}{2}} \sin\varphi - \delta F_A, \tag{53}$$

from which the equations of motion (49)-(52) can be derived using the canonical equations $\partial_t F = \partial_\varphi H$ and $\partial_t \varphi = -\partial_F H$ (Marion 1970). The other constants of motion are the Manley-Rowe relations (Lashmore-Davies 1981)

$$F_W + F_L = c_1, \tag{54}$$

$$F_W + F_A = c_2. \tag{55}$$

Other dependent constants of motion can be derived from a combination of (54)-(55).

With the choice of the initial conditions $F_W = c_1$, $F_L = 0$ and $F_A = c_2 - c_1$, which implies $H = 0$ when $\delta = 0$, and corresponds to the decay process $W \Rightarrow L + A$ at $\tau = 0$, the solution of (49)-(52) for the whistler wave can be written in the form

$$F_W = a_0 - (a_0 - a_1)\mathrm{sn}^2 \left((a_2 - a_1)^{\frac{1}{2}}\tau, \left(\frac{a_0 - a_1}{a_2 - a_1}\right)^{\frac{1}{2}} \right), \tag{56}$$

where

$$a_0 = c_1, \tag{57}$$

$$a_1 = c_1 - \frac{(2c_1 - c_2 - \delta^2/4)}{2} + \frac{\sqrt{(2c_1 - c_2 - \delta^2/4)^2 + 4c_1 c_2 - 4c_1^2}}{2}, \tag{58}$$

$$a_2 = c_1 - \frac{(2c_1 - c_2 - \delta^2/4)}{2} - \frac{\sqrt{(2c_1 - c_2 - \delta^2/4)^2 + 4c_1 c_2 - 4c_1^2}}{2}. \tag{59}$$

The solutions for F_L and F_A are readily obtained from (54) and (55). If $c_2 - c_1 = 0$, a_2 becomes c_1 and the modulus of sn-function becomes one. Under these circumstances, the solution (56) yields solitary waves (Taniuti & Nishihara 1983). Similarly, the solutions corresponding to the fusion process $L + A \Rightarrow W$ at $\tau = 0$ can be determined with the initial solutions: $F_L = c_1$, $F_W = 0$ and $F_A = c_2$.

Figures 3a and 3b give examples of periodic wavetrain solutions, in the case of perfect frequency matching ($\delta = 0$) for initial conditions corresponding to $W \Rightarrow L + A$ and $L + A \Rightarrow W$ at $\tau = 0$, respectively. Note that in the decay process $W \Rightarrow L + A$ the direction of energy transfer reverses when the pump wave (W) is depleted, as shown in Fig. 3a. In contrast, in the fusion process $L + A \Rightarrow W$ the direction of energy transfer reverses when one of the daughter waves is depleted, as shown in Fig. 3b. Note that the amplitude of wave modulations in the fusion process is a function of the initial amplitude of the daughter waves. In this case the pump is only partially depleted, which indicates that in general the fusion parametric instability is less efficient than the decay parametric instability.

(a)

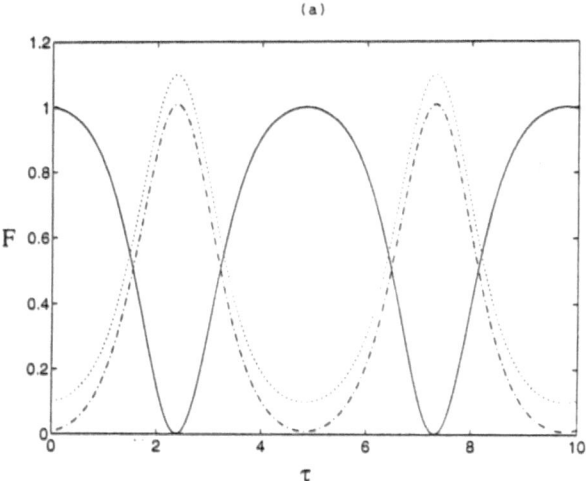

(b)

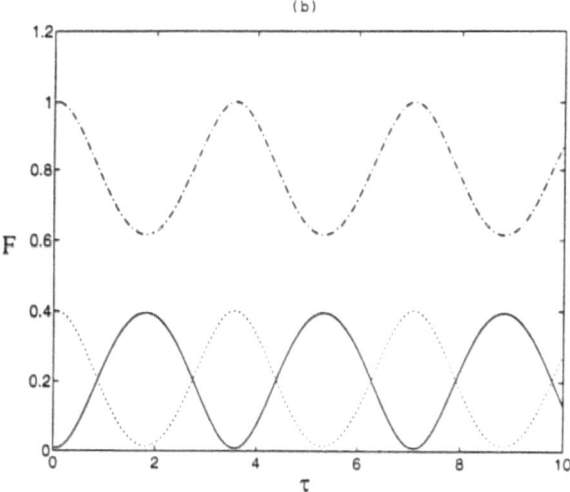

Figure 3. Nonlinear waveforms for the case of non-dissipation ($\nu = 0$) and perfect frequency matching ($\delta = 0$) for (a) $W \rightleftharpoons L + A$ with $F_W(0) = 1, F_A(0) = 0.1$ and $F_L(0) = 0.01$; (b) $L + A \rightleftharpoons W$ with $F_L(0) = 1, F_A(0) = 0.4$ and $F_W(0) = 0.01$. Solid curve (W), dot-dashed curve (L), and dotted (A).

Figures 4a and 4b give examples of periodic wavetrain solutions, in the case of finite frequency mismatching ($\delta \neq 0$). A comparison of Figs. 3 and 4 indicates that the frequency mismatch reduces the efficiency of energy transfer. When $\delta = 0$, it follows from (54), (55) and (58) that for the decay process $W \Rightarrow L+A$, $(F_L)_{max} = a_1 = c_1$ and $(F_A)_{max} = c_2 - c_1$ when the pump is fully depleted (i.e., $(F_W)_{min} = 0$). However, when $\delta \neq 0$, $(F_L)_{max} < c_1$ and $(F_A)_{max} < c_2 - c_1$ when the pump reaches its minimum with $(F_W)_{min} > 0$. Thus, the frequency mismatch prevents the complete depletion of pump energy.

2.4.3. Dissipative Case

Approximate analytical solutions of (49)-(52) including dissipation can be derived using the method developed by Weiland & Wilhelmsson (1977). This method is valid when

$$(F_W F_A F_L)^{\frac{1}{2}} \sin \varphi \simeq 0, \qquad (60)$$

at $\tau = 0$. The dissipative solution may be expressed in terms of nondissipative solutions (56)-(59). The constants of motion of the Manley-Rowe relations (54)-(55) become dependent on τ, namely

$$F_W + F_L = m_1(\tau), \qquad (61)$$

$$F_W + F_A = m_2(\tau). \qquad (62)$$

The resulting solution is still given by (56)-(59), with c_1 and c_2 now replaced respectively by $m_1(\tau)$ and $m_2(\tau)$ (Weiland & Wilhelmsson 1977).

Figures 5a and 5b give examples of numerical solutions of (49)-(52) with dissipation and without frequency mismatching. Since whistler and Alfvén waves are electromagnetic, in general they are much less damped than the electrostatic Langmuir waves in the absence of wave coupling. However, Figs. 5a and 5b demonstrate that due to nonlinear mode coupling the whistler wave, which is the mode with the highest frequency in $W \rightleftharpoons L + A$, follows the damping pattern of the Langmuir wave; whereas the dissipative effect of Alfvén wave is not much influenced by the nonlinear coupling.

Figures 6a and 6b give examples of numerical solutions of (49)-(52) with both dissipation and frequency mismatching. The same features observed in Figs. 5a and 5b are seen in Figs. 6a and 6b. In addition, the effect of partial pump depletion, due to finite frequency mismatching, can also be identified in Figs. 6a and 6b.

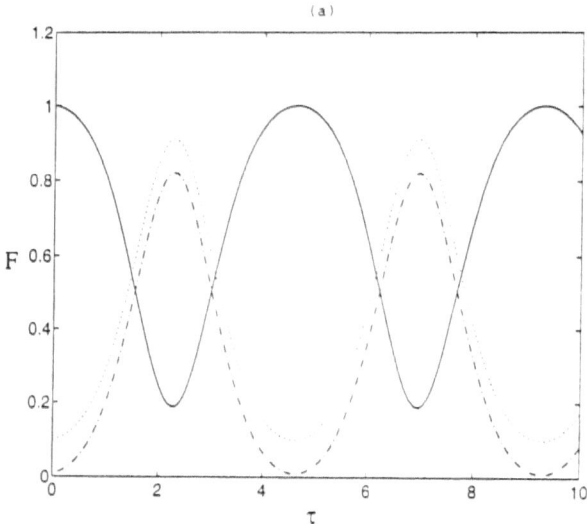

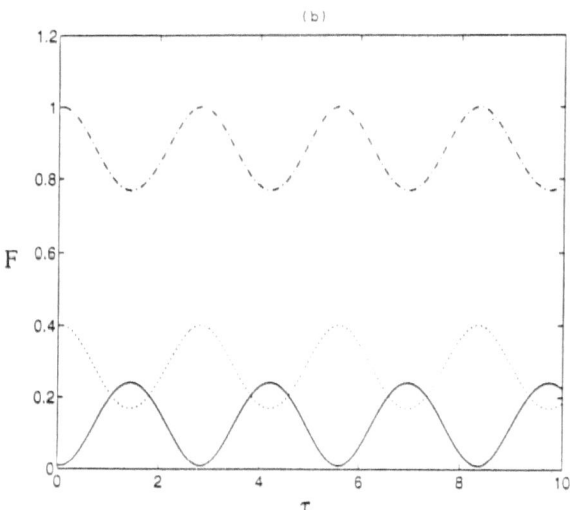

Figure 4. Nonlinear waveforms for the case of non-dissipation ($\nu = 0$) and finite frequency mismatching ($\delta = 1$) for (a) $W \rightleftharpoons L + A$ with $F_W(0) = 1, F_A(0) = 0.1$ and $F_L(0) = 0.01$; (b) $L + A \rightleftharpoons W$ with $F_L(0) = 1, F_A(0) = 0.4$ and $F_W(0) = 0.01$. Solid curve (W), dot-dashed curve (L), and dotted (A).

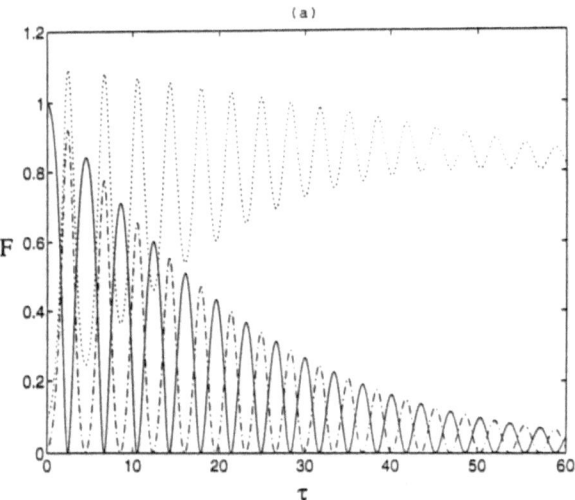

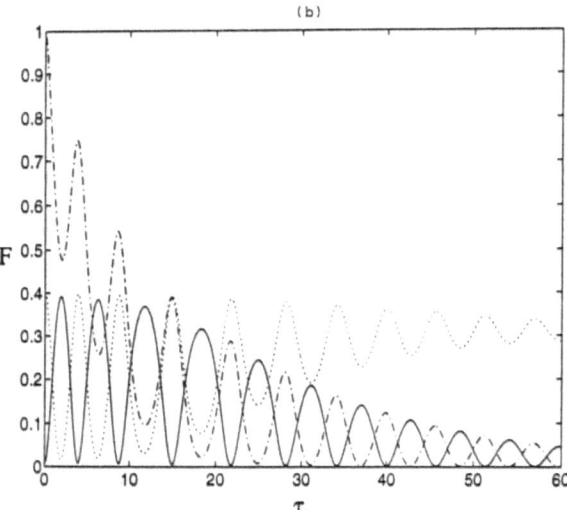

Figure 5. Nonlinear waveforms for the case of dissipation ($\nu_W = \nu_A = 0.005$, $\nu_L = 0.1$) and perfect frequency matching ($\delta = 0$) for (a) $W \rightleftharpoons L + A$ with $F_W(0) = 1, F_A(0) = 0.1$ and $F_L(0) = 0.01$; (b) $L + A \rightleftharpoons W$ with $F_L(0) = 1, F_A(0) = 0.4$ and $F_W(0) = 0.01$. Solid curve (W), dot-dashed curve (L), and dotted (A).

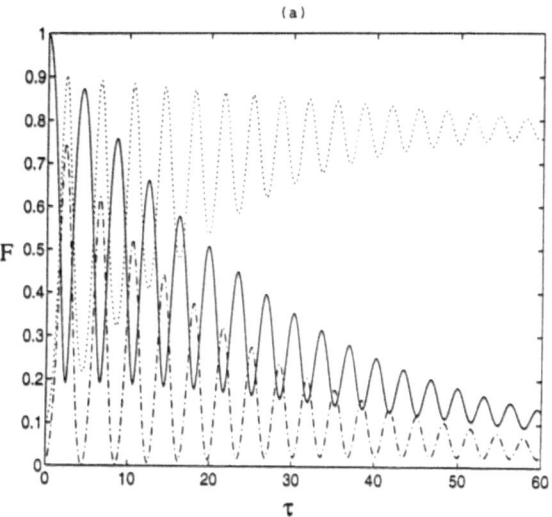

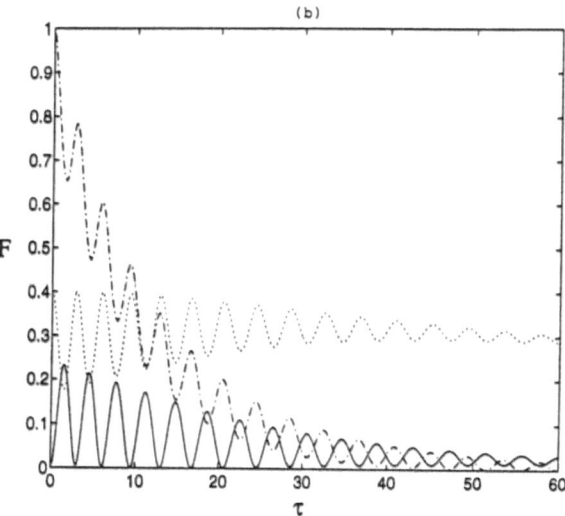

Figure 6. Nonlinear waveforms for the case of dissipation ($\nu_W = \nu_A = 0.005$, $\nu_L = 0.1$) and finite frequency mismatching ($\delta = 1$) for (a) $W \rightleftharpoons L + A$ with $F_W(0) = 1, F_A(0) = 0.1$ and $F_L(0) = 0.01$; (b) $L + A \rightleftharpoons W$ with $F_L(0) = 1, F_A(0) = 0.4$ and $F_W(0) = 0.01$. Solid curve (W), dot-dashed curve (L), and dotted (A).

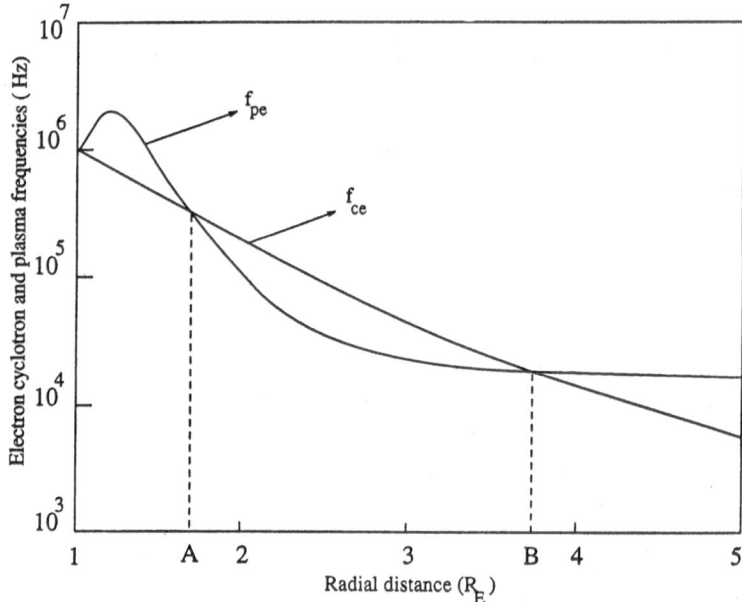

Figure 7. Typical profiles of electron cyclotron and plasma frequencies as a function of the radial distance in the Earth's magnetosphere. The source region of auroral LAW events is located in the density depletion region ($f_{pe} \leq fce$) between A and B. R_E is the Earth radius.

2.5. DISCUSSION

The above nonlinear theory shows that the three-wave process $W \rightleftharpoons L + A$ provides an efficient mechanism for dynamic (spatial and temporal) energy transfer between high-frequency whistler and Langmuir waves via the ponderomotive coupling to low-frequency fast magnetosonic waves. The present theory can be straightforwardly extended to the process $L \rightleftharpoons W + A$ (Chian et al. 1994a,b), where due to the wave helicity conservation the low-frequency mode is the left-hand circularly polarized shear Alfvén wave. Hence, it is likely that the mode-mode coupling processes $W \rightleftharpoons L + A$ and $L \rightleftharpoons W + A$ take place during the auroral LAW events, as detected by Boehm et al. (1990). These parametric processes can only operate within the auroral density depletion region where $f_{pe} \leq f_{ce}$ (Chian et al. (1994a,b). Figure 7 shows typical profiles of electron cyclotron and electron plasma frequencies, as a function of the radial distance, in the Earth's magnetosphere. According to our theory, LAW events occur in the auroral depletion region located between the two points A and B in Fig. 7.

The present section shows that as the consequence of nonlinear effects, space-temporal modulations of wave amplitude appear in the waveforms of auroral plasma waves. The modulated wave packets are represented by (1)-(3), where the wave modulations $\mathcal{E}_{W,L,A}(z,t)$ are given by the solutions of (49)-(52). Next, we discuss the application of this nonlinear theory to the observation of fine structures of auroral whistler-mode emissions and modulated Langmuir waves in the auroral plasma.

In the auroral acceleration region several types of radio emissions, such as X and O-mode auroral kilometric radiation (AKR) and auroral whistler-mode radiation, are excited (Gurnett et al. 1983). Fine structures of both extraordinary (X)-mode and ordinary (O)-mode have been identified in the AKR data (Gurnet et al. 1979; Benson et al. 1988). It was suggested by Grabbe et al. (1980) and Grabbe (1982) that a three-wave coupling involving electrostatic ion-cyclotron wave can explain both the generation and fine structures of the X-mode AKR. The auroral whistler waves consist of various components: hiss, saucers and leaked-AKR. Chian et al. (1994b) proposed that the processes $L \rightleftharpoons W \pm A$ may be a viable source mechanism for auroral whistler waves in the vicinity of the electron plasma frequency. These processes can account for the sharp peak near f_{pe} in the auroral whistler wave spectrum reported by Gurnett et al. (1983) as well as the intense spiky component of leaked-AKR observed by Oya et al. (1985). Benson et al. (1988) suggested that fine structures might be a common feature of auroral generated radio emissions, including the whistler-mode. In fact, Oya et al. (1985) pointed out that the leaked-AKR component of auroral whistler waves contains highly variable fine structures. The nonlinear theory of $W \rightleftharpoons L + A$ and $L \rightleftharpoons W + A$ gives a natural explanation for fine structures of auroral whistler waves near f_{pe}. The amplitude modulation of auroral whistler waves evolves naturally from the nonlinear effects of three-wave coupling, leading to the fast-varying fine structures of wave fields.

Amplitude modulations and localized structures of intense Langmuir waves have also been seen in the auroral plasma (Boehm et al. 1984; Ergun et al. 1991). Two types of large electric field signatures of auroral Langmuir waves were identified by Boehm et al. (1984): individual pulses and pulse trains. Boehm et al. (1984) interpreted these pulse trains as Langmuir solitons. According to the nonlinear theory presented in this section amplitude modulations of Langmuir waves, due to parametric coupling to whistler and Alfvén waves, produce waveforms for the envelope Langmuir electric field of either pulse train type or soliton type. Observational evidence of nearly 100% amplitude modulations of Langmuir waves was given by Ergun et al. (1991). They suggested that the low-frequency modulations is related to either lower hybrid or ion Bernstein waves. We showed that low-frequency Alfvén waves may also contribute to the amplitude modulation of auroral Langmuir waves through the processes $W \rightleftharpoons L + A$ and $L \rightleftharpoons W + A$.

2.6. SUMMARY

In this section, we discussed a coherent nonlinear theory of auroral LAW events in the planetary magnetosphere. We demonstrated that large-amplitude auroral whistler and Langmuir waves, driven by field-aligned electron beams, can nonlinearly convert into each other via ponderomotive coupling to either fast magnetosonic waves ($W \rightleftharpoons L + A$) or shear Alfvén waves ($L \rightleftharpoons W + A$).

By dividing the physical variables in the two-fluid equations into two time-scales, we were able to derive a system of three nonlinearly coupled wave equations (18)-(20) governing both resonant and non-resonant three-wave interactions involving whistler, Alfvén and Langmuir waves. Detailed analytical and numerical solutions for the resonant ($W \rightleftharpoons L + A$) interactions were presented. The influence of pump depletion, dissipation and frequency mismatch on the mode-mode coupling was studied. These results show that, in terms of energy transfer from the pump wave to the daughter waves, the decay parametric instabilities $W \Rightarrow L + A$ and $L \Rightarrow W + A$ are more efficient than the fusion parametric instabilities $L + A \Rightarrow W$ and $W + A \Rightarrow L$. In general, the frequency mismatch diminishes the efficiency of energy transfer. In the presence of dissipation, for both decay and fusion processes, the two high-frequency waves (whistler and Langmuir waves) follow similar space-temporal damping profiles but the damping pattern of the low-frequency Alfvén waves is not much affected by the nonlinear wave dynamics.

The present analysis shows that space-temporal modulations of wave amplitude can evolve directly from the nonlinear effect of mode-mode coupling. The nonlinear amplitude modulations produced by $W \rightleftharpoons L + A$ and $L \rightleftharpoons W + A$ can account for the fine structures of auroral whistler-mode emission near f_{pe} as well as the modulated Langmuir wave packets in the auroral plasma.

3.   Nonlinear Three-Wave Process in the Solar Wind

3.1. INTRODUCTION

Recent observations by the Voyager-1 (Gurnett et al. 1981), ISEE-3 (Lin et al. 1986), AMPTE (Koons et al. 1987), Ulysses (Kellogg et al. 1992; Canu et al. 1993; Thiessen & Kellogg 1993) and Galileo (Gurnett et al. 1993) spacecrafts have provided interesting data for the study of nonlinear plasma processes associated with Langmuir waves in the interplanetary medium and planetary foreshocks. These high-frequency electron plasma waves are generated by a bump-on-tail instability arising from the interaction of a stream of energetic electrons with the solar wind plasma. A proper understanding of the nonlinear dynamics of Langmuir waves is essential for obtaining an accurate picture of the generation mechanism of interplanetary radio emissions and plasma turbulence phenomenon upstream of planetary bow shocks.

Langmuir waves detected in the interplanetary medium and planetary foreshocks are usually very spiky and associated with large-amplitude fluctuations. The observed features of Langmuir waves suggest that nonlinear processes such as soliton collapse or mode-mode coupling are operative. Soliton collapse via the nucleation mechanism occurs when the Langmuir waves are trapped in the pre-existing density cavities (Cairns & Robinson 1992a). Nucleated wave collapse is a strong turbulence phenomenon that requires relatively high threshold field strengths. In contrast, three-wave mode-mode coupling is a weak turbulence phenomenon that appears at relatively low field strengths, hence easily excitable in space plasmas (see e.g., Chian et al. 1994 a, b). In the electrostatic Langmuir decay process, $L \rightarrow L + S$, a beam-excited Langmuir wave (L) decays into a backward-propagating Langmuir wave and a forward-propagating ion-acoustic wave (S); in the electromagnetic Langmuir decay process, $L \rightarrow T + S$, a Langmuir wave decays into a perpendicular-propagating electromagnetic wave (T) and a forward-propagating ion-acoustic wave.

The stability theory of nonlinear conversion of Langmuir waves into electromagnetic and Langmuir waves via ponderomotive coupling to ion-acoustic waves, using the fixed-phase approximation, has been developed and applied to type-III interplanetary radio emissions and stimulated electromagnetic emissions in active experiments in space (Chian & Alves 1988; Chian 1991; Rizzato & Chian 1992; Glanz et al. 1993). The dynamics and efficiency of type-III solar radio bursts were studied by Robinson et al. (1994) for various three-wave processes excited by Langmuir waves, using the random-phase approximation. The aim of this section is to discuss a coherent nonlinear theory of three-wave processes, $L \rightleftharpoons L + S$ and $L \rightleftharpoons T + S$, based on the fixed-phase approximation. This formalism allows us to derive analytical expressions for the modulated Langmuir wave packets, which are useful for interpreting space observations. In addition, we determine a formula for the nonlinear modulation period which is shown to depend on the wave amplitude.

3.2. Theory

The set of nonlinearly coupled wave equations governing the electro-static Langmuir decay process, $L \rightleftharpoons L + S$, can be obtained from the electrostatic Zakharov equations (Zakharov 1972; Chian 1991)

$$(\partial_t^2 + \nu_L \partial_t - \gamma_e v_{th}^2 \nabla^2 + \omega_{pe}^2) \mathbf{E} = -\frac{\omega_{pe}^2}{n_0} n \mathbf{E}, \tag{63}$$

$$(\partial_t^2 + \nu_S \partial_t - v_S^2 \partial_x^2) n = \frac{\varepsilon_0}{2m_i} \partial_x^2 \langle E^2 \rangle, \tag{64}$$

where $\mathbf{E}$ is the Langmuir electric field, n is the ion density fluctua-tion, $v_S = [K(T_e + T_i)/m_i]^{1/2}$ is the ion-acoustic velocity, $\nu_L(\nu_S)$ is the damping frequency of the Langmuir (ion-acoustic) wave and the angular brackets denote averaging over the fast timescale. ¿From (63) and (64), we have for $L \rightleftharpoons L + S$

$$(\partial_t^2 + \nu_L \partial_t - \gamma_e v_{th}^2 \nabla^2 + \omega_{pe}^2) \mathbf{E}_0 = -\frac{\omega_{pe}^2}{n_0} n \mathbf{E}_L, \tag{65}$$

$$(\partial_t^2 + \nu_L \partial_t - \gamma_e v_{th}^2 \nabla^2 + \omega_{pe}^2) \mathbf{E}_L = -\frac{\omega_{pe}^2}{n_0} n \mathbf{E}_0, \tag{66}$$

$$(\partial_t^2 + \nu_S \partial_t - v_S^2 \nabla^2) n = \frac{\varepsilon_0}{2m_i} \nabla^2 \langle \mathbf{E}_0 \cdot \mathbf{E}_L \rangle, \tag{67}$$

where $\mathbf{E}_0$ ($\mathbf{E}_L$) is the pump (daughter) Langmuir electric field.

The set of nonlinearly coupled wave equations governing the elec-tromagnetic Langmuir decay process, $L \rightleftharpoons T + S$, can be obtained from the generalized Zakharov equations (Rizzato & Chian 1992; Glanz et al. 1993; Chian & Rizzato 1994), with (63) replaced by

$$(\partial_t^2 + \nu_e \partial_t + c^2 \nabla \times \nabla \times - \gamma_e v_{th}^2 \nabla(\nabla \cdot) + \omega_{pe}^2) \mathbf{E} = -\frac{\omega_{pe}^2}{n_0} n \mathbf{E}, \tag{68}$$

where ν_e denotes the damping frequency of the high-frequency wave. ¿From (68) and (64), we have for $L \rightleftharpoons T + S$

$$(\partial_t^2 + \nu_L \partial_t - \gamma_e v_{th}^2 \nabla^2 + \omega_{pe}^2) \mathbf{E}_0 = -\frac{\omega_{pe}^2}{n_0} n \mathbf{E}_T, \tag{69}$$

$$(\partial_t^2 + \nu_T \partial_t - c^2 \nabla^2 + \omega_{pe}^2) \mathbf{E}_T = -\frac{\omega_{pe}^2}{n_0} n \mathbf{E}_0, \tag{70}$$

$$(\partial_t^2 + \nu_S \partial_t - v_S^2 \nabla^2) n = \frac{\varepsilon_0}{2m_i} \nabla^2 \langle \mathbf{E}_0 \cdot \mathbf{E}_T \rangle, \tag{71}$$

where $\mathbf{E}_T$ (ν_T) is the electric field (damping frequency) of the electro-magnetic wave. In writing (69) and (70) we have selected the longitudi-nal and transverse parts, respectively, of the high-frequency generalized Zakharov equation. In general, the electrostatic and electromagnetic fields are coupled in the wave operator of the high-frequency general-ized Zakharov equation. For the purely growing case, Rizzato & Chian (1992) showed that if $W \ll \delta_L/\omega_{pe}$, where $W = \varepsilon_0 E_0^2/n_0 K T_e$ and $\delta_L \approx \omega_L - \omega_0$ is a detuning parameter (ω_0 and ω_L are the frequency of the pump Langmuir wave and the linear frequency of the daughter Langmuir wave, respectively), the induced high-frequency electrostatic field is negligible. For simplicity, we ignore here the coupling of electro-static and electromagnetic fields in the wave operator.

Nonlinear wave coupling leads to energy transfer among the interact-ing waves. Hence, in general the wave fields become variable in space and time as seen in the previous section. The resulting nonlinearly amplitude-modulated wave packets can be represented by $\mathbf{E}_\alpha(\mathbf{r}, t) = \frac{1}{2}\mathcal{E}_\alpha(\mathbf{r}, t) \exp i\theta_\alpha + c.c.$, where $\mathcal{E}_\alpha(\mathbf{r}, t)$ is a slowly varying complex enve-lope such that $|\partial_r^2 \mathcal{E}_\alpha| \ll |k_\alpha \partial_r \mathcal{E}_\alpha|$ and $|\partial_t^2 \mathcal{E}_\alpha| \ll |\omega_\alpha \partial_t \mathcal{E}_\alpha|$, $\theta_\alpha = \mathbf{k}_\alpha \cdot \mathbf{r} - \omega_\alpha t$ is a fast-varying phase and α refers to each interacting wave. The slowly varying amplitude approximation adopted implies weak nonlin-earity and is equivalent to ignoring the field component propagating in the opposite direction. We treat only resonant modes with linear wave dispersion relations given by $\omega_{0,L}^2 = \omega_{pe}^2 + \gamma_e v_{th}^2 k_{0,L}^2$, $\omega_T^2 = \omega_{pe}^2 + c^2 k_T^2$ and $\omega_S^2 = c_S^2 k_S^2$. Resonant three-wave interactions occur provided the following phase matching conditions are met: $\omega_0 \approx \omega_L + \omega_S$ and $\mathbf{k}_0 = \mathbf{k}_L + \mathbf{k}_S$ for the electrostatic Langmuir decay process; $\omega_0 \approx \omega_T + \omega_S$ and $\mathbf{k}_0 = \mathbf{k}_T + \mathbf{k}_S$ for the electromagnetic Langmuir decay process. We assume perfect wave vector matching, but imperfect wave frequency matching.

In order to interpret the temporal electric field variations observed in space, we focus on the temporal rather than the spatial wave dynamics. An introduction of the modulation representation into (65)-(67) gives, for the electrostatic Langmuir decay process,

$$(\partial_t + \nu_L/2)\mathcal{E}_0 = (c_{LS}/2\omega_0)\mathcal{E}_L\mathcal{E}_S \exp i\Delta t, \tag{72}$$

$$(\partial_t + \nu_L/2)\mathcal{E}_L = -(c_{0S}/2\omega_L)\mathcal{E}_0\mathcal{E}_S^* \exp -i\Delta t, \tag{73}$$

$$(\partial_t + \nu_S/2)\mathcal{E}_S = -(c_{0L}/2\omega_S)\mathcal{E}_0\mathcal{E}_L^* \exp -i\Delta t, \tag{74}$$

where the nonlinear coupling coefficients are given by $c_{LS} = c_{0S} = ek_S/2m_e$ and $c_{0L} = (m_e/2m_i)c_{LS}$, $\Delta = \omega_0 - \omega_L - \omega_S$ is the frequency mismatch. Note that in writing (72)-(74) we used the linearized Pois-son's equation $E_S = ien/\varepsilon_0 k_S$ and the approximation $\omega_0 \sim \omega_L \sim \omega_T \sim$

ω_{pe}. Similarly, for the electromagnetic Langmuir decay process, we have

$$(\partial_t + \nu_L/2)\mathcal{E}_0 = (c_{TS}/2\omega_0)\mathcal{E}_T\mathcal{E}_S \exp i\Delta t, \tag{75}$$

$$(\partial_t + \nu_T/2)\mathcal{E}_T = -(c_{0S}/2\omega_T)\mathcal{E}_0\mathcal{E}_S^* \exp -i\Delta t, \tag{76}$$

$$(\partial_t + \nu_S/2)\mathcal{E}_S = -(c_{0T}/2\omega_S)\mathcal{E}_0\mathcal{E}_T^* \exp -i\Delta t, \tag{77}$$

where $c_{TS} = c_{0S} = ek_S/2m_e$, $c_{0T} = (m_e/2m_i)c_{TS}$, $\Delta = \omega_0 - \omega_T - \omega_S$.

In the linear theory, $\partial_t\mathcal{E}_0 = 0$, a Fourier analysis of coupled wave equations (73)-(74) and (76)-(77) gives nonlinear dispersion relations which yield

$$\Gamma = (c_{0S}c_{0L}/\omega_S\omega_{pe})^{1/2}| \mathcal{E}_0 | / 2,$$

for the growth rate of either the electrostatic or electromagnetic Langmuir decay instability.

In the nonlinear theory, the Langmuir pump depletion is taken into account. The nonlinear solution of the set of equations (72)-(74) and (75)-(77) is facilitated by introducing the real variables F_α and ϕ_α, as defined by $\mathcal{E}_\alpha = \beta_\alpha F_\alpha^{1/2} \exp i\phi_\alpha$. A substitution of this polar representation into equations (72)-(74) and (75)-(77), respectively, leads identically to

$$\dot{F}_0 = 2(F_0F_1F_2)^{1/2} \cos\phi - \nu_0'F_0, \tag{78}$$

$$\dot{F}_1 = -2(F_0F_1F_2)^{1/2} \cos\phi - \nu_1'F_1, \tag{79}$$

$$\dot{F}_2 = -2(F_0F_1F_2)^{1/2} \cos\phi - \nu_2'F_2, \tag{80}$$

$$\dot{\phi} = [(F_0F_1/F_2)^{1/2} + (F_0F_2/F_1)^{1/2} - (F_2F_1/F_0)^{1/2}] \sin\phi + \delta, \tag{81}$$

where $\phi \equiv \phi_0 - \phi_1 - \phi_2$, $\delta \equiv \Delta/\omega_{pe}$ and $\nu_\alpha'' = \nu_\alpha'/\omega_{pe}$; the subscript 0 denotes the pump Langmuir wave, 1 denotes the daughter Langmuir (electromagnetic) wave for $L \rightleftharpoons L + S$ ($L \rightleftharpoons T + S$), and 2 denotes the ion-acoustic wave; the dot denotes differentiation with respect to $\tau = \omega_{pe}t$. For $L \rightleftharpoons L + S$, the parameters β_α are given by $\beta_0 = (4/ek_S)(2m_em_i\omega_L\omega_S)^{1/2}$, $\beta_1 = (4/ek_S)(2m_em_i\omega_0\omega_S)^{1/2}$ and $\beta_2 = (4m_e/ek_S)(\omega_0\omega_L)^{1/2}\exp i\delta t$. For $L \rightleftharpoons T + S$, β_α are given by $\beta_0 = (4/e\,k_S)(2\,m_e\,m_i\,\omega_T\,\omega_S)^{1/2}$, $\beta_1 = (4/e\,k_S)(2\,m_e\,m_i\,\omega_0\,\omega_S)^{1/2}$ and $\beta_2 = (4\,m_e/e\,k_S)(\omega_0\,\omega_T)^{1/2}\exp i\,\delta\,t$.

In the absence of dissipation, the following energy conservation laws known as the Manley-Rowe relations (54)-(55) can be derived from (78)-(81): $c_1 = F_0 + F_1$, $c_2 = F_0 + F_2$, and $H = 2(F_0F_1F_2)^{1/2}\sin\phi - \delta F_2$, where c_1 and c_2 are constants of motion and H is the Hamiltonian of the system. A specification of $F_0(0) = c_1$, $F_1(0) = 0$ and $F_2(0) = c_2 - c_1$

as initial conditions yields the following cnoidal wave solution for the pump Langmuir wave

$$F_0(\tau) = a_0 + (a_0 - a_1)\text{sn}^2[(a_2 - a_1)^{1/2}\tau, ((a_0 - a_1)/(a_2 - a_1))^{1/2}], \quad (82)$$

where $a_0 = c_1$ and $a_{1,2} = (1/2)[c_2 + \delta^2 \pm ((c_2 + \delta^2)^2 - 4c_1\delta^2)^{1/2}]$, with a_0 and a_1 being the maxima and minima of $F_0(t)$, respectively. The period of nonlinear modulation is given by

$$P = 2K[\sqrt{(a_0 - a_1)/(a_2 - a_1)}]/\sqrt{a_2 - a_1}, \quad (83)$$

where K is the elliptic integral of the first kind. Equation (83) indicates that the period of nonlinear modulation is a function of the wave amplitude.

3.3. DISCUSSION

Temporal amplitude modulations of Langmuir waves were first detected in space by the Voyager-1 spacecraft in the Jupiter's foreshock (Gurnett et al. 1981). These intense Langmuir waves are driven by an electron beam streaming into the solar wind from the Jovian bow shock. Gurnett et al. (1981) interpreted the Voyager-1 observations in terms of parametric decay ($L \rightarrow L + S$) and spatial collapse of Langmuir waves. Later, a strong turbulence analysis carried out by Cairns and Robinson (1992a) demonstrated that the Jovian Langmuir waves measured by Voyager-1 cannot collapse via nucleation mechanism because the waves are disrupted before collapse takes place. Instead, they showed that the modulations are direct evidence of $L \rightarrow L + S$ as suggested by Gurnett et al. (1981) (Cairns & Robinson, 1992b). Thiessen & Kellogg (1993) showed that $L \rightarrow L + S$ is commonly seen in the Ulysses's data in the Jovian foreshock. In particular, low-frequency modulations of Langmuir wave packets like those previously reported by Gurnett et al. (1981) can be seen in many events. Thiessen & Kellogg (1993) proposed the electrostatic Langmuir decay as a viable mechanism for localizing Langmuir waves and increasing their energy density toward the collapse threshold. Observational evidence of $L \rightarrow L + S$ and $L \rightarrow T + S$ in the Earth's foreshock was obtained by Gurnett & Frank (1975), Anderson et al. (1981) and Koons et al. (1987).

Using the ISEE-3 data, Lin et al. (1986) presented clear evidence for nonlinear wave-wave interactions, $L \rightarrow L + S$ and $L \rightarrow T + S$, in association with type-III radio bursts in the interplanetary medium. Nonlinearly modulated Langmuir waves, in connection with an interplanetary shock, were detected by the Ulysses spacecraft in the solar wind (Kellogg et al. 1992); the observed Langmuir events seem to have the characteristics of nucleated collapse. Temporally amplitude-modulated Langmuir wave packets were also observed by the Galileo spacecraft in the solar wind (Gurnett et al. 1993). These bursts of Langmuir waves are produced by energetic electrons ejected from a solar flare and associated with a type-III radio event. Gurnett et al. (1993) showed that the observed Langmuir wave intensities are too low for strong turbulence effects such as wave collapse to be important; instead, they concluded that $L \rightarrow L + S$ provides the most obvious explanation of their observations.

The high resolution digital waveforms of the modulated Langmuir wave packets seen in the Jovian foreshock (Gurnett et al. 1981) and interplanetary medium (Gurnett et al. 1993) contain fine structures with durations down to a few milliseconds. The observed waveforms are highly coherent and consist mainly of isolated wave packets and beat-type waveforms. The beat-type waveshapes are suggestive of a beat between two quasi-monochromatic waves of slightly different frequency, both near the local ω_{pe}, but of comparable amplitude. According to the present theory, the beat-type waveforms can be explained in terms of beating between two Langmuir waves $(L \rightleftharpoons L + S)$ or between a Langmuir wave and an electromagnetic wave $(L \rightleftharpoons T + S)$. In the presence of beam-excited Langmuir turbulence, both $L \rightleftharpoons L + S$ and $L \rightleftharpoons T + S$ are likely to occur (Chian 1991; Rizzato & Chian 1992; Robinson et al. 1994). Hence, either process can contribute to produce the observed Langmuir wave packets. The fact that the two high-frequency beat waves have comparable amplitudes can be explained by the Manley-Rowe relations derived in this paper, namely, $|E_0(0)|^2/|E_{Lmax}|^2 = \omega_0/\omega_L$ for $L \rightleftharpoons L + S$ and $|E_0(0)|^2/|E_{Tmax}|^2 = \omega_0/\omega_T$ for $L \rightleftharpoons T + S$, where the initial conditions $E_L(0) = 0$ and $E_T(0) = 0$ are assumed, respectively. Evidently, the pump and daughter Langmuir or electromagnetic waves have comparable energy levels since $\omega_0 \sim \omega_L \sim \omega_T \sim \omega_{pe}$.

From another Manley-Rowe relation derived in this section, $|E_0(0)|^2/|E_{Smax}|^2 = \omega_0/\omega_S$, we see that the energy level of ion-acoustic waves is much lower than the Langmuir energy since $\omega_S \ll \omega_0$. This may explain why ion-acoustic waves associated with Langmuir turbulence are difficult to measure in the solar wind (Gurnett et al. 1993). The present theory can provide numerical estimates of the ion-acoustic field. In the interplanetary Langmuir events detected by Gurnett et al. (1993), most of the Langmuir field strengths are in the range from 0.01 to 1.0 mV m^{-1}. For plasma frequency $f_{pe} = 24$ kHz and ion-acoustic frequency $f_S = 400$ Hz, the ion-acoustic field $|E_S| = (f_S/f_{pe})^{1/2}|E_L|$, is estimated to vary from 0.0013 to 0.13 mV m^{-1}.

The spectrum of Langmuir beat-type waveforms in the interplanetary medium observed by Gurnett et al. (1993) consists of two sharply defined peaks separated by a beat frequency of about 400 Hz. For both Langmuir decay processes, $L \rightleftharpoons L + S$ and $L \rightleftharpoons T + S$, the beat frequency is given by $f_0 - f_{L,T} \sim f_S$, namely, the ion-acoustic frequency. Hence, it follows from the present theory that the ion-acoustic frequency $f_S \sim 400$ Hz. This conclusion is in agreement with Cairns & Robinson (1992b) which shows that the beat-type Langmuir wave packets observed in the Jovian foreshock are modulated at the Doppler-shifted ion-acoustic frequency.

Note that, in addition to the beat-type wave packets, a variety of other types of modulated Langmuir waveforms are observed. Hence, it is likely that a variety of different modulation processes, each with its own characteristic modulation period, may operate. Equation (18) can be used to identify the nonlinear modulation time scales due to 3-wave processes discussed here, which are expected to be longer than the ion-acoustic period. Examples of this type of modulation in the Jovian foreshock can be seen in Gurnett et al. (1981). The waveforms of nonlinearly modulated Langmuir wave packets due to either $L \rightarrow L + S$ or $L \rightarrow T + S$ can be determined from (82). In the limit $(a_0 - a_1) \rightarrow 0$, the elliptic function in (82) reduces to a periodic sine-function. In general the modulated waveforms can be rather non-sinusoidal. In the limit when the modulus of cnoidal sn-function tends to unity, the modulation period tends to infinity and the solution becomes a sech-type solitary wave.

3.4. SUMMARY

In this section, we discussed a coherent nonlinear theory of conversion of Langmuir waves into either Langmuir waves or electromagnetic waves via ponderomotive coupling to ion-acoustic waves. We showed that the temporal modulation of Langmuir wave amplitude may evolve directly from the weakly nonlinear effects of either electrostatic or electromagnetic Langmuir decay processes. The resulting waveforms may explain certain characteristics of the highly structured waveforms of Langmuir waves observed in the interplanetary medium and Jovian foreshock. This theory can also be applied to the interpretation of the fine structures of Langmuir waves upstream of Terrestrial and Venusian bow shocks (Hospodarsky et al. 1991, 1992) as well as the fine structures of decimetric and decametric solar radio emissions (Wang & Li 1991; Barrow et al. 1994).

4.   Order and Chaos in Nonlinear Four-Wave Process in Cosmic Plasmas

4.1. INTRODUCTION

We discuss next the nonlinear dynamical behavior of the Langmuir stimulated modulation instability and investigate the analytical and numerical solutions of a set of four coupled wave equations, with quadratic nonlinearities, which describes the resonant parametric interaction of two wave triplets. We show that, in the absence of driving and dissipation, such a Hamiltonian system exhibits transition to chaos via the route of quasi-periodicity. This type of four-wave process with second-order nonlinear terms may find application in a variety of parametric processes in cosmic plasmas.

The electrostatic Langmuir parametric instability contains basically three distinct regimes (Bardwell & Goldman 1976; Weatherall et at. 1981; Goldman 1984): PDI (parametric decay instability), OTSI (oscillating two-stream instability), and SMI (stimulated modulation instability). In the PDI, a Langmuir pump wave decays into a Stokes Langmuir wave and an ion-acoustic daughter wave. In the OTSI, a Langmuir pump wave interactswith purely growing density perturbations to excite a pair of Stokes and anti-Stokes Langmuir waves. The SMI can be regarded as a four-wave decay instability involving the pump, Stokes, and anti-Stokes Langmuir waves, along with an ion-acoustic daughter wave whose real frequency is on the order of the ion-acoustic frequency (i.e., a *normal* plasma mode). A three-dimensional linear stability analysis (Bardwell & Goldman 1976) shows that the maximum growth rate of the SMI is comparable to the PDI and OTSI, albeit the wavenumber for which the maximum growth occurs is determined by a four-wave interaction wherein the anti-Stokes mode is resonant. The OTSI is very sensitive to the magnetic field; in contrast, the SMI and PDI are not much affected by the presence of a weak magnetic field (Weatherall et al. 1981).

The Langmuir SMJ is of interest for the study of type-III solar radio bursts (Bardwell & Goldman 1976; Weatherall et at. 1981; Goldman 1984; Chian & Alves 1988; Chian & Abalde 1995) such as discussed in the previous section and may play an important role during active experiments in space where intense Langmuir waves are excited (Chian 1991; Chian & Rizzato 1994b). Previous works of the Langmuir SMI were based on the linear stability analysis (Bardwell & Goldman 1976; Weatherall et al 1981). In this section, we develop a fully nonlinear theory to investigate the saturation state of the Langmuir SMI.

4.2. GOVERNING EQUATIONS

In the absence of dissipation, the one-dimensional nonlinear coupling of Langmuir and ion-acoustic waves is governed by the following electrostatic Zakharov equations (Zakharov 1972; Goldman 1984)

$$(\partial_t^2 - 3v_{th}^2\partial_x^2 + \omega_{pe}^2)E = -\frac{\omega_{pe}^2}{n_0}nE, \tag{84}$$

$$(\partial_t^2 - v_S^2\partial_x^2)n = \frac{\varepsilon_0}{2m_i}\partial_x^2\langle E^2\rangle, \tag{85}$$

To study the four-wave process $L_0 \rightleftharpoons L_- + L_+ + S$, we write the total Langmuir electric field as $E = E_0 + E_- + E_+$, with the subscript $\alpha = (0,-,+)$ referring to the pump, Stokes and anti-Stokes Langmuir waves, respectively. We adopt the modulation representation $E_\alpha(x,t) = (1/2)\mathcal{E}_\alpha(x,t)\exp i(k_\alpha x - \omega_\alpha t) + c.c.$, where $\mathcal{E}_\alpha(x,t)$ is a slowly varying complex envelope such that $|\partial_x^2\mathcal{E}_\alpha| \ll |k_\alpha\partial_x\mathcal{E}_\alpha|$ and $|\partial_t^2\mathcal{E}_\alpha| \ll |\omega_\alpha\partial_t\mathcal{E}_\alpha|$; and $\omega_\alpha^2 = \omega_{pe}^2 + 3v_{th}^2k_\alpha^2$. Likewise, we write $n(x,t) = (1/2)n\exp i(k_S x - \omega_S t)+c.c.$, with $\omega_S^2 = v_S^2k^2$. The four-wave process under study involves the coupling of two wave triplets which satisfy the following phase-matching conditions

$$\omega_\mp \approx \omega_0 \mp \omega_S, \qquad\qquad k_\mp = k_0 \mp k_S. \tag{86}$$

We shall assume perfect wave-vector matching, but imperfect frequency matching.

An introduction of the modulation representation into (84)-(85) gives the following non-autonomous complex dynamical system

$$\partial_t\mathcal{E}_0 = c_-\omega_-(\mathcal{E}_S\mathcal{E}_- \exp i\delta_-t - r\mathcal{E}_S^*\mathcal{E}_+ \exp i\delta_+t), \tag{87}$$

$$\partial_t\mathcal{E}_S = -c_-(m_e\omega_p^2/2m_i\omega_S)(\mathcal{E}_0\mathcal{E}_-^* \exp -i\delta_-t + r\mathcal{E}_0^*\mathcal{E}_+ \exp i\delta_+t), \tag{88}$$

$$\partial_t\mathcal{E}_- = -c_-\omega_-\mathcal{E}_0\mathcal{E}_S^* \exp -i\delta_-t, \tag{89}$$

$$\partial_t\mathcal{E}_+ = c_+\omega_+\mathcal{E}_0\mathcal{E}_S \exp -i\delta_+t, \tag{90}$$

where $\mathcal{E}_S$ is the ion-acoustic electric field; the coupling coefficients are given by $c_- = (ek_S)/(4m_e\omega_0\omega_-)$ and $c_+ = (ek_S)/(4m_e\omega_0\omega_+)$; $\delta_- = \omega_0 - \omega_S - \omega_-$ and $\delta_+ = \omega_0 + \omega_S - \omega_+$ are the parameters of linear frequency mismatch for each of two wave triplets.

The linear stability analysis of the Langmuir stimulated modulation instability can be performed by assuming $\partial_t\mathcal{E}_0 = 0$ and $|\mathcal{E}_0| \gg |\mathcal{E}_\pm|$, $|\mathcal{E}_S|$. A Fourier analysis of (88)-(90) then yields a nonlinear dispersion relation whose solution gives a growth rate $\Gamma = (\sqrt{3}/2)[(c_Sc_0\omega_0|\mathcal{E}_0|^2)/(2\omega_+\omega_-)]^{1/3}$

(a)

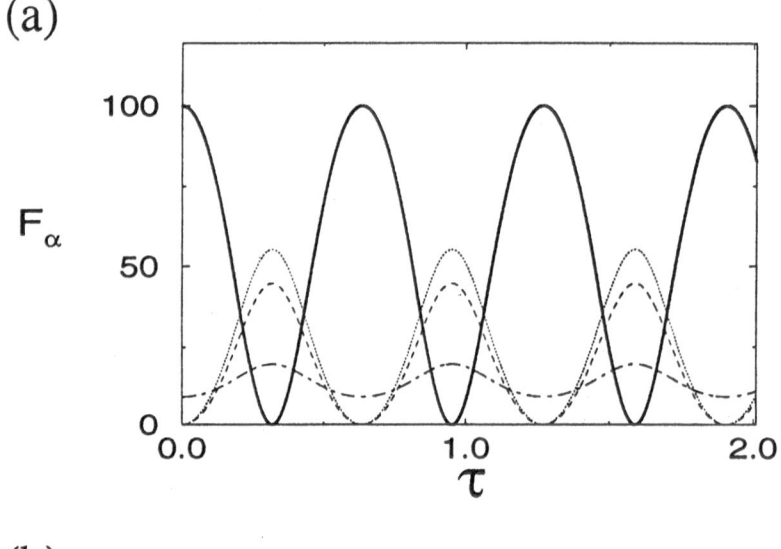

(b)

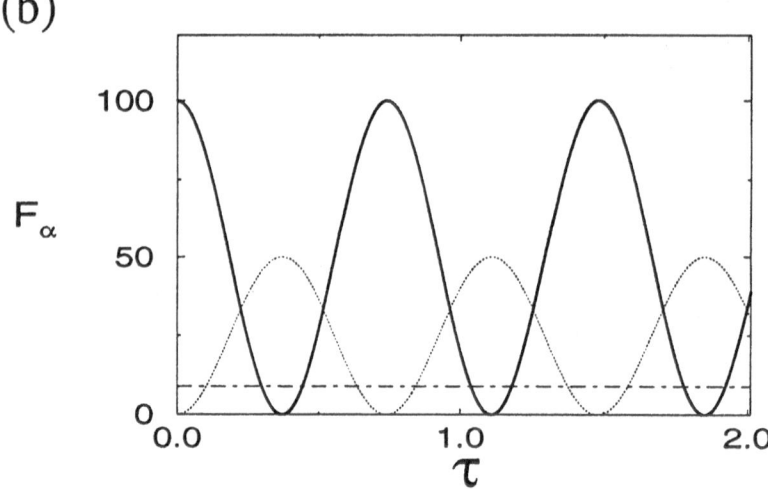

Figure 8. The plot of periodic $F_\alpha(\tau)$ for $\delta'_- = \delta'_+ = 0$, $r = 0.9$(a) and $r = 1.0$(b); $F_1(0) = 100.01$, $F_2(0) = 10$, and $F_3(0) = F_4(0) = 0$. In (a) the solid curve is $F_1(\tau)$, the dot-dashed curve is $F_2(\tau)$, the dotted curve is $F_3(\tau)$, and the long dashed curve is $F_4(\tau)$. In (b) the dotted curve represents both $F_3(\tau)$ and $F_4(\tau)$.

The analysis of nonlinear solutions of (87)-(90) is simplified by defining $\mathcal{E}_\alpha = \beta_\alpha F_\alpha^{1/2} \exp i\, \phi_\alpha$, where the amplitude F_α and phase ϕ_α are real variables, β_α are complex normalization constants. Substituting this polar representation into (87)-(90) leads to the following autonomous six degree-of-freedom system

$$\dot{F}_1 = 2(F_1 F_2 F_3)^{1/2} \cos \phi_- - 2r(F_1 F_2 F_4)^{1/2} \cos \phi_+, \qquad (91)$$

$$\dot{F}_2 = -2(F_1 F_2 F_3)^{1/2} \cos \phi_- - 2r(F_1 F_2 F_4)^{1/2} \cos \phi_+, \qquad (92)$$

$$\dot{F}_3 = -2(F_1 F_2 F_3)^{1/2} \cos \phi_-, \qquad (93)$$

$$\dot{F}_4 = 2r(F_1 F_2 F_4)^{1/2} \cos \phi_+, \qquad (94)$$

$$\dot{\phi}_- = (1/2)(H + \delta'_- F_3 + \delta'_+ F_4)(1/F_2 - 1/F_1) + (F_1 F_2/F_3)^{1/2} \sin \phi_- - \delta'_-, \qquad (95)$$

$$\dot{\phi}_+ = (1/2)(H + \delta'_- F_3 + \delta'_+ F_4)(-1/F_2 - 1/F_1) - r(F_1 F_2/F_4)^{1/2} \sin \phi_+ - \delta'_+, \qquad (96)$$

where H is the Hamiltonian of the system

$$H = 2(F_1 F_2)^{1/2}(F_3^{1/2} \sin \phi_- - r F_4^{1/2} \sin \phi_+) - \delta'_- F_3 - \delta'_+ F_4, \qquad (97)$$

and
$$\phi_- = \phi_1 - \phi_2 - \phi_3,$$
$$\phi_+ = \phi_1 + \phi_2 - \phi_4,$$
$$\delta'_\mp = \delta_\mp / \omega_{pe},$$
$$\beta_1 = (1/c_-)(2m_i \omega_2 / m_e \omega_3)^{1/2},$$
$$\beta_2 = (\omega_{pe}/c_-)(\omega_1 \omega_3)^{1/2},$$
$$\beta_3 = (1/c_-)(2m_i \omega_2 / m_e \omega_1)^{1/2} \exp i\delta'_- \tau,$$
$$\beta_4 = (1/c_-)(2m_i \omega_2 / m_e \omega_1)^{1/2} \exp i\delta'_+ \tau.$$

The dot denotes differentiation with respect to $\tau = \omega_{pe} t$; the subscript 1 denotes the pump Langmuir wave, 2 the ion-acoustic daughter wave, 3 the Stokes Langmuir wave and 4 the anti-Stokes Langmuir wave. The ratio r, which measures the relative coupling strengths of the anti-Stokes and Stokes modes, is defined as

$$r = c_+/c_-. \qquad (98)$$

In addition to the Hamiltonian given by (97), the set of equations (91)-(96) admits other conservation laws known as the Manley-Rowe relations (54)-(55):

$$F_1 + F_3 + F_4 = c_1, \qquad (99)$$

$$F_2 - F_3 + F_4 = c_2. \qquad (100)$$

When the anti-Stokes mode F_4 is off-resonant, $c_+ \to 0$ and $r \to 0$, the system given by (91)-(96) degenerates to a three-wave decay process $L_0 \rightleftharpoons L_- + S$. Alternatively, when the Stokes mode F_3 is off-resonant, $c_- \to 0$ and $r \to \infty$, the system degenerates to a three-wave fusion process $L_0 + S \rightleftharpoons L_+$.

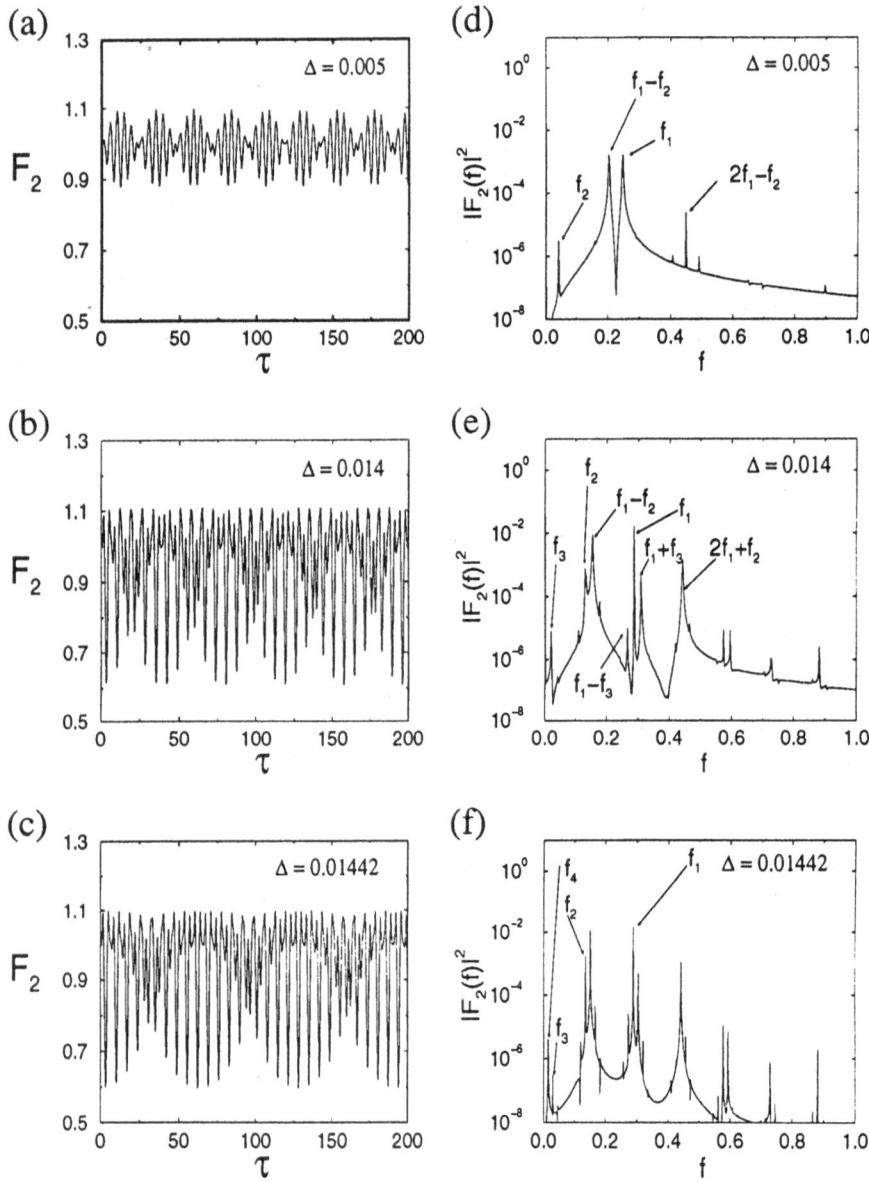

Figure 9. The plot of quasi-periodic $F_2(\tau)$ and $|F_2(f)|^2$ for $r = 1$ and three increasing values of Δ; the initial conditions are $F_1(0) = 100.01$, $F_2(0) = 1$, and $F_3(0) = F_4(0) = 0$.

4.3. Regular Solutions

If we specify $F_3(0) = F_4(0) = 0$ as initial conditions, we have $H = 0$ and $F_4(\tau) = r^2 F_3(\tau)$. In the absence of frequency mismatch, $\delta'_- = \delta'_+ = 0$, we can obtain periodic analytical solutions of (91)-(96). The solution for $F_2(\tau)$ is

$$F_2(\tau) = c_2 + (1 - r^2)a_1 \text{cn}^2\{(a_1 - a_2)^{1/2}\tau, [a_1/(a_1 - a_2)]^{1/2}\}, \quad (101)$$

where $a_1 = c_1/(1+r^2)$ and $a_2 = c_2/(r^2 - 1)$; the solutions for $F_{1,3,4}$ can be determined from (99)-(101). The periodic saturated states of the Langmuir SMI are illustrated in Fig. 8. Fig. 8a shows a plot of $F_\alpha(\tau)$ for $r = 0.9$ and $\delta'_- = \delta'_+ = 0$; note that the value of r was arbitrarily chosen in order to stress the distinct behaviors of F_3 and F_4. In the limit $r = 1$, we have $F_3(\tau) = F_4(\tau)$, and it follows from (100) that $F_2 = c_2 = $ constant. The modulus of the cnoidal function in (101), $[a_1/(a_1 - a_2)]^{1/2}$, becomes zero if $r = 1$ since $a_2 = \infty$ and the elliptic function becomes a sinusoidal function; Fig. 8b shows a graph of $F_\alpha(\tau)$ for $r = 1$ and $\delta'_- = \delta'_+ = 0$.

4.4. Transition from Order to Chaos

Let us now turn our attention to the transition from the regular to chaotic states of the nonlinear solutions of (91)-(96). Nonlinear mode coupling provides a good physical example for the study of chaotic dynamics. It is known that the onset of turbulence in a dissipative medium does not require a large number of modes and disordered motions can arise even in a system containing three or even two coupled non-conservative modes (Vyshikind & Rabinovich 1976); when the pump amplitude is increased in a three-wave process, two of which are para-metrically induced, a strange attractor appears (Pikovskii et al. 1978). The analysis of the instability saturation by nonlinear three-wave coupling, whereby one wave is linearly unstable and two waves are linearly damped, shows that as the damping rate is increased period-doubling bifurcations occur which leads to chaotic solutions (Wersinger et al. 1980a, b). The above model was found to undergo transitions to chaos via intermittence at certain tangent bifurcations (Meunier et al. 1982).

All the aforementioned works were directed to the nonconservative systems. Recently, the Hamiltonian dynamics of nonresonant nonlinear coupling of Langmuir and ion-acoustic waves for a system with infi-nite degree-of-freedom, based on the relativistic Zakharov equations, was studied (de Oliveira et al. 1995). We investigate here the low-dimensional Hamiltonian dynamics of resonant four-wave coupling of

(a)

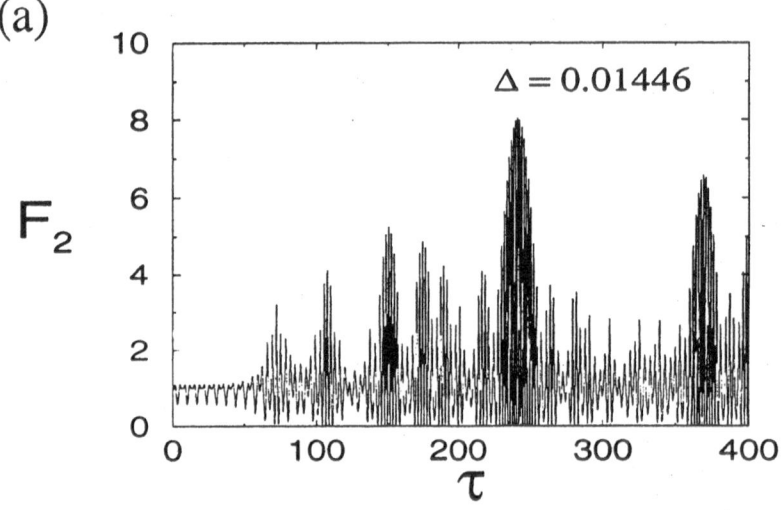

(b)

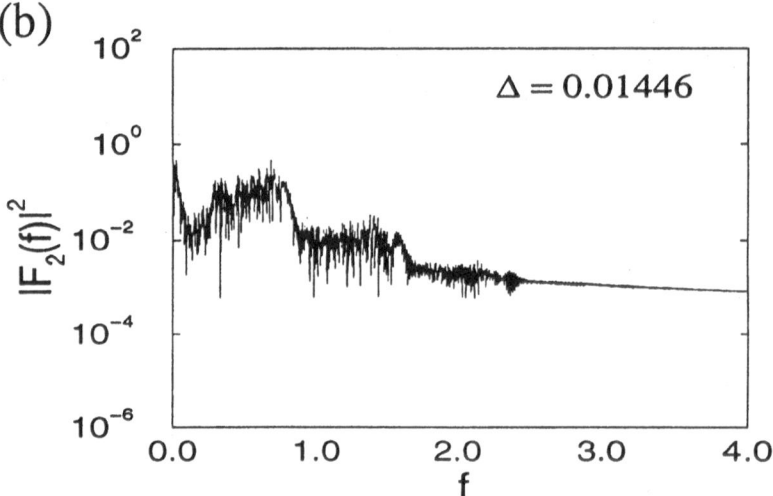

Figure 10. The plot of chaotic $F_2(\tau)$ and $|F_2(f)|^2$ for $r = 1$, and $\Delta = 0.01446$; the initial conditions are as given in Fig. 9.

Langmuir and ion-acoustic waves described by (91)-(96). In the special case $\delta'_- = \delta'_+ = 0$, it has been proved that the system of Eqs. (91)-(96) is integrable and the solutions are regular (Walters & Lewak 1977; Romeiras 1983). We introduce a parameter $\Delta = \delta'_- - \delta'_+$ which is a measure of the relative linear frequency mismatch of the two wave triplets. When $\Delta \neq 0$ the system becomes nonintegrable and chaotic solutions may appear. For the numerical solutions, we take the initial conditions $F_3(0) = F_4(0) = 0$ so that $H = 0$, and for simplicity we assume $r = 1$. When $\Delta = 0$, there is a kind of symmetry in the system and we have $F_3(\tau) = F_4(\tau)$ and $F_2 = $ constant, as seen in Fig. 8. For $\Delta \neq 0$ this symmetry is broken, thus $F_3(\tau) \neq F_4(\tau)$ and $F_2(\tau)$ is no longer constant. To solve (91)-(96) numerically, we use their complex form (Walters & Lewak 1977) and set $\delta'_+ = 0$ and $\Delta = \delta'_-$.

The numerical solutions show that the finite value of Δ introduces new frequencies to the system. For a fixed initial condition, as Δ is increased, the periodic solutions become quasi-periodic and evolve to chaos. Figure 9 shows the plot of $F_2(\tau)$ and its power spectrum for three increasing values of Δ, indicating three distinct quasi-periodic orbits residing on two-torus ($\Delta = 0.005$), three-torus ($\Delta = 0.014$) and four-torus ($\Delta = 0.01442$), respectively. The four-torus state shown in Figs. 9c and 9f is unstable, so that a slight increase of Δ destroys this torus and turns the orbit into a chaotic one, as displayed in Fig. 10 (with $\Delta = 0.01446$). We have also identified windows of regular solutions within the chaotic regions due to phase locking, as exemplified in Fig. 11 (with $\Delta = 0.1245$). Figure 12 shows the behavior of the maximum Lyapunov exponent λ for five different values of Δ, which confirms that the periodic ($\Delta = 0$, Fig. 8), quasi-periodic four-torus ($\Delta = 0.01442$, Figs. 9c and 9f) and phase-locked ($\Delta = 0.1245$, Fig. 11) orbits are in the regular states since $\lambda \to 0$ as $\tau \to \infty$; and the orbits for $\Delta = 0.01446$ (Fig. 10) and $\Delta = 1.5$ are chaotic since λ tends to a finite positive value as $\tau \to \infty$ (Lichtenberg & Lieberman 1983). Note from Fig. 12 that although the overall dynamics of the system is sensitive to Δ, in the chaotic regime the degree of stochasticity of the system, measured by the asymptotic value of λ, is rather insensitive to Δ.

4.5. SUMMARY

In this section, we showed that the nonlinear saturation of the Langmuir stimulated modulation instability is governed by a set of four coupled wave equations with quadratic nonlinearity. The saturated state of the Langmuir SMI can be either regular or chaotic. As the frequency mismatch parameter Δ is varied, this Hamiltonian system can evolve from the regular state to chaotic state via the route of quasi-periodicity. It is anticipated that the behavior of this type of resonant four-wave process will also take place in a variety of other nonlinear mode coupling processes in cosmic plasmas.

(a)

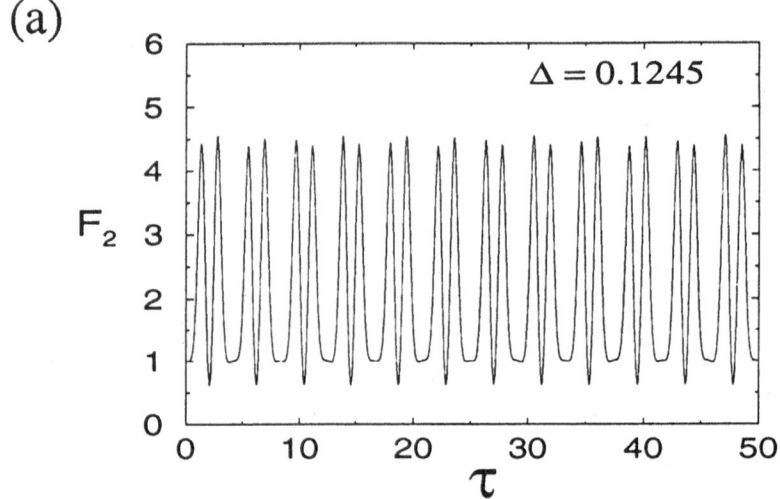

(b)

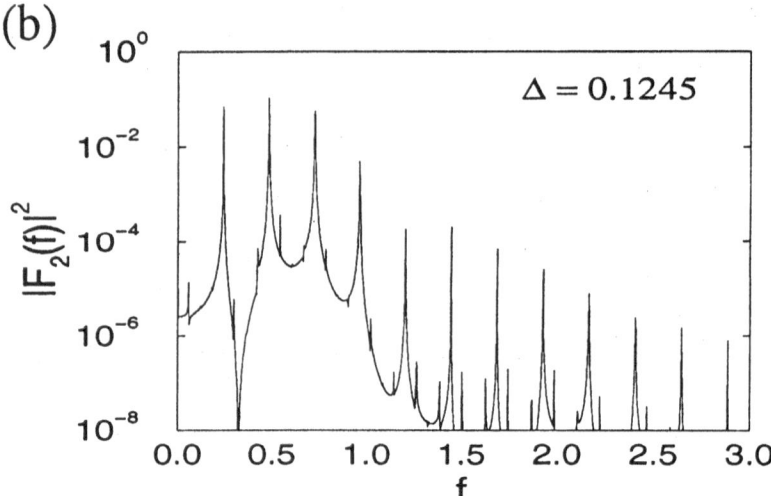

Figure 11. The plot of phase-locked $F_2(\tau)$ and $|F_2(f)|^2$ for $r = 1$, and $\Delta = 0.1245$; the initial conditions are as given in Fig. 9.

5.   Order and Chaos in Pulsar and AGN Emissions

5.1. INTRODUCTION

Variability of emissions from pulsars and active galactic nuclei (AGN) has been observed in a wealth of different forms. Temporal fluctuations of pulsar dispersion measure have been detected in the Crab pulsar (Rankin and Roberts 1971) and various other pulsars (e.g., Cordes et al. 1990; Phillips and Wolszczan 1991). A number of binary pulsars

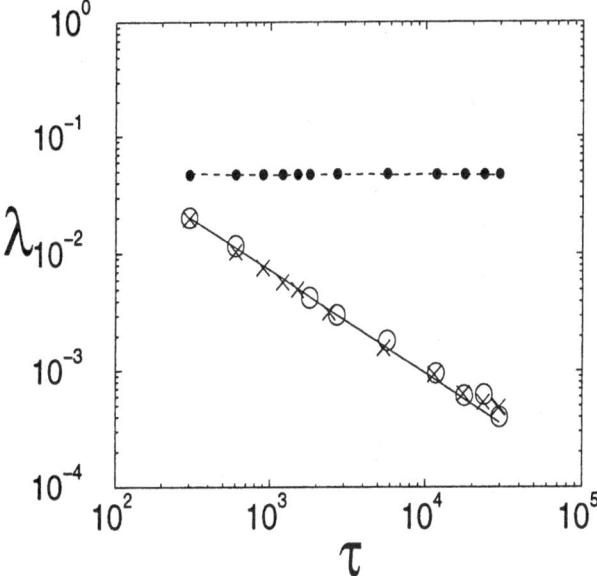

Figure 12. The behavior of the maximum Lyapunov exponent λ for $r = 1$ and $\Delta = 0$ (solid curve), $\Delta = 0.01442$ (empty circle), $\Delta = 0.1245$ (x-curve), $\Delta = 0.01446$ (dashed curve) and $\Delta = 1.5$ (filled circle); the initial conditions are as given in Fig. 9.

are known to undergo eclipses (e.g., Fruchter et al. 1988; Manchester et al. 1991; Johnston et al. 1992). Several forms of variations such as flaring, flickering, extreme scattering events, low-frequency variability and intra-day variability have been identified in AGN (e.g., Hundstead 1972; Quirrenbach et al. 1989; Kunieda et al. 1990).

The study of variability in pulsars and AGN can provide major clues to the physical mechanisms operating near the source region as well as along the propagation path of electromagnetic waves. For example, the microstructures of pulsar radio pulses (with timescales ranging from milliseconds to microseconds) and the intra-day radio variability of AGN may be caused by the nonlinear modulation of radiation as it comes out of the source region (Chian and Kennel 1983; Chian 1992; Gangadhara et al. 1993). Moreover, a recent analysis by Wu and Chian (1995) of a number of radio pulsars presented evidence that the observed temporal variations of pulsar dispersion measure may be due to the influence of the wave intensity on the dispersion relation. In view of the extremely high values of the observed radio luminosity of pulsars and AGN, the electromagnetic waves in the vicinity of source region may obey a nonlinear dispersion relation which depends on the radiation amplitude.

In addition to the plasma effects exemplified above, a great variety of other plasma mechanisms can also contribute to the variability in pulsars and AGN. Krishan and Witta (1994) classified these mechanisms into two categories: (1) variability induced by the fluctuation of plasma densities; (2) variability induced by the modulation of electromagnetic fields. Fluctuations in plasma density can be produced, for example, by the Langmuir turbulence (Zakharov 1972) or Alfvén turbulence (Chian and Oliveira 1994). Nonlinear modulation of radiation may lead to filamentation or self-focusing of the electromagnetic field with associated plasma density fluctuations (Bingham 1990; Chian and Rizzato 1994).

In this section, we discuss the role played by Langmuir turbulence in the variability in pulsars and AGN. Langmuir turbulence is easily excited by particle beams or electromagnetic waves in cosmic plasmas as mentioned earlier. It has been observed by spacecrafts upstream and downstream of interplanetary shocks, in type-III solar radio events, upstream of Earth's and Jovian bow shocks, in the Earth's auroral ionosphere-magnetosphere, and during active experiments in space.

5.2. RELATIVISTIC ZAKHAROV EQUATIONS

In the vicinity of the source regions of the pulsar and AGN emissions, the field can be sufficiently strong to induce relativistic mass variations of plasma particles (Chian 1982; Chian and Kennel 1983; Chian 1992; Wu and Chian 1995). Hence, a proper description of physical processes in pulsars and AGN should take into account the relativistic effect. In this section, we study the nonlinear dynamical behavior of the weakly relativistic Zakharov equations (Berezhiani et al. 1980; Bingham 1990; de Oliveira et al. 1995)

$$i\partial_t \mathcal{E} + \partial_x^2 \mathcal{E} = n\mathcal{E} - \alpha|\mathcal{E}|^2\mathcal{E}, \tag{102}$$

$$\partial_t^2 n - \partial_x^2 n = \partial_x^2 |\mathcal{E}|^2, \tag{103}$$

where $\mathcal{E} = \mathcal{E}(x,t)$ is the slowly varying amplitude of the Langmuir wave electric field, $n = n(x,t)$ is the low-frequency density fluctuations associated with the ion-acoustic perturbations, and the parameter α is a measure of the relativistic effect induced by the wave intensity as defined by

$$\alpha = \frac{3T_e}{2m_e c^2}\left(1 - \frac{8k_0^2 c^2}{3\omega_0^2}\right), \tag{104}$$

where ω_0 (k_0) is the frequency (wavenumber) of the high-frequency Langmuir wave with the dispersion relation $\omega_0^2 = \omega_{pe}^2 + 3v_{th}^2 k_0^2$. The physical quantities are normalized as $E/(8\ sqrt\pi m_e n_0 T_e/3m_i) \rightarrow E$, $(m_i n)/(m_e n_0) \rightarrow n$, $(m_e/m_i)\omega_{pe}t \rightarrow t$, $\sqrt{m_e/m_i}(\omega_{pe}/v_{th})x \rightarrow x$.

5.3. TRANSITION FROM ORDER TO CHAOS

The nonrelativistic Zakharov equations ($\alpha = 0$) reduce to a single equation called the nonlinear Schrödinger equation, in the subsonic limit $\partial_t^2 << \partial_x^2$, which admits regular (periodic and solitary) wave solutions in the absence of driving and dissipation. A soliton gas model of Langmuir turbulence was developed by Shen and Nicholson (1987) by solving numerically the nonrelativistic Zakharov equations and nonlinear Schrödinger equation, respectively, with zero driver and dissipation. The chaotic nature of Langmuir turbulence was investigated by Doolen et al. (1983) and Moon and Goldman (1984) by solving numerically the nonrelativistic Zakharov and nonlinear Schrödinger equations, repectively, including driver and dissipation. Doolen et al. (1983) showed that above the modulation threshold, the system evolves into regular patterns of cavitons which become temporally chaotic for stronger driving. Moon and Goldman (1984) showed that as the driver strength is increased, the system exhibits a transition from intermittence to a two-torus to chaos.

The relativistic Zakharov equations (102) and (103) also admit analyticperiodic and solitary stationary wave solutions (Berezhiani et al., 1980). In order to determine the dynamical behavior of (102) and (103), we solve them numerically by assuming periodic boundary conditions, with L=2π/k defining the basic lengthscale, where k is the wavenumber of the low-frequency field. The dynamical variables $\mathcal{E}$ and n are expanded into Fourier series

$$\mathcal{E}(x, l) = \sum_{N=-\infty}^{\infty} \mathcal{E}_N(t)c^{iNkx}, \tag{105}$$

$$n(x, t) = \sum_{N=-\infty}^{\infty} n_N(t)e^{iNkx}, \tag{106}$$

and the dipole approximation $k_0 = 0$ is made. A number of Fourier modes ranging from $N = 32$ to $N = 128$ for each dynamical variable is used. Nonlinear terms are evaluated with a Fast Fourier Transform subroutine and the set of temporal equations is advanced in time with a predictor-corrector algorithm (de Oliveira et al., 1995). Numerical precision is tested by monitoring the temporal evolution of the conserved quantity, namely, the Hamiltonian of the system

$$H = \int_0^L \left[|\partial_x \mathcal{E}|^2 - \alpha|\mathcal{E}|^4 + n|\mathcal{E}|^2 + \frac{1}{2}\left(n^2 + v^2\right) \right] dx, \tag{107}$$

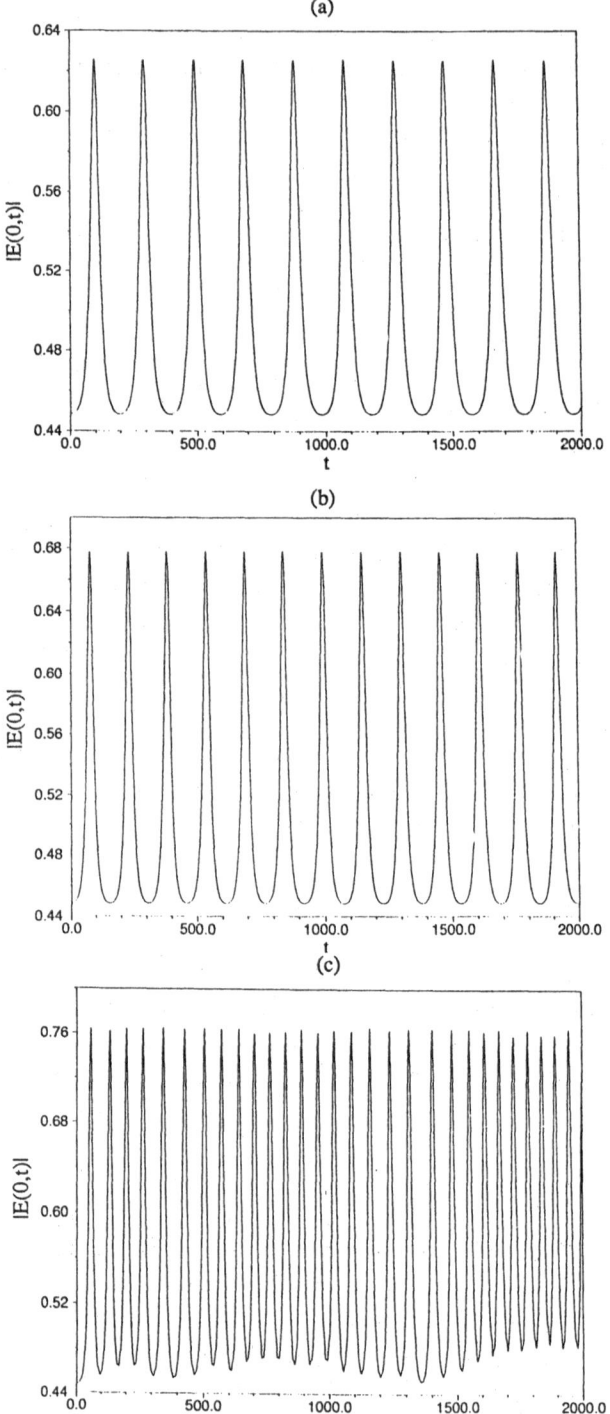

Figure 13. Time series of the electric field for $\rho_* = 0.2$ and $k = 0.62$: (a) $\alpha = 0.001$, (b) $\alpha = 0.03$, and (c) $\alpha = 0.1$.

with $\partial_t n = -\partial_x v$, where v is an auxiliary variable. Our numerical simulations demonstrate that the relativistic Langmuir turbulence can nonlinearly evolve into either *regular* (periodic or quasi-periodic) or *chaotic* regimes, depending on the values of ρ_* (the initial energy of the Langmuir field), k (the wavenumber of the low-frequency field) and α (the relativistic parameter). Fig. 13 illustrates three examples of periodic (Fig. 13a), quasi-periodic (Fig. 13b), and chaotic (Fig. 13c) time series of the electric field $|\mathcal{E}(0,t)|$. The corresponding time series of the plasma density $n(0,t)$ and the power spectra ($|\rho_0(f)|^2$) are displayed in Figs. 14 and 15, respectively. The dynamical variability in Figs. 13-15 is obtained by fixing ρ_* and k, but varying the values of α. Alternatively, the transition from the periodic to quasi-periodic to chaotic regimes can be observed by fixing ρ_* and α, but varying k. The resulting time series of $\mathcal{E}$ and n and the corresponding power spectra are similar to Figs. 13-15. Figures 15a and 15b show that the power spectra of the regular regime consist of a combination of N-number of commensurate (periodic) or incommensurate (quasi-periodic) discrete frequencies. In contrast, the spectrum of the chaotic regime (Fig. 15c) is broadband and continuous, which indicates the onset of turbulence. In addition, we have characterized the regular and chaotic behaviors of the time series in Figs. 13 and 14 by calculating the maximum Lyapunov exponents. Note that the quasi-periodic time series in Figs. 13b and 14b appear 'periodic' due to the timescale displayed; their quasi-periodic features become evident in the power spectrum in Fig. 15b.

5.4. VARIABILITY IN PULSARS AND AGN

The radio wave emitted by a pulsar or AGN is scattered when it traverses a turbulent plasma. In particular, Langmuir waves (turbulence) can effectively stimulate scattering processes such as the stimulated Raman scattering (SRS) or stimulated Compton scattering (SCS). The SRS occurs when a strong electromagnetic wave is scattered off a weakly damped Langmuir wave; whereas, the SCS occurs when a strong electromagnetic wave is scattered off a heavily damped Langmuir wave. The variability in pulsars and AGN can be a consequence of nonlinear scattering processes involving the Langmuir turbulence since they are triggered only above a certain threshold flux and the transmitted flux is a function of various parameters such as wave frequency, scattering angle, density, temperature and magnetic field.

The role of Langmuir turbulence in eclipsing binary pulsars has been studied by numerous papers. It is believed that pulsar eclipses are caused by the plasma surrounding the companion star, either in the wind from the companion or the companion magnetosphere; or

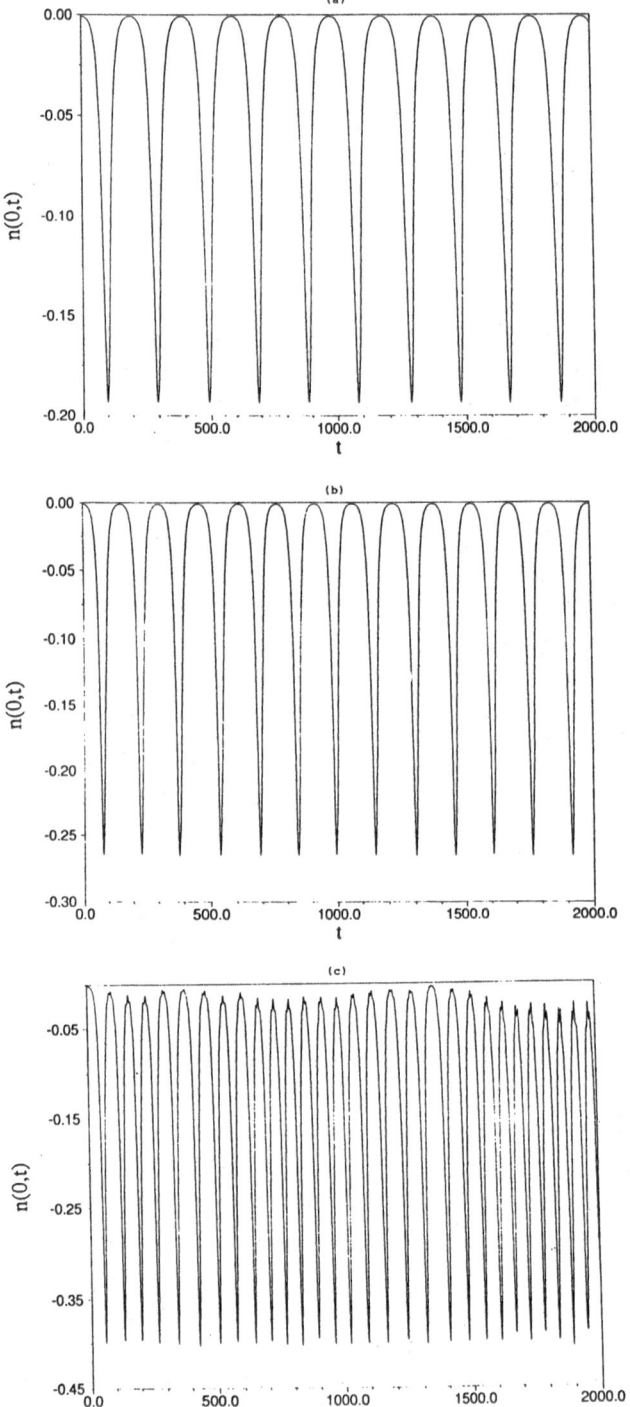

Figure 14. Time series of the plasma density for $\rho_* = 0.2$ and $k = 0.62$: (a) $\alpha = 0.001$, (b) $\alpha = 0.03$, and (c) $\alpha = 0.1$.

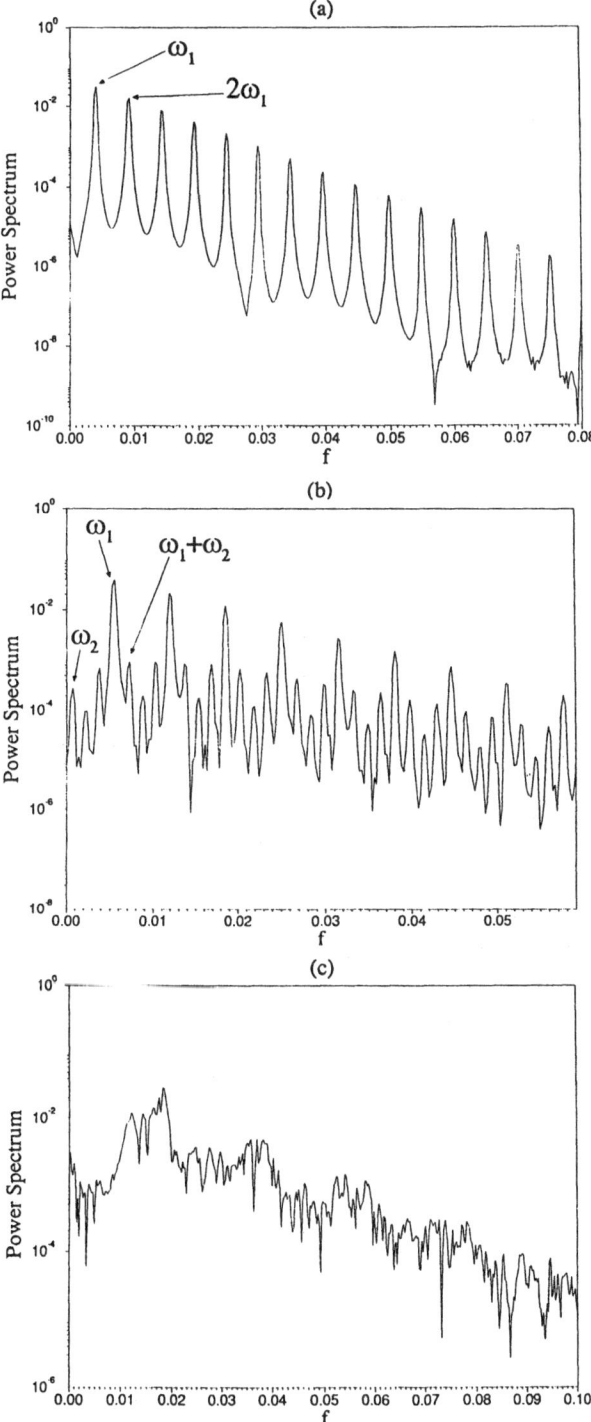

Figure 15. Power spectra for $\rho_* = 0.2$ and $k = 0.62$: (a) $\alpha = 0.001$, (b) $\alpha = 0.03$, and (c) $\alpha = 0.1$.

by material entrained in the pulsar wind. Gedalin and Eichler (1993) suggested that the high level of Langmuir turbulence required for the SRS to operate in PSR 1957+20 and PSR 1744-20A can be excited by the penetration of the relativistic electron-positron pulsar wind into the eclipsing plasma through two-stream instability. Melrose (1994) derived a pair of kinetic equations for three-wave interactions between photons and Langmuir waves; this study showed that the photon-induced Langmuir turbulence is capable of producing the observed eclipse of PSR 1957+20. Luo and Melrose (1994, 1995) analyzed the effect of ambient magnetic field on three-wave scattering processes; they concluded that the inclusion of magnetic field improves the agreement between the theory of SRS and observation of PSR 1957+20. Thompson et al. (1994) showed that the observational properties of PSR 1744-24A favor an eclipse model based on the SRS. In addition, they concluded that the radio brightness temperature of PSR 1718-19 is high enough to generate Langmuir turbulence via the SRS.

The role of Langmuir turbulence in the AGN has been discussed by several papers. Colgate et al. (1970) calculated electromagnetic emissions from the coalescence of Langmuir waves as a source of the radio and infrared radiations in quasars. Baker et al. (1988) constructed a model of astrophysical jets, such as the BL Lacertae objects, in which a relativistic electron beam injected from the central engine into the jet plasma drives Langmuir turbulence which leads to the collective emission of electromagnetic waves. In this model, a spatially varying electrostatic field of wavelength λ_L due to Langmuir turbulence acts as a wiggler to produce a forward-beamed free-electron-laser emission at $\lambda_{em} \sim \lambda_L/(2\gamma_b^2)$, where γ_b is the Lorentz factor of the electron beam. Sol et al. (1989) proposed a two-flow model for extragalactic radio jets: one flow is a beam of relativistic electrons and positrons coming from the inner most part of the accretion disc with $\gamma_b \sim 10$; the second flow is a classical or mildly relativistic wind of electrons and protons coming out from all parts of the accretion disc. They showed that in the strong magnetic field region the Langmuir turbulence is well described by the one-dimensional undriven Zakharov equations; and the fluctuations of Langmuir waves may be an origin of the AGN variability. Krishan and Witta (1990) developed a model for the production of the continuum emission of the AGN via a sequence of the SRS processes. They argued that cyclotron radiation due to nonrelativistic electrons provides a source of photons which can beat with each other to excite Langmuir waves through the forward SRS. Langmuir waves in turn accelerate electrons to high energies with $\gamma_b \sim 10^3$ to 10^4 and at the same time set up a quasi-stationary magnetic field via the magnetic modulation instabilities. The electron beams then scatter off the mag-

netic wiggler to generate free-electron-laser emission in the AGN via
the SRS. According to Krishan and Witta (1990), since the frequency of this emission is proportional to the plasma density, γ-rays can
be emitted in the high-density regions near the central engine and X-
rays through radio waves emitted at larger distances. Weatherall and
Benford (1991) showed that the emission produced by the scattering
of energetic electron beams off solitary wave structures in the regions
of Langmuir turbulence can be more efficient than synchrotron emission; such radiation may reduce the overall energy requirements for
central engines of the AGN. Lesch and Pohl (1992) suggested a model
for intra-day variability in the AGN in which relativistic electrons are
accelerated in magnetic reconnection zones in accretion discs; the variable nonthermal emission is due to coherent radiation arising from the
interaction of relativistic electron beams with the Langmuir turbulence.
This model can account for the AGN variability in radio, optical and
X-rays. Coppi et al. (1993) studied the observational signatures of the
SCS in the AGN and concluded that the coherent emission with high
brightness temperature can be modified and escape via the SCS. They
suggested that the SRS can compete with the SCS to cause similar
effects. Krishan and Witta (1994) showed that the intra-day variability
in the AGN may be induced by either Langmuir or electromagnetic
modulation instabilities, and a shift from the SRS to SCS or vice versa
may cause micro-variability with timescales of a few microseconds to
a few milliseconds. Melrose (1994) concluded that the induced plasma
emission may be the origin of the variability of low-frequency emission
from the AGN, and radiation scattering off Langmuir waves should
dominate over the SCS.

From the above discussions, it is evident that the Langmuir turbulence *de facto* plays an important role in the variability in pulsars and
AGN. Cordes et al. (1990) performed a detailed analysis of micropulse
periodicities of five radio pulsars. They found the quasiperiods of periodic microstructures vary over a 10:1 range about an average period of
0.5-5 ms. Since the pulsar radio emission may be generated by the Langmuir turbulence driven by particle beams in the pulsar magnetosphere,
periodic and quasi-periodic variabilities of pulsar fine structures might
be related to the periodic and quasi-periodic regimes of the Langmuir
turbulence in the pulsar magnetosphere. Our numerical simulations of
the relativistic Zakharov equations confirm the existence of such periodic and quasi-periodic regimes, as seen in the graphs (a) and (b) of
Figs. 13-15, respectively. Note that (102) and (103) are derived for an
electron-ion plasma; however, regular and chaotic regimes of the Langmuir turbulence are expected to be present also in electron-positron
plasmas in the pulsar and AGN magnetospheres. Zhuravlev and Popov

(1990) showed that the pulsar microstructures of PSR 0809+74 on 10-100 μsec timescales may be well described by a model of deterministic chaos and concluded that the formation of a chaotic time series of pulsar micropulses is due to a nonlinear dynamical system. Romani et al. (1992) obtained evidence of deterministic chaos in the radio intensity fluctuations of PSR 0823+26. The observations of Zhuravlev and Popov (1990) and Romani et al. (1992) are in good agreement with our theoretical model based on the nonlinear dynamical system described the relativistic Zakharov equations (102) and (103). The chaotic regime of the Langmuir turbulence identified by our numerical simulations, as seen in the graphs (c) of Figs. 13-15, may be the origin of the chaotic structures of pulsar radio emissions detected in PSR 0809+74 and PSR 0823+26.

5.5. SUMMARY

The results of this section suggest that the periodic, quasi-periodic and chaotic behaviors in the variability of pulsars and AGN may be the manifestation of the corresponding regular and chaotic dynamics of the Langmuir turbulence present either in the neighborhood of the source region or along the propagation path of emissions. It is likely that a careful analysis of the time series of the pulsar and AGN emissions will reveal more details of these fascinating nonlinear dynamical features of the system.

6.   Concluding Remarks

In this paper, we showed through four physical examples that nonlinear wave-wave interactions indeed play an important role in astrophysical and space plasmas. Since a great variety of wave modes can be excited in plasmas, it is likely that the various types of nonlinear processes discussed in this paper may also apply to other plasma modes. Further advances in the concepts of deterministic chaos and nonlinear dynamics, such as exemplified in this paper, will provide good insight to the complex behavior of plasmas in the cosmos.

7.   Acknowledgements

The author wishes to thank J.R. Abalde, M.V. Alves, G.I. de Oliveira, S.R. Lopes, L.P.L. Oliveira and F.B. Rizzato for collaborations, and CNPq for support.

References

Anderson, R.R., Parks, G.K., Eastman, T.E., Gurnett, D.A. and Frank, L.A.: 1981, *J. Geophys. Res.* **86**, 4493.

Baker, D.N., Borovsky, J.E., Benford, G. and Eilek, J.A.: 1988, *ApJ*, **326**, 110.

Bardwell, S. and M. V. Goldman: 1976, *ApJ* **209**, 912.

Barrow, C.H., Zarka, P. and Aubier, M.G.: 1994, *A&A* **286**, 597.

Benson, R.F., Desch, M.D., Hunsucker, R.D., and Romick, G.J.: 1988, *J. Geophys. Res.* **93**, 277.

Berezhiani, V.I., Tsintsadze, N.L. and Tskhakaya, D.D.: 1980, *J. Plasma Phys.* **24**, 15.

Bingham, B.: 1990, *Physica Scripta* **T30**, 24.

Boehm, M.H., Carlson, C.W., McFadden, J. and Mozer F.S.: 1984, *Geophys. Res. Lett.* **11**, 511.

Boehm, M.H., Carlson, C.W., McFadden, J.P., Clemmons, J.H. and Mozer, F. S.: 1990, *J. Geophys. Res.* **95**, 12157.

Cairns, I.H. and Robinson, P.A.: 1992b, *Geophys. Res. Lett.* **19**, 2187.

Cairns, I.H. and Robinson, P.A.: 1992a, *Geophys. Res. Lett.* **19**, 2069.

Canu, P., Conilleau-Wehrlin, N., de Villedary, C., Kellogg, P.J., Harvey, C.C. and MacDowall, R.J.: 1993, *Planet. Space Sci.* **41**, 811.

Chian, A.C.-L.: 1982, *A&A* **112**, 391.

Chian, A.C.-L.: 1991, *Planet. Sp. Sci.* **39**, 1217.

Chian, A.C.-L.: 1992, in *Proc. IAU Colloquium 128 on The Magnetospheric Structure and Emission Mechanisms of Radio Pulsars*, eds. T.H. Hankins, J.M. Rankin and J.A. Gil, Pedagogical University Press: Zielona Góra, 356.

Chian, A.C.-L. and Kennel, C.F.: 1983, *Ap. Space Sci.* **97**, 9.

Chian, A.C.-L. and Alves, M.V.: 1988, *ApJ Lett.* **330**, L77.

Chian, A.C.-L., Lopes, S.R. and Alves, M.V.: 1994a, *A&A* **288**, 981.

Chian, A.C.-L., Lopes, S.R. and Alves, M.V.: 1994b, *A&A* **290**, L13.

Chian, A.C.-L. and Oliveira, L.P.L.: 1994, *A&A* **286**, L1.

Chian, A.C.-L. and Rizzato, F.B.: 1994, *J. Plasma Phys.* **51**, 61.

Chian, A.C.-L. and Abalde, J.R.: 1995, *A&A* **298**, L9.

Colgate, S.A., Lee, E.P. and Rosenbluth, M.N.: 1970, *ApJ*, **162**, 649.

Coppi, P., Blandford, R.D. and Rees, M.J.: 1993, *MNRAS* **262**, 603.

Cordes, J.M., Weisberg, J M. and Hankins, T.H.: 1990, *Astr. J.* **100**, 1882.

de Oliveira, G.I., Rizzato, F.B. and Chian, A.C.-L.: 1995, *Phys. Rev. E* **52**, 2025.

Doolen, G.D., Dubois, D.F., Rose, H A. and Hafizi, B.: 1983, *Phys. Rev. Lett.* **51**, 335.

Ergun, R.E., Carlson, C.W., McFadden, J.P., Clemmons, J.H. and Boehm, M.H.: 1991, *Geophys. Res. Lett.* **18** , 1177.

Fruchter, A.S., Stinebring, D.R. and Taylor, J.H.: 1988, *Nature* **333**, 237.

Gangadhara, R.T., Krishan, V. and Shukla, P.K.: 1993, *MNRAS* **262**, 151.

Gedalin, M. and Eichler, D.: 1993, *ApJ* **406**, 629.

Glanz, J., Goldman. M.V., Newman, D.L. and McKinstrie, C.J.: 1993, *Phys. Fluids* **5**, 1101.

Goldman, M.V.: 1984, *Rev. Modern Phys.* **56**, 709.

Grabbe, C.: 1982, *Geophys. Res. Lett.* **9**, 155.

Grabbe, C., Papadopoulos K. and Palmadesso P.: 1980, *J. Geophys. Res.* **85**, 3337.

Gurnett, D.A. and Frank, L.A.: 1975, *Solar Phys.* **45**, 477.

Gurnett, D.A., Kurth, W.S. and Scarf, F.L.: 1979, *Nature* **280**, 767.

Gurnett, D.A., Maggs, J.E., Gallagher, D.I., Kurth, W.S., and Scarf, F.L.: 1981, *J. Geophys.Res.* **86**, 8833.

Gurnett D.A., Shawhan S.D., Shaw R.R.: 1983, *J. Geophys. Res.* **88**, 329.

Gurnett, D.A., Hospodarsky, G.B., Kurth, W.S., Williams, D.J., and Bolton, S. J.: 1993, *J. Geophys. Res.* **98**, 5631.

Hospodarsky, G.B., Gurnett, D.A., Kurth, W.S. and Bolton, S.J.: 1991, *EOS* **72**, No. 44, 390.

Hospodarsky, G.B., Gurnett, D.A., Kurth W.S., Bolton, S.J., Kivelson, M. and Strangeway, R.: 1992, *EOS* **73**, No. 14, 241.

Hunstead, R.W.: 1972, *Ap. Lett.* **12**, 193.

Johnston, S., Manchester, R.N., Lyne, A.G., Bailes, M., Kaspin, V.M., Qiao, G.J. and D'Amico, N.: 1992, *ApJ* **387**, L37.

Kellogg, J.P., Goetz, K., Howard, R.L. and Monson, S.J.: 1992, *Geophys. Res. Lett.* **19**, 1303.

Koons, H.C., Roeder, J.L., Bauer, O.H., Haerendel, G., Treumann, R., Anderson, R.R., Gurnett, D.A. and Holzworth, R.H.: 1987, *J. Geophys. Res.* **92**, 5865.

Krishan, V. and Wiita, P.J.: 1990, *MNRAS* **246**, 597.

Krishan, V. and Wiita, P.J.: 1994, *ApJ* **423**, 172.

Kunieda, H., Turner, T.J., Awaki, H., Koyama, K., Mushotzky, R. and Tsusaka, Y.: 1990, *Nature* **345**, 786.

Lashmore-Davies, C.N.: 1981, *Plasma Physics and Nuclear Fusion Research*, ed. R.D. Gill, Academic Press: London, 319.

Lesch, H. and Pohl, M.: 1992, *A&A* **254**, 29.

Lichtenberg, A.J. and Lieberman, M.A.: 1983, *Regular and Stochastic Motion*, Springer-Verlag: New York, 280.

Lin, R.P., Levedahl, W.K., Lotko, W., Gurnett, D.A., and Scarf, F.L.: 1986, *ApJ* **308**, 954.

Luo, Q. and Melrose, D.B.: 1994, *Solar Phys.* **154**, 187.

Luo, Q. and Melrose, D.B.: 1995, *ApJ* **452**, 346.

Manchester, R.N., Lyne, R.G., Robinson, C., D'Amico, N., Bailes, M., and Lim, J.: 1991, *Nature* **352**, 219.

Marion J.B.: 1970, *Classical Dynamics of Particles and Systems*, Academic Press: New York.

Melrose, D.B.: 1994, *J. Plasma Phys.* **51**, 13.

Meunier, C., Bussac, M.N. and Laval, G.: 1982, *Physica* **4D**, 236.

Moon, H.T. and Goldman, M.V.: 1984, *Phys. Rev. Lett.* **53**, 1821.

Oya H., Morioka A. and Obara T.: 1985, *J. Geomagn. Geoelect.* **37**, 237.

Phillips, J.A. and Wolszczan, A.: 1991, *ApJ Lett.* **382**, L27.

Pikovskiĭ, A.S., Rabinovich, M.I. and Trakhtengerts, V.Yu., *Sov. Phys. JEPT* **47**, 715.

Quirrenbach, A., Witzel, A., Krichbaum, T., Hummel, G.A., Alberdi, A. and Schalin-ski, C.: 1989, *Nature* **337**, 442.

Rankin, J.M. and Roberts, J.A.: 1971, *Proc. IAU Symp. 46 on The Crab Nebula*, eds. R.D. Davies and F.G. Smith, 114.

Rizzato, F.B. and Chian, A.C.-L.: 1992, *J. Plasma Phys.* **48**, 71.

Robinson, P.A., Cairns, I.H. and Willes, A.J.: 1994, *ApJ* **422**, 870.

Romani, R.W., Rankin, J.M. and Backer, D.C.: 1992, in *Proc. IAU Colloquium 128 on The Magnetospheric Structure and Emission Mechanisms of Radio Pulsars*, eds. T.K. Hankins, J.M. Rankin and J.A. Gil, Pedagogical University Press: Zielona Góra, 326.

Romeiras, F.J.: 1983, *Phys. Lett. A* **93**, 227.

Shen, M.-M. and Nicholson, D.R.: 1987, *Phys. Fluids* **30**, 1096.

Sjölund, A. and Stenflo, L.: 1967, *Zeitschrift Physik* **204**, 211.

Sol, H., Pelletier, G. and Asséo, E.: 1989, *MNRAS* **237**, 411.

Taniuti, T. and Nishihara, K.: 1983, *Nonlinear Waves*, Pitman, 148.

Thiessen, J.P. and Kellogg, P.J.: 1993, *Planet. Space Sci.* **41**, 823.

Thompson, C., Blandford, R.D., Evans, C.R. and Phinney, E.S.: 1994, *ApJ* **422**, 304.

Vyshikind S. Ya. and Rabinovich, M.I.: 1976, *Sov. Phys. JEPT* **44**, 292.

Walters, D. and Lewak, G.: 1977, *J. Plasma Phys.* **18**, 525.

Wang, D.-Y., and Li, D.-Y.: 1991, *Solar Phys.* **135**, 393.
Weatherall, J.C., Goldman, M.V. and Nicholson, D.R.: 1981, *ApJ* **246**, 306.
Weatherall, J.C. and Benford, G.: 1991, *ApJ* **378**, 543.
Weiland, J. and Wilhelmsson, H.: 1977, *Coherent Non-Linear Interaction of Waves in Plasmas*, Pergamon Press.
Wersinger, J.M., Finn, J.M. and Ott, E.: 1980a, *Phys. Rev. Lett.* **44**, 453.
Wersinger, J.M., Finn, J.M. and Ott, E.: 1980b, *Phys. Fluids* **23**, 1142.
Wilhelmsson, H.: 1969, *J. Plasma Phys.* **3**, 215.
Wu, X. and Chian, A.C.-L.: 1995, *ApJ* **443**, 261.
Zakharov, V.E.: 1972, *Soviet Phys. JETP* **35**, 908.
Zhuravlev, V.I. and Popov, M.V.: 1990, *Sov. Astr.* **34**, 377.

ANNOUNCEMENT

CALL FOR PAPERS
Causality and Locality in Modern Physics and Astronomy:
Open Questions and Possible Solutions.
To be held at York University, Toronto, Canada
August 25-29,1997

The scientific program will include invited and contributed lectures and poster presentations concerned with the topic of locality and its implications for modern physics and astronomy. These include:
• Non-locality-results of recent optical experiments.• Classical models of particle spin.• Quantum measurement, gravitation and locality.
• Neutrino oscillations-experimental and theoretical implications for neutrino mass. • Superluminal waves-theory and experiment.
• Non-locality vs Special Relativity.• $E_n=0$ solutions of Maxwell's equations.
•Longtitudinal solutions of Maxwell's equations • Related topics.

The International Organising Committee is:

• D.V.Ahluwalia, Physics Division, Los Alamos National Laboratory, USA.
•V.Dvoeglazov, University of Zacatecas, Mexico.
• M.W.Evans, Visiting Professor, ISI, Calcutta and York University, Canada.•
A.Garuccio, University of Bari, Italy.• G.Hunter, Chemistry , York University, Canada.•
S.Jeffers, Physics and Astronomy, York University , Canada.•
M.Meszaros, TKI, Hungary.•M. Novikov, Russian Academy of Sciences
• S. Roy, Indian Statistical Institute, Calcutta, India.
• T.Van Flandern, University of Maryland, USA
.• A. van der Merwe, University of Denver, USA.
• J-P. Vigier, University of Pierre and Marie Curie, Paris, France.

Please submit one page abstracts to Dr.S.Jeffers, Department of Physics and Astronomy, York University, 4700 Keele St, North York, Ontario, M3J 1P3,Canada.

Telephone Enquiries: 416-736-2100 ext.33851
Fax: 416-736-5516
E-mail: vigier@yorku.ca
Web Site with On-Line Registration
URL: http://www.vigier.yorku.ca/VigierII

LIST OF FORTHCOMING PAPERS

L.A. Dávalos-Orozco: Rayleigh-Taylor Instability of a Two-Fluid Layer under a General Rotation Field and a Horizontal Magnetic Field (Received 18 April, 1996; accepted October 9, 1996.)

S.N. Goderya, K.C. Leung and E.G. Schmidt: Photometric Investigation of the Short Period Eclipsing Binary Star V719 Her (Received May 8, 1996; accepted October 9, 1996.)

L.S. Lyubimkov: Observational Manifestations of Early Mixing in B- and O-Type Stars (Received October 3, 1996; accepted October 9, 1996.)

Ranjna Bakaya and R.R. Rausaria: Microwave Emission From Solar Flares and Comparison with Observations (Received January 19, 1996; accepted October 11, 1996.)

Q. Zeng: Methanol $7_2-8_1A^+$ Emission and Some Maser Series in Orion KL (Received April 10, 1996; accepted October 22, 1996.)

J.L. Han and X.Z. Zhang: A Possible Association of a Long-Period Pulsar with a Supernova Remnant? (Received April 23, 1996; accepted October 22, 1996.)

Young-Jong Sohn, Yong-Ik Byun and Mun-Suk Chun: Color Gradients in Four King Type Globular Clusters: NGC 2298, NGC 6402, NGC 6934, and NGC 7089 (Received June 27, 1996; accepted October 22, 1996.)

Huang Guang-Li, Qin Zhi-Hai and Yao Qi-Jun: A Model for Solar Radio Pulsations at Short Centimetric Band (Received August 13, 1996; accepted October 22, 1996)

W. Köhnlein: Cross-Correlation of Solar Wind Parameters with Sunspots ('Long-Term Variations') at 1 AU During Cycles 21 and 22 (Received May 31, 1996; accepted November 13, 1996)

Astrophysics and Space Science **242**: 299, 1996.